HANDBOOK OF HOUSING SYSTEMS FOR DESIGNERS AND DEVELOPERS

HANDBOOK OF HOUSING SYSTEMS FOR DESIGNERS AND DEVELOPERS

LAURENCE STEPHAN CUTLER

SHERRIE STEPHENS CUTLER

VNR VAN NOSTRAND REINHOLD COMPANY
NEW YORK CINCINNATI TORONTO LONDON MELBOURNE

Van Nostrand Reinhold Company Regional Offices:
New York Cincinnati Chicago Millbrae Dallas

Van Nostrand Reinhold Company International Offices:
London Toronto Melbourne

Library of Congress Catalog Card Number: 73-7629
ISBN: 0-442-21820-6

Manufactured in the United States of America

Published by Van Nostrand Reinhold Company
450 West 33rd Street, New York, N.Y. 10001

Published simultaneously in Canada by Van Nostrand Reinhold Ltd.

15 14 13 12 11 10 9 8 7 6 5 4 3 2 1

Library of Congress Cataloging in Publication Data

Cutler, Laurence S.
Handbook of housing systems for designers and developers.

Bibliography: p.
1. Industrialized building—Handbooks, manuals, etc. 2. Prefabricated houses—Handbooks, manuals, etc. 3. Housing—Handbooks, manuals, etc.

I. Cutler, Sherrie Stephens, joint author.

II. Title

TH1000.C87 693.97 73-7629
ISBN 0-442-21820-6

[redacted]

To
MAXIMILIAN and ZACHARY

One man's system is another man's component.
(Courtesy Ezra D. Ehrenkrantz)

FOREWORD

When I first met Laurence and Sherrie Cutler, we were jointly engaged in the whirlwind activity known as the "In-Cities Experimental Housing Research and Development Project" program, undertaken by the Department of Housing and Urban Development in late 1968 during the Johnson administration. The Cutlers, already deeply involved with the study of industrialized building both here and abroad, became even more involved as housing systems were an integral part of that project.

Then came Operation BREAKTHROUGH, undertaken by HUD in the Nixon administration. The Cutlers took the lead in organizing one of the teams, involving United States and European systems and producers, and their team reached the semifinals—no mean achievement for a small, young design office.

Quite aside from, but closely related to, these activities has been their continuing efforts in the evolution of systems for the efficient, flexible design and production of housing by industrialized processes. Additionally, for the past several years, we have collaborated in organizing special one-week summer sessions in industrialized housing at M. I. T.

The Cutlers have drawn upon the sum of this experience to produce this book, whose aim is to provide insights into the gathering of fundamental data, the design of housing systems, and organization for the carrying out of such projects. It should be of assistance to the person entering this rapidly developing and changing field, as well as to the practitioner already in it.

Albert G. H. Dietz
Professor of Building Engineering, M.I.T.
Cambridge, Massachusetts

ACKNOWLEDGMENTS

It is with great pleasure that we acknowledge the assistance of those people and institutions who have helped with this book.

Some of the research which acted as a basis for this book was undertaken by our professional firm, ECODESIGN, INC., of Cambridge, Massachusetts, for the U. S. Department of Housing and Urban Development under the "In-Cities Low Cost Housing Experiment" in 1968. Additional technical studies and the basic philosophies expounded herein were developed during a research project, SYSTEM ECOLOGIC—A Transitional Building System, which was supported by a grant from the National Endowment for the Arts in 1970 and again in 1972.

We are also indebted to Daphne Allen Rice of ECODESIGN for her untiring efforts and assistance during the preparation of this manuscript.

Finally, we are grateful to Michael Hamilton and Alberta Gordon for their enduring patience.

Laurence Stephan Cutler
Sherrie Stephens Cutler

Cambridge, Massachusetts

CONTENTS

A PROPHECY

by H. G. Wells

Quoted from his "Anticipations" published in 1902.

"I find it incredible that there will not be a sweeping revolution in the methods of building during the next century. The erection of a house wall, come to think of it, is an astonishingly tedious and complex business; the final result is exceedingly unsatisfactory. It has been my lot recently to follow in detail the process of building a private dwelling-house, and the solemn succession of deliberate, respectable, perfectly satisfied men who have contributed each so many days of his life to this accumulation of weak compromises, has enormously intensified my constitutional amazement at my fellow-creatures. The chief ingredient in this particular house-wall is the common brick, burned earth, and but one step from the handfuls of clay of the ancestral mud hut, small in size and permeable to damp. Slowly, day by day, the walls grew tediously up, to a melody of tinkling trowels."

"Everything in this was hand work, the laying of the bricks, the dabbling of the plaster, the smoothing of the paper; it is a house built of hands—and some I saw were bleeding hands—just as in the days of the pyramids when the only engines were living men. The whole confection is now undergoing incalculable chemical reactions between its several parts. Lime, mortar, and microscopical organisms are producing undesigned chromatic effects in the paper and plaster; the plaster, having methods of expansion and contraction of its own, crinkles and cracks; the skirting, having absorbed moisture and now drying again, opens its joints; the rough-cast coquettes with the frost and open chinks and crannies for the humbler creation. I fail to see the necessity of (and, accordingly, I resent bitterly) all these coral-reef methods. Better walls than this, and better and less life-wasting ways of making them, are surely possible."

1 INTRODUCTION

SYSTEMS CLASSIFICATION

The term "system ecologic" is simply the name given to the transitional industrialized building system we have developed under a research grant from the National Endowment for the Arts, a federal agency. Ecology, of course, means the totality of the relationship between organisms and their environment. Ecologic relates to the logic of ecology and emphasizes the coherent relationship between the built environment and the natural environment—particularly relating to the "eco" or habitat of man. Exploratory work has been undertaken to study the possibilities of applying past research as a basis for the development of a transitional building system, the prime objective being to develop SYSTEM ECOLOGIC, a design technique, a kit, capable of performing immediately within the present range of building codes, union constraints, and economic considerations which, collectively, have retarded the U. S. building industry.

Although this discussion is devoted primarily to the development of a particular transitional technology, available for the immediate production of housing units, it is necessary to frame the concept with a short description of building systems in general.

There are many technologies, but no one of them is applicable to all cases. They are each dependent upon and influenced by the many other aspects of the housing situation, such as: land use, density, volume, site conditions, user-needs, continuity, codes, and labor.

Generally, industrialized systems building may be defined as *the total integration of all subsystems and components into an overall process fully utilizing industrialized production, transportation, and assembly techniques.* This integration is achieved through the exploitation of underlying organizational principles. Industrialized housing systems may be classified in a variety of ways, of which the following are the major categories:

1-1
MONOLITHIC UNITS (BOXES) Factory-produced and preassembled volumetric elements with a high amount of finish and a minimum amount of erection time. *Engineering News-Record*

Monolithic Units (Boxes)

Monolithic units are usually factory-produced and preassembled volumetric elements with a high degree of finish and a minimum amount of required site erection time (primarily for site utility connections). For the most part though, these units are boxes and they may be four- or six-sided room sized or complete dwellings. They may be further categorized as a function of their relative degree of self-containment and weight.

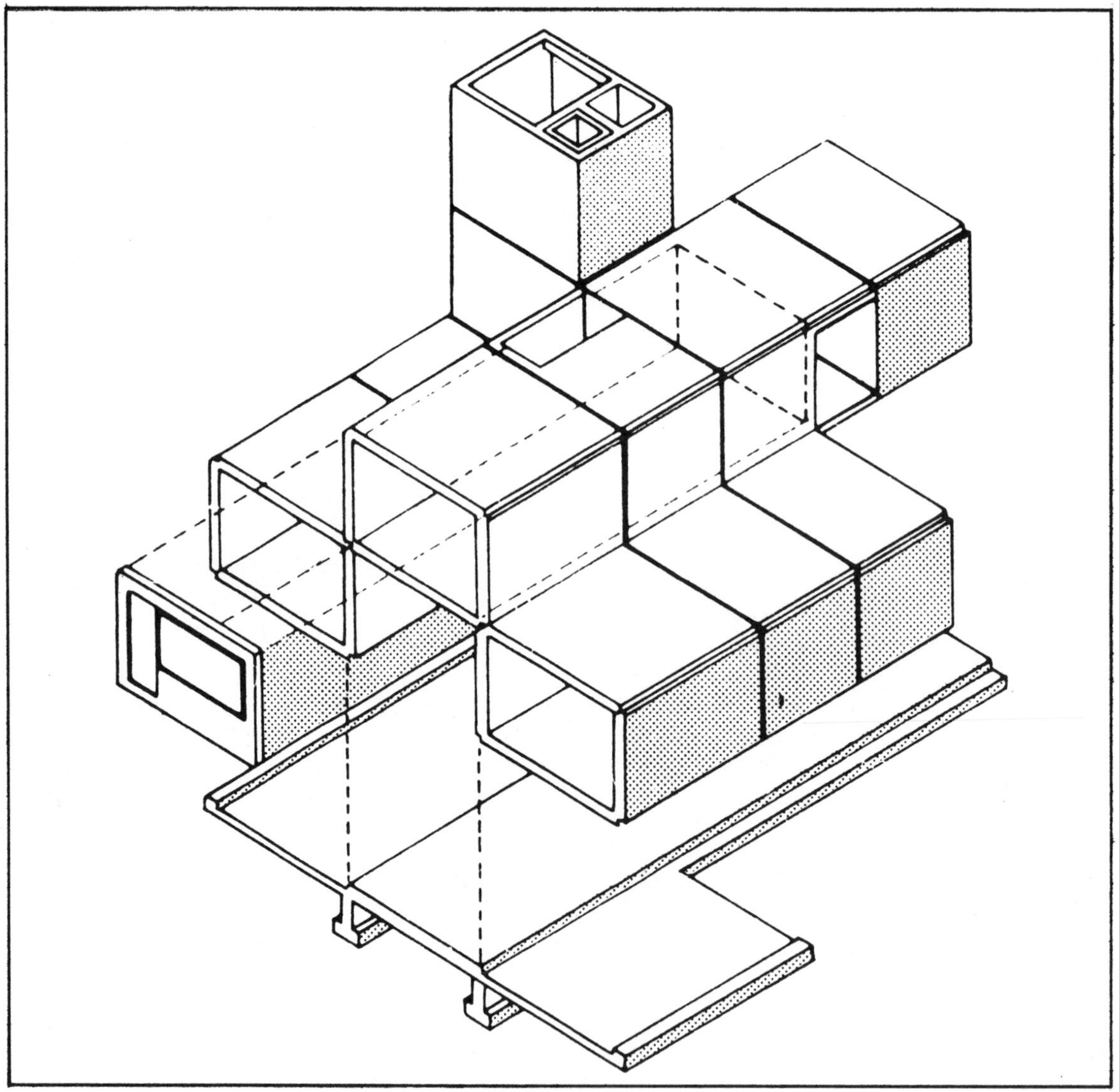

1. Lightweight Boxes—Mobile and Sectional Homes: Totally self-contained units which can retain their mobility or be permanently installed and grouped horizontally or stacked with the addition of a demountable frame or strengthened wall as a framework. Mobile homes are presently fabricated wood stud construction, but some stressed skin construction is now being used. In most cases, modulars are completely preassembled and finished and require only site utility connections for occupancy. Full foundations are generally not required. The use of materials such as fiber reinforced plaster, lightweight aluminum and steel frames, honeycomb, stressed skins, can permit a reduction in labor for assembly of structural systems in this category. However, fire and structural codes and materials economies often limit these units to low-rise usage.
2. Heavyweight Boxes or Volumetric Components—Habitat '67-type (Moshe Safdie, Architect) room-size (or smaller) volumes of concrete, steel sandwich, or fiber-reinforced plastic, which can be grouped horizontally and/or stacked vertically (if wall bearing structure) and can be dry connected to form single- or multi-

1-2
GRUMMAN LIGHTWEIGHT BOXES Constructed of a thin skin structure with a steel frame, the units are lifted into final position within a steel superstructure. This prototype concept illustrates the flexibility of these boxes and also the redundancy between the floor of one unit and the roof of the other. *Grumman Allied Industries, Inc.*

1-3
SAN ANTONIO, TEXAS An experimental heavyweight box structure whereby short and long boxes are built in an on-site factory, lifted by crane into position between slip-formed concrete service cores. *H. B. Zachry System*
Photo by Tell-Pics

family attached or detached housing. In some cases, these volumes may be incorporated in traditional structural/mechanical space grids (infrastructure) to provide high-rise multifamily housing. This concept, originated in the 1920s by the Swiss architect Le Corbusier, has finally captured the imagination of every young architect and planner today.

Basically, boxes are restricted by weight and bulk in their travel radius from the plant (action radii) and are extremely uneconomical to handle, because of the amount of empty space being transported. This inefficiency can be cut by increasing the amount of finishing within the protected volume, such as with completely furnished trailers. Another economic drawback is the duplication of structural walls which usually occurs when the boxes are stacked next to each other. Some high-rise box systems overcome this by using three-sided boxes, slab sharing, or checkerboard stacking (see Shelley System). Monolithic boxes are generally "closed" systems and do not offer great flexibility in the design treatment or the massing.

Total Systems (Panels)

Total systems are usually large slabs, or otherwise panelized units, not made in the form of a box but often large enough to constitute entire walls, partitions, and floors, and substantial parts of floors and roofs. They are fabricated in a factory and assembled at the site. In some cases, the components of one manufacturer are incorporated in the subsystem of another manufacturer, if the two elements are compatible and dimensionally integrative.

1-4
TOTAL SYSTEMS (PANELS) Panelized units, often large enough to constitute entire walls, partitions, and floors.
Engineering News-Record

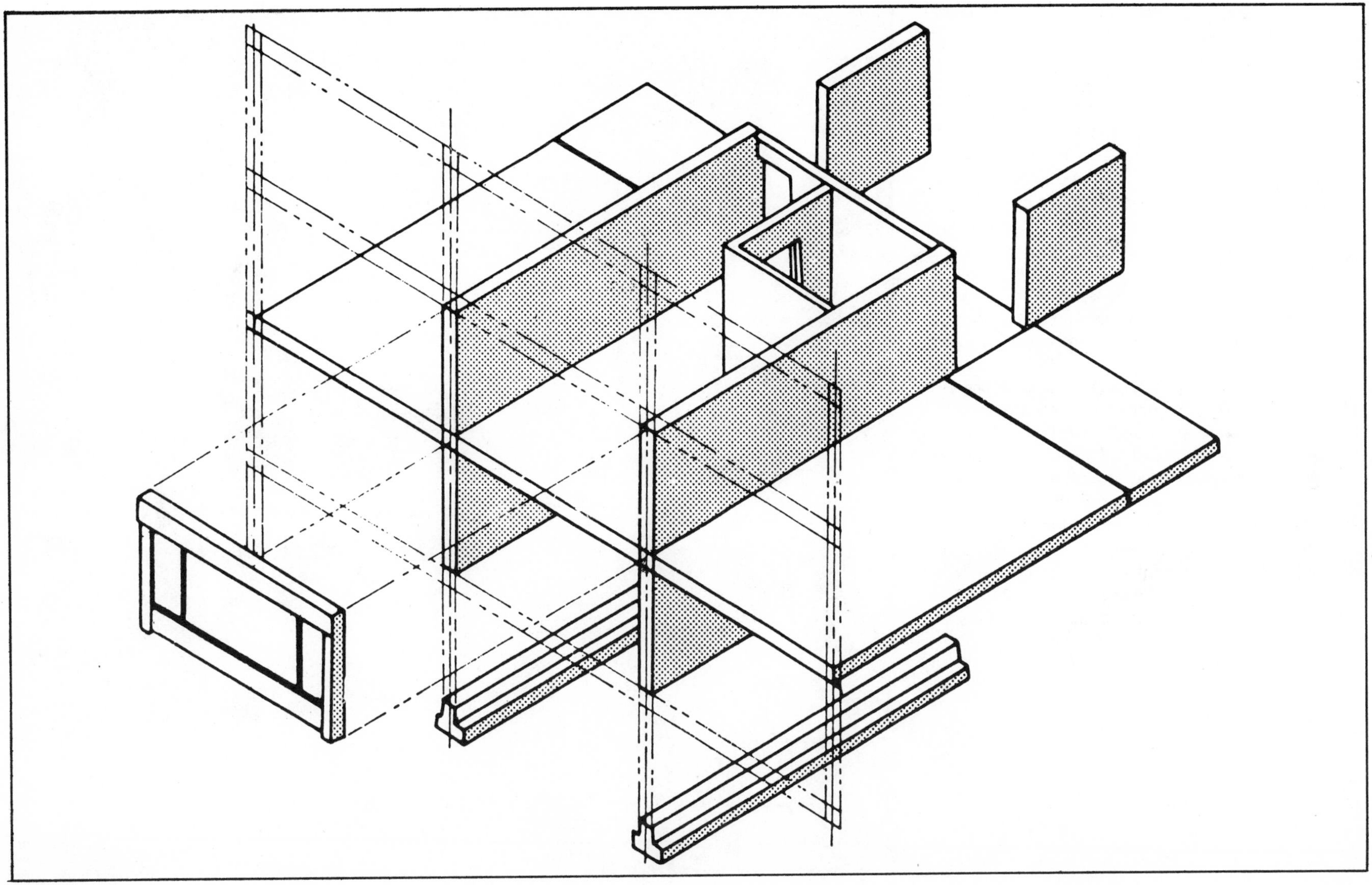

1-5
LIVERPOOL, ENGLAND These maisonettes are constructed of an *open production* casting method which permits flexibility in panel size and in the incorporation of doors, windows, ducts, ventilation flues, holes, and conduit runs for service. *Unit Camus Ltd., Liverpool, U.K. Photo by John Mill*

Total systems can be further analyzed according to whether they are:

1. Open Production—Mold sizes and incorporations are modified to specifications per project of 300 dwelling units or more. These systems are mostly concrete slabs and panels incorporating almost any specified finish, and are usually either room-size or in manageable pieces.
2. Closed Production—The components produced are usually bearing panels and other pieces comprising what amounts to a set number of inalterable pieces. The system is based on a small number of standardized components which, when combined, can offer many solutions to a given problem.

Generally, heavy panels require a huge aggregated market (300 units minimum) to justify going to systems building at all and are often limited to buildings of over two to three stories in order to be efficient. However, they do offer great design flexibility, and in large scale projects, they exhibit 20–30% cost savings. Bulk and production space is controlled as they can be cast, stored, and shipped vertically. The panel joint details are critical for rigidity and continuity—"wet joints," structurally superior, take longer in erection than "dry joints."

1-6
THAMESMEAD, ENGLAND A specific number of standardized pieces are put together in a limited number of ways, making this a *closed production* system. *Riccardo Meregaglia, Balency MBM, Milan, Italy*

1-7
STRUCTURAL SYSTEMS (FRAMES) Generally, this classification includes the frame parts of the building, the beams and columns, fabricated off-site. *Engineering News-Record*

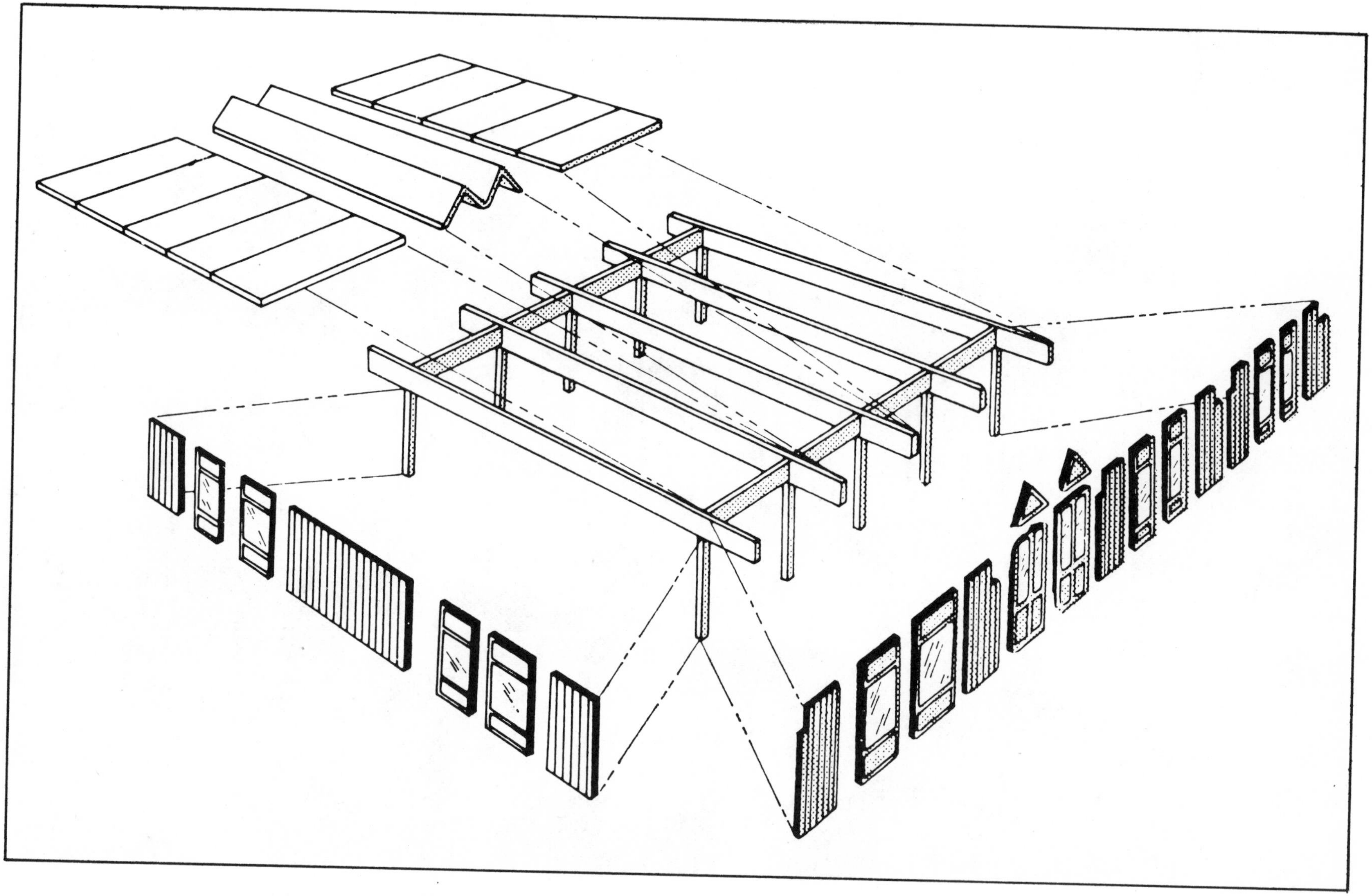

1-8
CAMBRIDGE, MASSACHUSETTS A heavyweight concrete frame of a closed production technique requires special equipment for handling the pieces. Rapid erection of the building skeleton is a positive asset for the developer, but the infill between the frame is time-consuming. *Componoform, Inc., Cambridge, Massachusetts*
Photo by Steve Hanson. 1970

Structural (Frames)

Structural systems generally constitute the frame parts of the building, such as beams and columns, fabricated off-site but assembled on-site. Into these frames are fitted infill units, such as walls, partitions, floors, ceilings, and roofs, also usually fabricated off-site and assembled on-site to the structural members. The primary advantage is the reduction of site work to strictly a component assembly. Transportation is economical primarily because the components are small and light, and for this reason, central factories can enjoy large market areas (action radii). However, the increase in joints and materials tends to complicate the total construction process, increases the cost, and does not insure good acoustical privacy between dwelling units. Heavy weights, such as steel (sophisticated) and concrete (common) frames are the two types of frames used most in present-day construction. They can both be lightened in weight—steel by punching holes in the members or by using higher strength steel, and concrete by utilizing lightweight aggregates in the mix—but in all cases the cost then increases. Materials such as aluminum and fiberglass are also widely in use, but fire protection methods and semi-exotic materials are both more costly.

1-9
SPECIAL CONSTRUCTION TECHNIQUES (ON SITE) A building technology where the building is able to build itself, utilizing special machinery and methods. *John Laing & Son, Ltd., London, England*

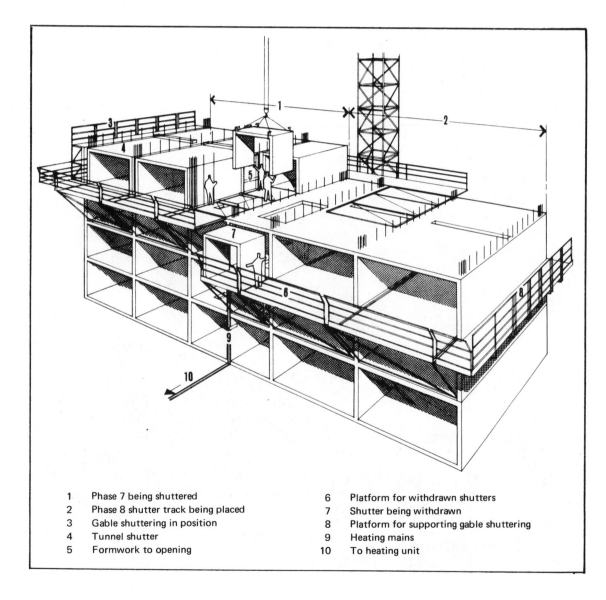

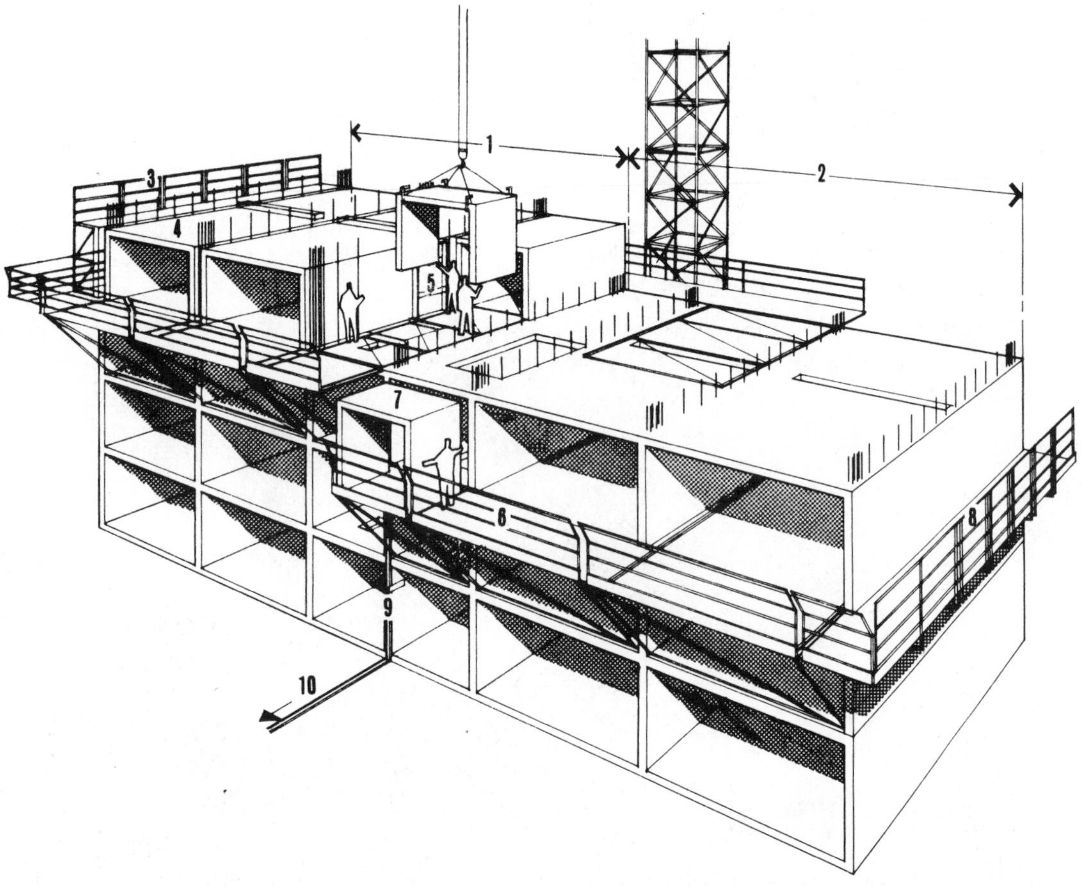

1	Phase 7 being shuttered	6	Platform for withdrawn shutters
2	Phase 8 shutter track being placed	7	Shutter being withdrawn
3	Gable shuttering in position	8	Platform for supporting gable shuttering
4	Tunnel shutter	9	Heating mains
5	Formwork to opening	10	To heating unit

Special Construction Techniques (On-Site)

Special construction techniques express the application of machine technology to the traditional craft procedures. The most successful of these include a total rationalization of all on-site jobs, procedures and materials movement. These techniques are characterized by on-site construction using special machinery and methods but are usually poured concrete. It would serve our purposes well to give a few examples of the many on-site systems in order to illustrate some of these techniques in more detail.

1. Lift Slab—Slabs are cast at grade, one on top of the other, with sleeves designed for slab connections to previously erected columns. After proper curing, they are lifted into position by lifting mechanisms attached to columns. Although this method eliminates difficult floor form work and casting is simplified by being at ground level, it does require sophisticated lifting mechanisms to insure all parts of slab being lifted at the same rate.
2. Tilt Up—This technique, like Lift Slab, is already widely used in the U. S. The walls are cast horizontally on the floor slabs and then tilted up into position. Need for wall formwork is eliminated, but lifting and jointing procedures limit its use to low structures.
3. Slip-Form—Short vertical steel forms are arranged in the intended configuration of the plan (this procedure is most often used for service cores only). Concrete is poured in the top of the forms while they rise vertically at a constant rate (conditioned by external temperature) which permits the concrete to reach initial set at the bottom of the form. Slabs are poured

1-10
TUNNELFORMS Inverted L-shaped metal formwork forms a tunnel when placed next to another on a slab. Walls and floors are cast in one single operation with methods of accelerating the concrete curing. *McKone Estates, Tallaght, Dublin, Ireland. 1970*

1-11
COMPONENTS The rationalization and application of modular coordination and assembly line techniques to traditional craft technology. *Engineering News-Record*

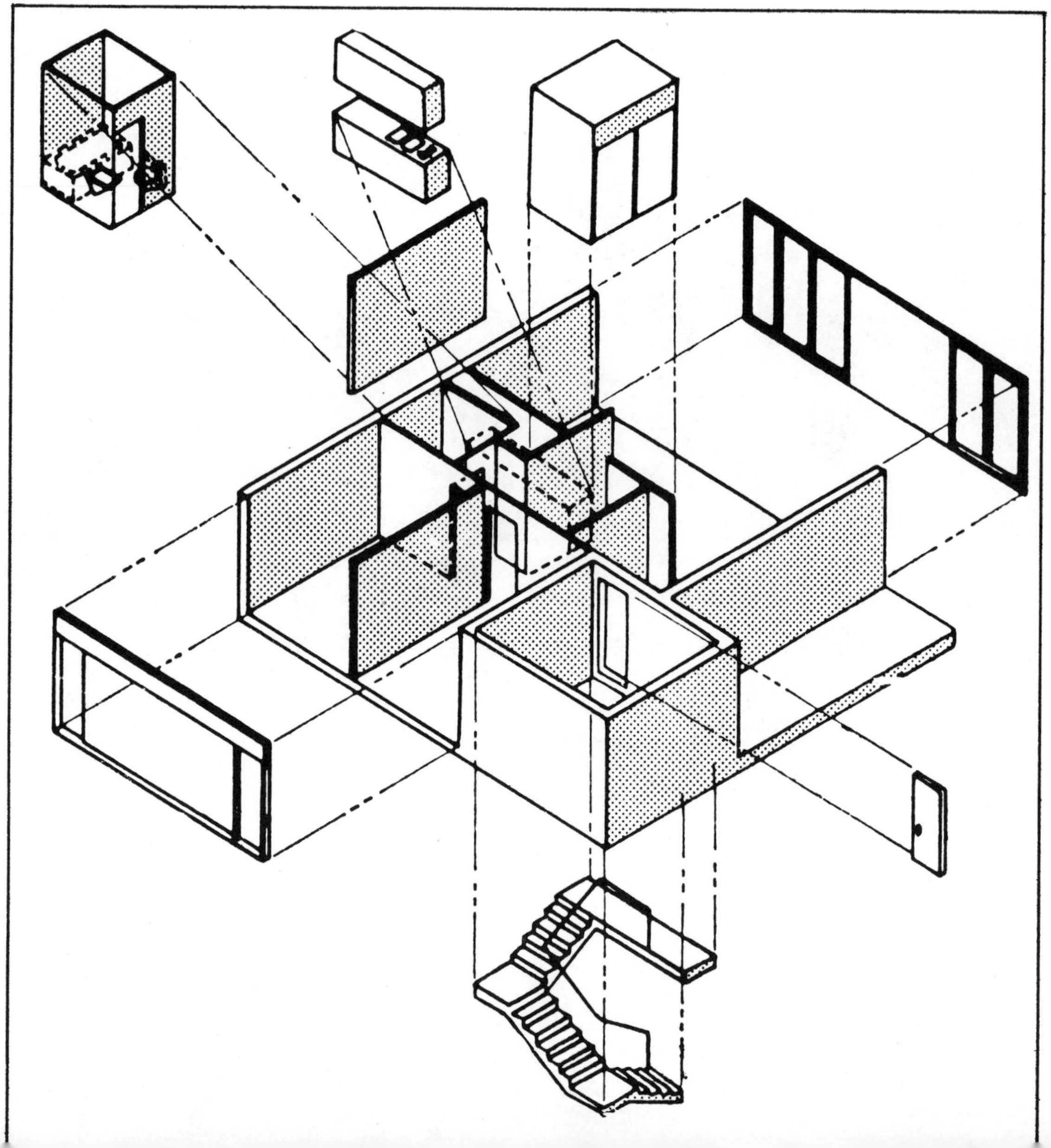

conventionally at each level. This method is primarily used for tall, self-contained, fairly windowless structures such as silos and elevator shafts.

4. Tunnelforms with Accelerated Curing—U-shaped forms are placed next to one another on a slab and floors are cast in one single operation, in one or two room widths. After casting, the formwork is removed along tracks and by crane to a new position for immediate re-use. Methods of accelerating the concrete curing can reduce the hardening process to only thirteen hours.
5. Push Up—Entire top floor is built first at ground level. It is then raised and the next and following stories are built beneath it until the building is complete. Cranes are eliminated, and virtually all work is done at ground level under cover. However, lifting equipment must be extremely heavy to lift an entire building in final stages. This method has not been extensively field-tested as yet.
6. All dry construction is the preselection of construction materials to eliminate the need for wet construction (i.e., concrete, plaster). This generally includes lightweight steel joists and studs or steel beams and columns, gypsum board, floor planks, etc. This method should be accompanied with highly organized materials handling but is beneficial in its relative independence of the weather. As with frames, extra consideration must be given to providing acoustic privacy between units.
7. Staggered Truss—This is a structural efficiency concept developed at M. I. T. The full height of the walls becomes a beam by actually making the walls story-height trusses. Further efficiencies are gained by staggering adjacent trusses on

1-12
UTILITY CORES These are inclusive of plumbing lines, fixtures, appliances, and heating units, and can be distributed as an entire unit or broken down into separate elements. *McKone Estates, Tallaght, Dublin, Ireland*

alternate floors, allowing floor planks to span only the short distance between the bottom cord of one truss and top cord of the next. This structural concept can be applied to a number of materials, but its primary benefits, outside of materials savings, are as a solution to special problems such as wide spans for air rights or in high-rise construction. Wide span spaces are, perhaps, more suitable for commercial than housing.

8. High-strength mortars and epoxy adhesives are used to bond brick and concrete block more strongly and consistently. This is particularly useful with brick because it allows the wall thickness of a two-story house to be decreased to one brick and makes possible prelaying of brick panels which can then be lifted into place.
9. Other special construction techniques include accelerated concrete curing methods, such as the use of heat and chemical accelerators.

Components

The industrialized production of materials and components is nothing more than the rationalization and application of modular coordination and assembly line techniques to traditional craft technology (windows, floors, panels, etc.). The exception to this is the grouping of units that had previously been produced and distributed separately (heart units or utility cores), the translation of a greater portion of production to the factory, and the use of new materials, especially molded plastics.

1-13
SERVICE CORES Containing component bathrooms with sanitary fittings now are commonplace in building construction with most plumbing manufacturers offering units. This fiberglass unit is particularly well designed, compact, and economical. *National Prefab House Industrial Company Ltd., Osaka, Japan, 1971*

Mechanical Units: The term "mechanical" as used here comprises plumbing, electrical components, and heating or air conditioning units. They may be assembled on the site in the traditional manner, or subassemblies may be fabricated off-site and connected together on-site. The degree of subassembly may vary from "plumbing trees" to completely preassembled combined bathrooms, kitchens, heating cores.

The prefabrication of mechanical units represents the most efficient form of prefabrication in that the units consist of a relatively small self-contained transportable section of work which is extremely skilled-labor intensive. These mechanical sections generally require several skilled trades, and money can easily be lost in this work on-site because of the somewhat delicate timing involved in the overlapping of trades. These timing and supervision requirements make them especially good candidates for factory assembly. Although many such units are already available, their wide usage has been slowed, usually because of uncertainties of union acceptance. A volume has, therefore, not been reached that can reflect their true competitive cost against traditional installations.

Performance Specifications (and Integrated Structural and Mechanical Systems)

Performance specifications can decree full compatibility of the subsystems. This is achieved by aggregating a market ample enough to guarantee the manufacturer a suitable exposure to warrant diversion from his stock products. Totally flexible, the design standards regulate the compatibility in such a way that one manufacturer is able to join with others and to offer competitive bids with higher

1-14
PREFABRICATION OF MECHANICAL UNITS The most efficient form of prefabrication in that units consist of relatively small self-contained transportable sections of work which are extremely skilled-labor intensive. *American-Standard*

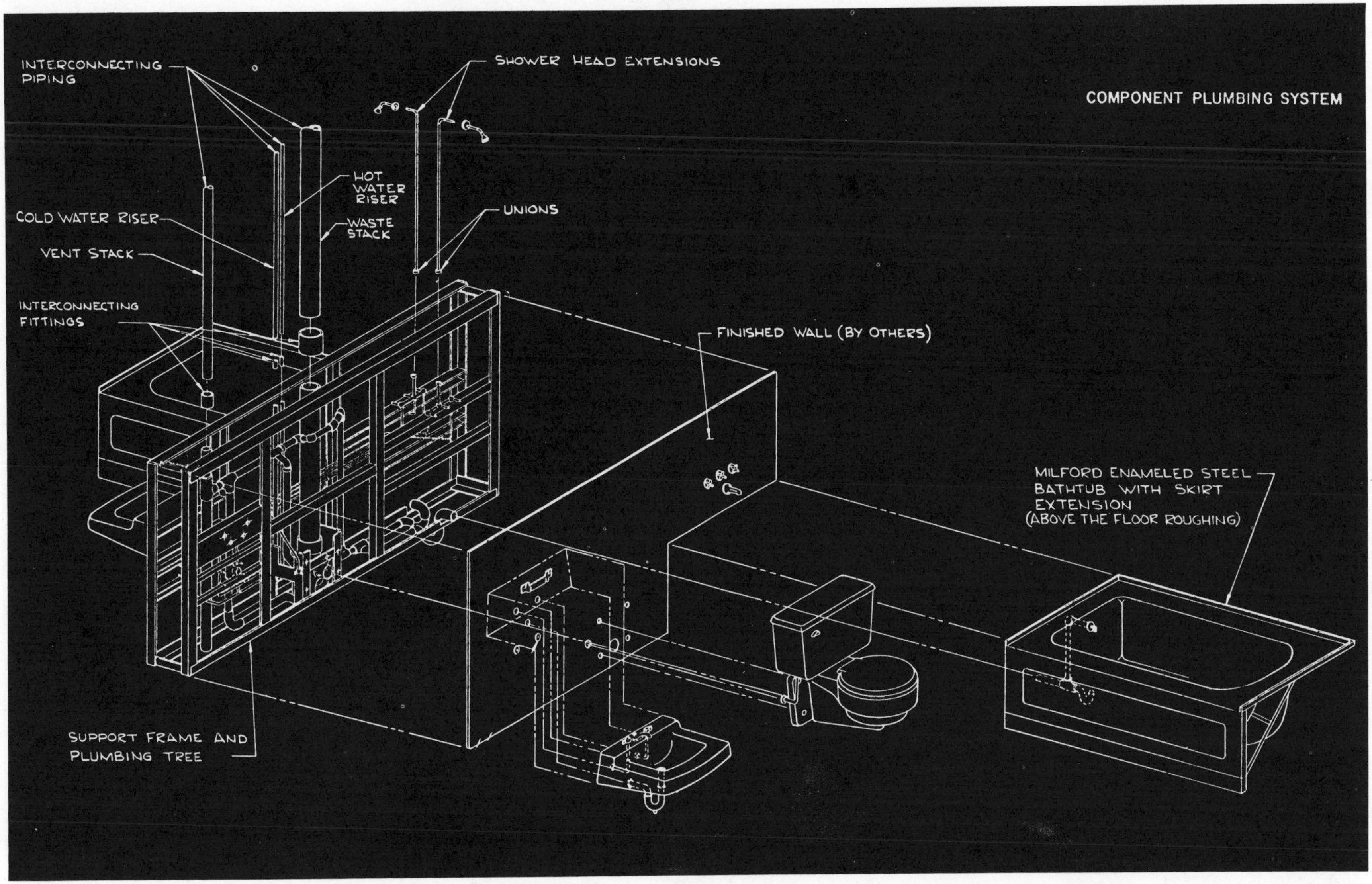

quality and integrative design. This is a uniquely "American" approach and is, perhaps, the only way that the lack of modular coordination among component manufacturers can be overcome in the U.S.

Instead of the usual set of complex, detailed, and protective specifications, performance specifications allow flexibility through mutual guarantees. Total systems can be developed in this way, using primarily variations on existing products. Performance specifications appear to be the best approach for developing integrated structural-mechanical systems which are carefully combined for maximum mutual efficiencies.

1-15 PRECAST STAIRS Easily stocked by a manufacturer, the design standards are such that one manufacturer's floor slab will integrate well with another's stair component. *Componoform, Inc., Cambridge, Massachusetts*
Photo by Steve Hanson

1-16
PERFORMANCE SPECIFICATIONS Probably the best approach for developing integrated structural-mechanical systems which are combined for maximum mutual efficiencies. *SCSD: The Project and The Schools, 1967. Educational Facilities Laboratories, Inc., New York, New York*
Photo by Rondal Partridge

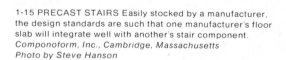

1-15

1-16

1-17

LOG CABIN. A late 19th century Maine logging camp illustrating the precut nature of the notching at building corners. This type of construction was first brought to these shores in 1638 by Swedish immigrants, and it is popular still today. *Downeast Magazine, Camden Maine, March 1973, Volume XIX, no. 6, page 50*

1-18

EXPOSITION UNIVERSELLE DE 1878. In 1878, these prefabricated industrial buildings were designed by M. Hardy, chief architect, as possible space enclosures. They were popular, systems-built, and easily constructed. *Ezra Ehrenkrantz, New York*

1-17

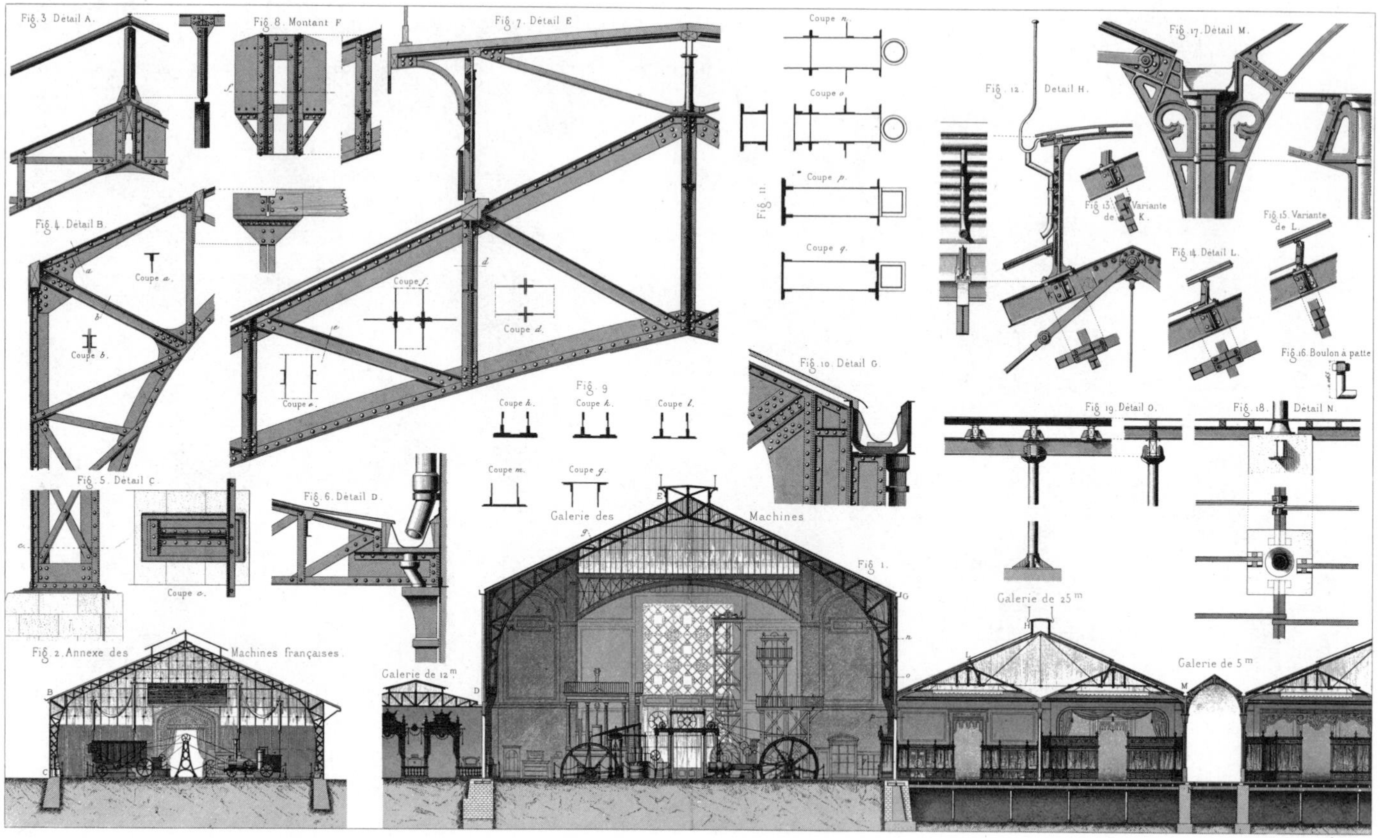

1-18

PREFABRICATION OF HOUSING

A Short History[1]

As early as 1624, the English brought a panelized house of wood to Cape Ann for use by a fishing fleet, and the house was subsequently disassembled, moved, and reassembled many times. Throughout the early years of American history, the log cabin provided the simple one-room dwelling unit constructed of precut logs, notched together at the corners. This type of precut construction was first brought to these shores around 1638 by Swedish immigrants. Later, many other nationalities, including Germans, Scots, Irish, Russian, contributed to the idiom making the log cabin as American as apple pie. New settlements provided a market for early prefabricators—the California Gold Rush of 1849 was a particularly lucrative market, as was the Union Army in the Civil War. In fact, railroad freight rates for portable houses date from around 1870.

Early in the 1900s, the "mail-order house" became popular on the frontiers. Sears, Roebuck Company claims it sold 110,000 houses in forty years. These were usually precut houses, but their production was important since it pioneered techniques for the production lines, standardization, and price packaging of the housing manufacturing industry.

In 1908, Thomas Edison proposed to pour an entire two- or three-story house of concrete, but this was found impractical and the idea was abandoned.

1. A Fact Book on Home Manufacturing, The Home Manufacturers Association, Washington, D. C.

In 1910, Walter Gropius, founder of the Bauhaus, proposed the "industrialization of housing," recommending "repetitive production of individual parts . . . made by machine to the same standard dimension and with provision for interchangeability of parts."

Finally, in the early part of this century, Roger Corbetta, a New York builder, conceived the idea of building a house using precast concrete panels, and this structural system resulted in several homes being built of precast hollow-core panels around 1917. However, the panels as a material were found unacceptable in the marketplace.

The construction industry actually began developing its present-day characteristics around 1930. With the establishment of F.H.A., it became possible to market homes in a mass volume in normal times of peace as buyers were able to buy homes on terms they could afford.

Early proponents of prefabrication concluded that the way to reduce construction costs was to deliver to the building site parts of the structure which fit and could be assembled without cutting or alteration, thus factory-fabricated housing systems.

The most exhaustive study of this subject was made by Albert Bemis and Associates of Boston, Massachusetts. The results of this study were presented in Volume III of the "Evolving House" published during the three-year period of 1933 to 1936. In this volume, entitled "Rational Design," Bemis suggests a typical module as the basis for design and develops a method for establishing standard assembly details and a simplified drafting technique in which all dimensions are referred to a modular grid. Regarding housing, Bemis states, "the reorganization that housing needs—and the redesign of structure here presented—is not a change of process. It does not suggest merely transferring to the shop what was previously done in the field. The parts of the house must be given the new forms and features required for versatility of design, economical mass production and ready-field erection." Albert Bemis died in 1936 and his heirs, wishing to see his work continued, organized the Modular Service Association to continue research in the field of modular standards. As a result of this effort, the American Standards Association initiated a project for the coordination of dimensions of building materials and equipment. This was a definite benchmark in the evolution toward prefabrication.

By 1940, there were about thirty firms manufacturing and selling prefabricated homes with approximately 10,000 units produced between 1935 and 1940. The most important single person at this time was Alfred Schwarz of Boston. Herr Schwarz was born in Vienna and educated as a mechanical engineer. In the thirties, he pioneered several new building insulation materials such as "Thermolit" for use in Belgium in the petroleum industry as protective cladding panels from Nazi firebombs. From 1941 to 1946, Herr Schwarz developed a housing system using a hollow-core precast panel filled with "Antagonit," a very practical and inexpensive insulation material. A factory was established and prototype units were built, but the moment for such sophisticated approaches had not yet arrived.

During World War II, home manufacturing met its severest test. The manufacturers were faced with the difficult task of providing emergency war housing that would meet three requirements: speed, flexibility, and the reduction of on-site labor. This provided the prefabrication industry with a great opportunity—quality was sacrificed

1-19
PRE-ENGINEERED METAL BUILDINGS These and various preassembled components for nonresidential use were developed to a great extent around 1950. This type of construction has made noticeable inroads into the total construction picture. *Butler Steel Products*

1-20
MOBILE HOMES account for 46% of *all* the single-family residential construction in the U.S. in 1971 and for 95% of the single-family residences constructed for under $15,000, according to *House & Home*, April 1971.
National Homes Corporation, Lafayette, Indiana

for quantity, and as a result of this effort, prefabrication gained a reputation as being "cheap" or "poor" construction.

Following the war, when there was a national demand for permanent private housing, the prefabrication industry had to work diligently to overcome the public's concept of a "prefab," which was based on what they had seen during the war.

Around 1950, other types of prefabrication started to make noticeable inroads—such as the pre-engineered metal buildings and various preassembled components for nonresidential use.

Within the past decade, the prefabrication industry has been growing at a rapid rate, both in the United States and abroad, but in very different areas. For instance, in the U.S. the sales of pre-engineered buildings have tripled; the mobile home industry is currently producing about 450,000 units annually; thousands of manufacturers are producing components for the industry; and, according to the Home Manufacturers Association, approximately 25% of all single-family dwelling units are prefabricated. It is important to remember that the greatest percentage of these are actually "trailer homes," reflecting the national schizophrenia—the American's desire for his own piece of real estate and his innate lust for mobility.

1-21
NATIONAL SCHIZOPHRENIA The American's desire for his own piece of real estate and his innate lust for mobility.
Explorer Motor Home Corporation

In Europe, where a very high percentage of housing is government-subsidized, the growth of prefabrication has been particularly rapid since about 1955. France is probably the furthest advanced in the extent of prefabrication, as 80% of all multifamily housing projects are industrialized.

Programs initiated in England by the London County Council after World War II give a typical picture of the European experience and may lend insights to our own. When special subsidies were offered for housing built by systems and, immediately, every man had a system, there appeared over 600 "housing systems" on the market overnight. During the following years, these hundreds of systems failed or were never developed until the present, when approximately 75% of systems housing in England is being built by only four giant systems contractors and the remaining 25% is built by another eight systems.

Now in the U.S., due to the tremendous promotional impact of Operation BREAKTHROUGH in 1969, a great number of the European systems groups have attempted to enter the American market—along with everyman's native system. It appears that most of these heavy European systems will be unsuccessful due to lack of promised market aggregation, the large capital investments required, and present saturation of precast concrete producers already in production. Operation BREAKTHROUGH has not followed through with either the financial backup or the market aggregation guarantees which were the heart of its early promise. These are an absolute prerequisite for the introduction of the heavy systems because of the high initial capital expenditure required by the production facilities.

The European experience has shown us several lessons. Even with substantial government subsidization of systems (which has been virtually unavailable in the U. S. despite Operation BREAKTHROUGH) the governments have often not been able to *maintain* the market required to financially support the heavy precast systems. *Many highly urbanized areas have been overbuilt, while needy markets in other areas are inaccessible to the "economic action radius" of highly capitalized existing plants.*

This is what has happened abroad, and one can already see signs of a similar pattern in the U. S. Many of the new American systems have already failed financially, a number of the European systems (Bison, Balency, Wates, Jesperson, etc.) have built plants here, but the market continuity is not as yet assured. The weeding-out period has begun, the "going public" stage is ending, and the real problem of solving the so-called housing problem lingers. Worse yet is the fact that all of the activity has been focused on the urban situation, when the crying need is in the dispersed, rural underdeveloped, and even suburban areas.

THE NATURE OF INDUSTRIALIZATION

The concept of prefabrication is originally an American pragmatic idea (not European), and although the word itself has long been considered a "dirty word" in the U. S. in reference to housing, virtually all buildings being erected today are utilizing some prefabricated components. However, to date, the reductions in initial building costs derived from prefabrication have not as yet met expectations, and the probability of further reduction in the near future is slight. Clearly, there is no revolution in the field of technology in sight, and what is more important, there is no need for a revolution. Technology has produced a tremendous volume of innovative materials, processes, and total systems; *the essential problem is that public acceptance and the present organization, attitudes, and structure of the building industry are nontechnical.*

The characteristic at the base of the "systems" approach to building is the translation from craft to machine production, which generally entails prefabrication in a centralized factory. Many advantages of factory and assembly line production tend to be the same for housing as for the other product industries.

Advantages

Quality Control: High-precision machinery imposes a schedule and a discipline. Higher tolerances, more accurate measurements, with less maintenance and material waste and greater consistency of finish, result.

Production Control: Programmed production, timed delivery and/or erection lessen the need for large stock inventoried left in the factory or on the site. Construction is faster on the site where a more efficient order of building sequences can be maintained.

Inventory Control: The tighter inventory controls possible in a factory setting, over small-piece building materials and components—piping, ducts, fans, windows, bricks, and tools—virtually eliminate the high rates of theft and vandalism on the site.

Labor Control: More extensive use of unskilled labor is possible in the factory because of

1-22
CRAFT TO MACHINE. Mass production generally entails quality control and inventory control, as well as labor control. (Note: In the most sophisticated plant, some acts must still be performed by hand.) *Camus (Great Britain) Ltd., Farnham, Surrey, England*

improved supervision (one foreman can adequately supervise more men) and because of the quality control inherent in the machinery.

"Wet" trades, in which labor is already in short supply, are eliminated as well as costly hours spent on skilled labor due to overlapping of plumbing, electrical, and finish carpentry trades.

A permanent site of employment guarantees full-time all-year jobs which provide a more successful continuous type of employment relationship for both employee and employer.

Climate Control: Factory conditions release certain areas from the "building season" limitations imposed by their climate conditions. A permanent labor force that can be employed when days lost to bad weather are minimized also reduces costs of training and initiation to jobs.

Problem Control: The detailed appraisal of constructional problems before work begins results in fewer delays in construction after commencement on the site.

1-23
OPERATION BREAKTHROUGH. Winner is this colonial-style townhouse row propagated by the building industry for the first time in response to the needs (and likes?) of the consumer and lured by the promise of large housing markets and the responsible administration of housing standards, codes, and regulations. *Stuart M. Heagy, V.P. Scholz Homes, Inc., 2001 N. Westwood, Toledo, Ohio 43607*

Deterrents

Machine production, however, requires large capital investments at preconstruction stages for tooling, factory installation, and transport (for central factory systems) and/or factory relocation (for on-site factory systems). At the present time, clear financing mechanisms for preconstruction investment do not exist for the building industry as they do for the automotive industry. In any case, these investments would have to be justified by large guaranteed markets in the form of large single projects (for on-site factories), coordinated smaller projects within economical transport distances (for heavy-component central factories—generally 100-mile action radius), or modular coordination for many small open-system projects on a nationwide scale (for light-component central factories). This guaranteed market must then be sustained over a long enough period of time to amortize these expenditures before profits can be calculated. Fragmented and localized subcontracting practices, administrative (code) and labor union policies, and professional attitudes still hinder the development of large markets on an area-wide or nationwide scale.

Management and Design for Industrialization

Industrialized building implies a new concept of rationalized architecture embodying a philosophy of planning and design, the use of advanced management techniques, organization of materials and manpower, forward thinking, and a progressive method of construction. It demands *a far more complex and challenging form of management* from the developer, the architect, and the contractor, than has been apparent before. It also involves *continuity of construction,* implying a steady flow of demand, standardization, and integration of the different stages of the whole production process.

In the operation of any system of industrialized building, all economical ways in which accuracy and control can be achieved must be employed, and specialized equipment and new techniques will evolve. Costly nonproductive works such as formwork, scaffolding, cleaning, and making good should and can be eliminated, and clear methods for prefinishing and packaging are included in a "total" system. Techniques of "dry" construction are used and overheads reduced by the use of computer programs for repetitive design, estimating, and general office operations.

Dimensional coordination or control is a prerequisite of successful industrialization. The situation where every designer is free to choose a series of dimensions which differ only marginally from those chosen by other designers is undesirable and costly. One inch one way or the other can make little difference in the usefulness of a living room, but it can mean the difference between the use of a standard versus a nonstandard element.

In order to achieve an optimum use of (1) men, (2) machines, (3), materials, (4) time, (5) allocated costs, it has been necessary to assess all the factors concerned with these five variables. A production control organization can forecast and analyze and ensure that projects run smoothly from financing and design through manufacture and erection.

Industrialized building is an overall conception of design, structure, and construction, integrated with production and the factor of demand, which must be mutually compatible. The architect is vitally involved as the use of an industrialized building system demands a rationalized approach not only to site organization, erection, and manufacturing process—but also within the architect's office. A balance must be achieved between design requirements and construction requirements. The design is developed with full collaboration between the architect and contractor responding to the system's guidelines.

Every design decision made must be weighed in the light of prefinishing, rapid fabrication, assembly, and organization. The architect should be aware of the function of the production control organization and how it can help him control the project. He should, for instance, realize the full implications of making design changes late in the predevelopment program.

Since construction of an industrialized building necessitates precise planning of the process of variation, the drawings needed by operatives can be simplified to show only the major factors, such as (1) where forms or panels go or cuts are made, (2) where the openings will be, and (3) how cladding and subsystems will be affixed.

A coordinated drawing system associated with standardization should be used in the architect's office to counteract the fragmentation of information resulting from requirements of the building team and the repetition of old methods of working drawings production. With a standardized drawing system incorporating reusable details and components and computerized specifications, extra time may be allocated to architectural and environmental research and design.

For the *contractor,* the *production control* is one of the most important sections of his organization. It is involved in preparing, either manually or on computers, program data for the use of every

1-24
REUSABLE DETAILS And Components devised at the birth of a system, these may be allocated (costwise) to architectural and environmental research and design. *Lambeth 1B Tower blocks by Wates, Ltd., London, England*

1-25
NEW ATTITUDES In design and approach are exemplified in the joint venture of the Rouse Company (Columbia, Maryland, new town developers) and Wates, Ltd. (contractors for the Lambreth project shown above) as Rouse-Wates, a team planning, designing, and marketing a complete residential community, utilizing an on-site factory.
Photo by Wates, Ltd.

member of the building team to the point where it is possible to have ***site operatives repeating the same general process each day, and everything can be coded or indicated by templates.*** No industrialized building system can be fully successful unless the follow-up to the erection of the superstructure in terms of components and subassemblies is effective with the general contractor's system. He is generally confined to the structural shell and must look to outside sources for his component supply.

The work and organization required of the architect's design team and the contractor's management and erection teams require a new philosophy allowing for full cooperation at an early design stage. The manufacturer, by the same token, has today to absorb this new attitude and supply factory-finished subassemblies which are larger and more complex than previously done on-site. In addition, the packaging and supplying of components is according to a more specific and rigorous schedule.

This new philosophy—a thinking awareness of the real meaning of industrialized building and the implications of its use—together with an understanding of what each member of the building team is trying to achieve, can result in more flexibility for negotiation, cost planning, and other techniques which are beneficial to client, architect, contractor, manufacturer, and user.

2 SYSTEMS EVALUATION

GOVERNMENT PROMOTION OF SYSTEMS

In-Cities Experimental Housing Project, 1968

"Potentially, reorganization and refinement of the construction process offers the only possibility of reducing the gap between housing need and supply. One of the ways that this can be accomplished is by providing the building industry with the physical means of producing higher quality accommodations at increased rates of production. In addition, cost benefit factors in the form of reduced construction and financing time, reduced material wastage, and the more extensive use of unskilled labor could extend the housing market downward toward the vast number of low–income families who are, at present, inadequately housed."

Such were the definitions and goals of the ECODESIGN research group for the HUD-sponsored "In-Cities Experimental Housing Research Project" which was completed in June 1968. The project was undertaken to find the means to produce the now famous "26 million dwelling units required over the next ten years [1968–1978]," as stated in the HUD Housing Act of 1968.

Challenged, as evident in the introduction, ECODESIGN and Professor A. G. H. Dietz of M.I.T. participated as the technology search subcontractors to the so-called Abt-DMJM joint venture (Contract #H-969). Simultaneous to the award of the HUD grant to our consortium, two identical Phase I grants were awarded, one to Building Systems Development, Inc. (and Kaiser Engineers), Contract #H-971, and another to Westinghouse Electric Corporation, Contract #H-970. A subsequent composite report (#69-8-R-5) was compiled in March 1969, Phase II, by Kaiser Engineers, consolidating the Phase I information provided by the three contractors, in a consistent but not very informative format. This four-volume composite report included user needs, constraints, technology, and city data. Because of naivété on the part of non-building trade editors as to what information was informative and comprehensible to the parties who would seek out this report, most of the technical analysis of the original "Technology Section" never reached print and, consequently,never reached the public. A vast amount of effort and money was spent for three separate research contracts to find methods to reduce the cost of housing and to realize the goals of the Housing Act of 1968 in the 67 or so Model Cities. It never happened, the information was scantily disseminated, the administration in Washington changed, and a new Secretary, George Romney, was appointed at the Department of Housing and Urban Development.

2-1
BOUCHARD The agrément system no. 1981 details of a single-family house system, illustrating some of the technical detail. *G. Blachere, Director, Agrément Cahiers du C.S.T.B., Paris, France*

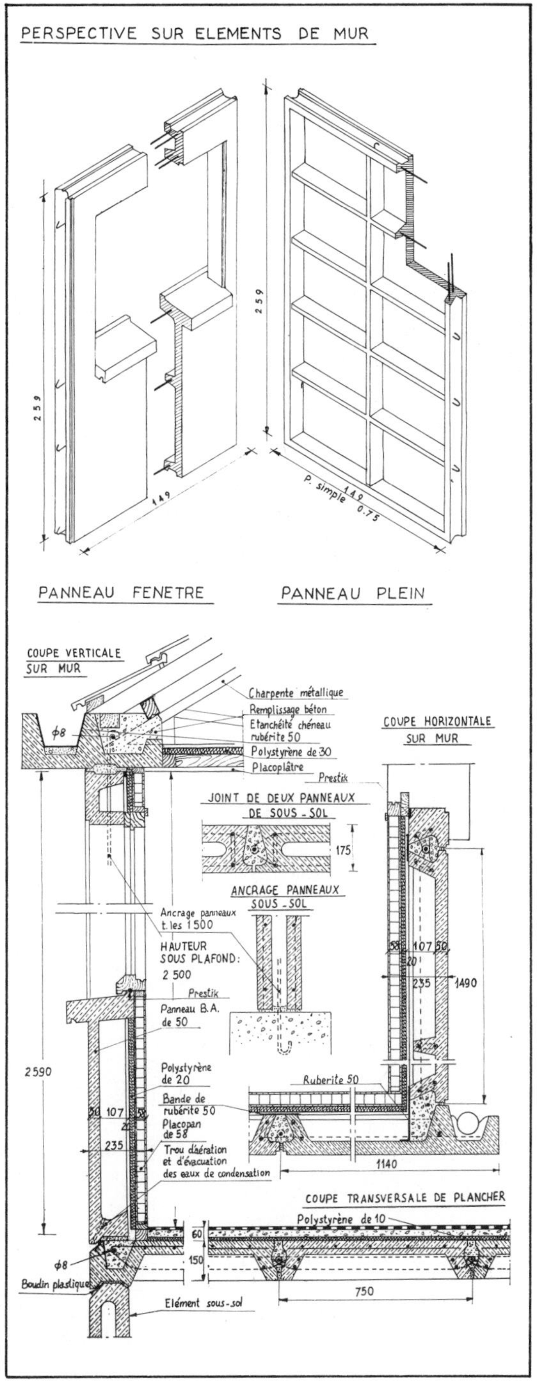

Operation BREAKTHROUGH, 1970

Secretary Romney, formerly a leader in the automobile industry, issued an RFP entitled "Operation BREAKTHROUGH" to replace, in effect, the In-Cities project (which produced only approximately 40 dwelling units in a small Florida project). BREAKTHROUGH oriented itself to mass-production *factory-type housing systems,* and Romney claimed that he would build houses as one builds cars. This was, perhaps, an oversimplified suggestion and the preconceived emphasis it gave to BREAKTHROUGH has proved self-defeating in that it stifled a clear view of the American housing situation.

With a fresh understanding of most of the industrialized building systems, ECODESIGN was in the position to select from our In-Cities sources an already operative system and accommodate more than the stated criteria which HUD had established for Operation BREAKTHROUGH. The In-Cities research had already indicated, for instance, that a large-scale housing project (about 1000–1500 units) was necessary to utilize a large panel system. In fact, the amortization of equipment and plant facilities for a system of this kind would require about 4000 units. The panel and heavy monolithic units also had severe transportation problems which limited them to areas which were not necessarily "BREAKTHROUGH City" territories.

After analysis of U. S. housing needs and consideration of HUD's stated criteria of flexibility, we felt the answer was an on-site system capable of building small numbers of units on various site sizes and in varying climates. The housing should be adaptable to different topography, land use, unit types, and physical access.

The French system, S.E.C.T.R.A., developed by M. Quentin of Chambery, an *in situ* tunnel-forming technique which integrated heating elements in the formwork and accelerated the concrete curing seemed to fill the bill. The system also lent itself to structuring a rigorous, self-imposed three-part proposal to HUD, suggesting one project in the East (Prototype A: Combination Cluster) with a mixed-income group, certain code confrontations such as PVC pipes, and a local contractor finishing the shell, on a flat site. The second project, 2000 miles away in the Midwest (Prototype B: Walk-Up Cluster) on a sloping site utilized self-help labor, and the third (Prototype C: High-Rise Spine) on an infill site on the West Coast utilized full community participation.

This project proposal could construct about 450 units with total experimental criteria, testing all kinds of architectural and social constraints, for less than any factory system. The assembled consortium was unique in being headed by an architecture firm and including a contractor (Gilbane Building Company, Providence, Rhode Island, one of the largest general contractors in the nation, according to *Engineering News-Record*) and a proven system, SECTRA, represented by John Laing Construction Ltd., London, England (the oldest and largest user of housing systems in the United Kingdom). Purposely not included were the product-oriented brand-names, as earlier experience had indicated that they were primarily lending their names for the ride and were not willing to research or contribute in any meaningful way until a sure thing became evident.

HUD, after an amazingly all-encompassing objective and well-intentioned review, selected 37 out of over 600 consortiums, including our entry SECTRA AMERICA. Then, from this 37, the final 22 winners were selected on what appeared to be a

2-2
SECTRA AMERICA The *only finalist consortium* with the prime contractors being architects (ECODESIGN) rather than the usual industrial giants, utilizing a system from France capable of low-, medium-, and high rise construction.
Project by John Laing & Son, Ltd.

2-3
OPERATION BREAKTHROUGH Loser. This 12′ wide steel frame module oddly combines modern construction technology with French mansard roof, Swiss-style simulated wall panels, Wild West gas lamp, and the ever-present chain link fencing bringing the whole esthetic into its proper perspective. *U.S. Steel Corporation*

political basis. Of the 15 semifinalists dropped, seven, or about half, were from New England. This eliminated all entries from politically unpopular New England, including SECTRA AMERICA.

It is hard to understand how all the New England entries could be eliminated as this was certainly the primary area for study and experimentation in the prefab housing field prior to HUD's interest, and the proposals represented some of the few American systems that were well along into their development.

The winners[1] were Levitt, Boise-Cascade, National Homes, and 20 other mighty magnates of industry, almost all product-oriented. None of them appeared to need the funds for research and prototype backing or even the guaranteed market or consequent publicity originally promised by HUD. Perhaps the government representatives felt the program needed to receive this financial security and publicity rather than giving the type of support required to develop new housing innovative potential. The point is, most of the BREAKTHROUGH winners would have undertaken an industrialized housing program with or without BREAKTHROUGH.

The essence of the housing problem on the national scale must be mass housing through interpretive experiments oriented towards industrialization of existing housing sources (contractors, developers, architects, engineers, etc.).

It seems obvious now that BREAKTHROUGH did have additional and justifiable (even if somewhat political) selection criteria which the winners group indicated and fulfilled—that is, to promote the housing market as big business in order to initiate the interest of the major industrialist (perhaps to fill a gap hopefully to be left by the cessation of the Indochina war). If this was the approach the government wished to take, then it should have been clearly indicated as one of the selection criteria. Rather than giving hope to the small independent entries, unnecessary expense in competing against the industrial giants might have been avoided. The bankruptcy of as many potentially useful systems and companies as BREAKTHROUGH created, is a sad record, when the need is so great.

Perhaps if the criterion for financial backing had been clearly stated with a *requirement* for management and program control in the hands of personnel with different previous experience in the building fields, then perhaps the experience and the financial backing could have worked together instead of competing. Selection criteria must be stated in complete sincerity and then carried through if they are to work.

THE ERA OF THE "GENERAL SYSTEMS" STUDY (1967–1972)

The response to inquiries during the previously discussed In-Cities Technology Study in 1967 reflected the first real blush of enthusiasm for sys-

1. Aluminum Company of America; Ball Brothers Research Corporation; Henry C. Beck Company; Boise Cascade Corporation; Christiana Western Structures; Descon/Concordia; Forest City Enterprises; General Electric Company; Hercules, Inc.; Home Building Corporation; Keene Corporation; Levitt Technology Corporation; Material Systems Corporation; Module Communities Inc.; National Homes Corporation; Pemtom, Inc.; Republic Steel Corporation; Rouse-Wates; Scholz Homes, Inc.; Shelley System; Stirling Homex Corporation; TRW Systems Group.

DEVELOPMENT OF SYSTEMS IN RELATION TO MODULES

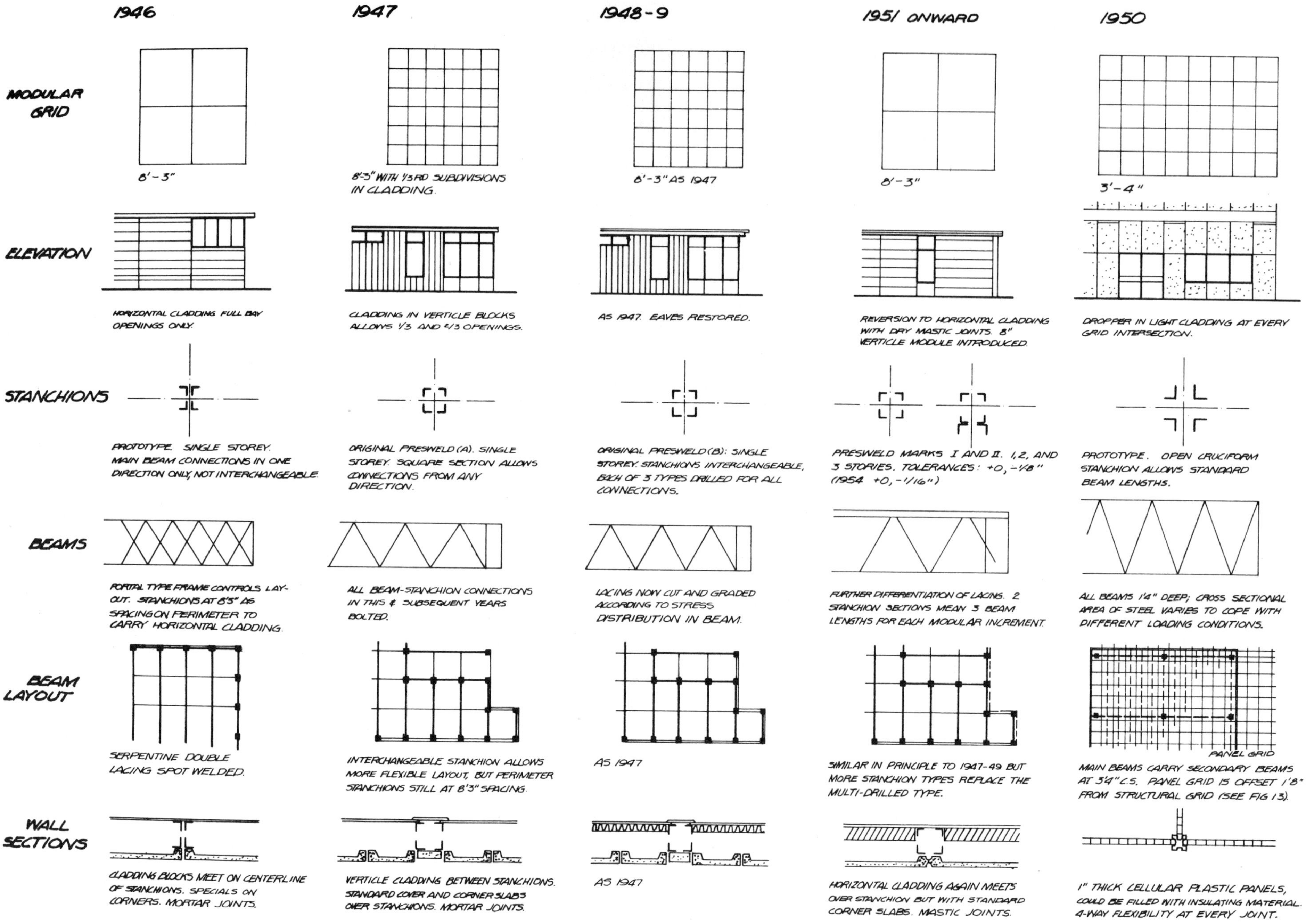
1946
1947
1948-9
1951 ONWARD
1950
MODULAR GRID
8'-3"
8'-3" WITH 1/3RD SUBDIVISIONS IN CLADDING.
8'-3" AS 1947
8'-3"
3'-4"
ELEVATION
HORIZONTAL CLADDING. FULL BAY OPENINGS ONLY.
CLADDING IN VERTICLE BLOCKS ALLOWS 1/3 AND 2/3 OPENINGS.
AS 1947. EAVES RESTORED.
REVERSION TO HORIZONTAL CLADDING WITH DRY MASTIC JOINTS. 8" VERTICLE MODULE INTRODUCED.
DROPPER IN LIGHT CLADDING AT EVERY GRID INTERSECTION.
STANCHIONS
PROTOTYPE. SINGLE STOREY. MAIN BEAM CONNECTIONS IN ONE DIRECTION ONLY, NOT INTERCHANGEABLE.
ORIGINAL PRESWELD (A). SINGLE STOREY. SQUARE SECTION ALLOWS CONNECTIONS FROM ANY DIRECTION.
ORIGINAL PRESWELD (B): SINGLE STOREY. STANCHIONS INTERCHANGEABLE, EACH OF 3 TYPES DRILLED FOR ALL CONNECTIONS.
PRESWELD MARKS I AND II. 1, 2, AND 3 STORIES. TOLERANCES: +0, -1/8" (1954 +0, -1/16")
PROTOTYPE. OPEN CRUCIFORM STANCHION ALLOWS STANDARD BEAM LENGTHS.
BEAMS
PORTAL TYPE FRAME CONTROLS LAY-OUT. STANCHIONS AT 8'3" AS SPACING ON PERIMETER TO CARRY HORIZONTAL CLADDING.
ALL BEAM-STANCHION CONNECTIONS IN THIS & SUBSEQUENT YEARS BOLTED.
LACING NOW CUT AND GRADED ACCORDING TO STRESS DISTRIBUTION IN BEAM.
FURTHER DIFFERENTIATION OF LACING. 2 STANCHION SECTIONS MEAN 3 BEAM LENGTHS FOR EACH MODULAR INCREMENT.
ALL BEAMS 1'4" DEEP; CROSS SECTIONAL AREA OF STEEL VARIES TO COPE WITH DIFFERENT LOADING CONDITIONS.
BEAM LAYOUT
SERPENTINE DOUBLE LACING SPOT WELDED.
INTERCHANGEABLE STANCHION ALLOWS MORE FLEXIBLE LAYOUT, BUT PERIMETER STANCHIONS STILL AT 8'3" SPACING.
AS 1947
SIMILAR IN PRINCIPLE TO 1947-49 BUT MORE STANCHION TYPES REPLACE THE MULTI-DRILLED TYPE.
PANEL GRID
MAIN BEAMS CARRY SECONDARY BEAMS AT 3'4" C.S. PANEL GRID IS OFFSET 1'8" FROM STRUCTURAL GRID (SEE FIG 13).
WALL SECTIONS
CLADDING BLOCKS MEET ON CENTERLINE OF STANCHIONS. SPECIALS ON CORNERS. MORTAR JOINTS.
VERTICLE CLADDING BETWEEN STANCHIONS. STANDARD COVER AND CORNER SLABS OVER STANCHIONS. MORTAR JOINTS.
AS 1947
HORIZONTAL CLADDING AGAIN MEETS OVER STANCHION BUT WITH STANDARD CORNER SLABS. MASTIC JOINTS.
1" THICK CELLULAR PLASTIC PANELS, COULD BE FILLED WITH INSULATING MATERIAL. 4-WAY FLEXIBILITY AT EVERY JOINT.

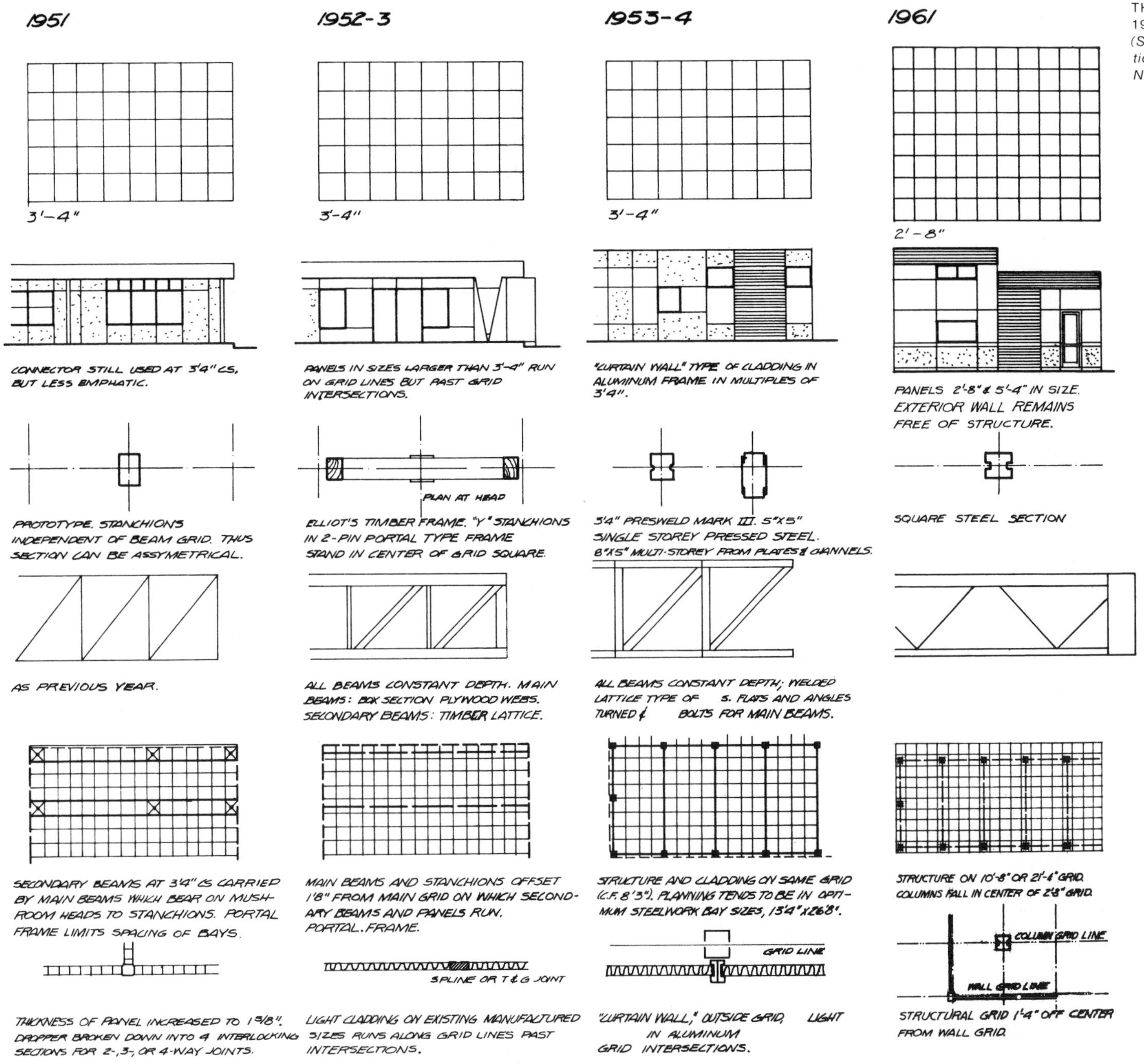
1951
1952-3
1953-4
1961
3'-4"
3'-4"
3'-4"
2'-8"
CONNECTOR STILL USED AT 3'4" CS, BUT LESS EMPHATIC.
PANELS IN SIZES LARGER THAN 3'-4" RUN ON GRID LINES BUT PAST GRID INTERSECTIONS.
"CURTAIN WALL" TYPE OF CLADDING IN ALUMINUM FRAME IN MULTIPLES OF 3'4".
PANELS 2'-8" & 5'-4" IN SIZE. EXTERIOR WALL REMAINS FREE OF STRUCTURE.
PLAN AT HEAD
PROTOTYPE. STANCHIONS INDEPENDENT OF BEAM GRID. THUS SECTION CAN BE ASSYMETRICAL.
ELLIOT'S TIMBER FRAME. "Y" STANCHIONS IN 2-PIN PORTAL TYPE FRAME STAND IN CENTER OF GRID SQUARE.
3'4" PRESWELD MARK III. 5"X5" SINGLE STOREY PRESSED STEEL. 8"X5" MULTI-STOREY FROM PLATES & CHANNELS.
SQUARE STEEL SECTION
AS PREVIOUS YEAR.
ALL BEAMS CONSTANT DEPTH. MAIN BEAMS: BOX SECTION PLYWOOD WEBS. SECONDARY BEAMS: TIMBER LATTICE.
ALL BEAMS CONSTANT DEPTH; WELDED LATTICE TYPE OF S. FLATS AND ANGLES TURNED ₵ BOLTS FOR MAIN BEAMS.
SECONDARY BEAMS AT 3'4" CS CARRIED BY MAIN BEAMS WHICH BEAR ON MUSHROOM HEADS TO STANCHIONS. PORTAL FRAME LIMITS SPACING OF BAYS.
MAIN BEAMS AND STANCHIONS OFFSET 1'8" FROM MAIN GRID ON WHICH SECONDARY BEAMS AND PANELS RUN. PORTAL. FRAME.
STRUCTURE AND CLADDING ON SAME GRID (C.F. 8'3"). PLANNING TENDS TO BE IN OPTIMUM STEELWORK BAY SIZES, 13'4"X26'8".
STRUCTURE ON 10'-8" OR 21'-4" GRID. COLUMNS FALL IN CENTER OF 2'8" GRID.
SPLINE OR T & G JOINT
GRID LINE
COLUMN GRID LINE
WALL GRID LINE
THICKNESS OF PANEL INCREASED TO 1 3/8". DROPPER BROKEN DOWN INTO 4 INTERLOCKING SECTIONS FOR 2-, 3-, OR 4-WAY JOINTS.
LIGHT CLADDING ON EXISTING MANUFACTURED SIZES RUNS ALONG GRID LINES PAST INTERSECTIONS.
"CURTAIN WALL," OUTSIDE GRID, LIGHT IN ALUMINUM GRID INTERSECTIONS.
STRUCTURAL GRID 1'4" OFF CENTER FROM WALL GRID.

2-4
THE ERA OF THE "GENERAL SYSTEMS" STUDY (USA 1967-1972). *School Construction Systems Development (SCSD) Report #2, British Prefabricated School Construction, School Planning Lab., Stanford University, E.F.L., New York*

tems in the U.S. In spite of a fairly complex and complete questionnaire which was sent to systems people by the ECODESIGN Technology Search staff, it was painstakingly answered by literally hundreds of manufacturers in the most minute detail. Yet, after a three-year period, the building industry was still halted in the systems search and research stage. Like anything else which requires enormous capital expenditure, the "study" seems cheap and safe, and groups that should be developing housing projects by now have produced shelves of studies and proposals instead. A name mentioned in magazines as a manufacturer of building systems can easily expect over 100 multipaged questionnaire forms and requests for literature, instead of requests for bids on projects.

The questions come from a variety of interested parties—often, engineering or architectural firms doing a "general systems" study for a public company interested in investing potentials or builders or developers interested in finding a system suitable to their purposes. It is doubtful that many of these questionnaires are returned now. The forms tend to be too time-consuming an effort to end up in a file instead of a job possibility.

There are two fallacies in this "general systems" study approach. First, it is an unnecessary and wasteful duplication of effort to be constantly starting from scratch, but this is normally because the "researcher" has had little or no previous experience with systems. The client is often paying for an education and a gathering of information which the consultant should already have in hand. The second fallacy is the "general systems" study for a specific problem. If the researcher was already aware of what alternatives existed, then ideally the client developer-builder-investor would come to him for services as a "Systems Broker." If the broker was already knowledgeable in all aspects of systems, the client could then describe his intended utilization and economic and physical needs to form a ***personal set of selection criteria*** and have a limited number of recommendations made on that basis which were realistic, well-examined, alternatives for his particular situation. A constant restudy of irrelevant alternatives would not then be necessary.

The questionnaire approach was feasible before 1967 and before Secretary Romney's Operation BREAKTHROUGH. During Secretary Weaver's In-Cities study, the systems groups were still the wall flowers of the construction industry waiting for someone to take an interest. ECODESIGN, as part of its technology search task for a prime contractor, conducted an extensive general study which covered in detail over 360 systems here and abroad. Each of the systems responded in full and exposed their technical assets all with the hopes that their systems would be "discovered." This youthful enthusiasm has now waned, and groups have become suspicious and secretive with their "privileged" information.

The following section presents the methodology forms and surveys utilized for information and data retrieval, cataloguing, categorizing, analyzing, selecting, evaluating, and final selection of systems alternatives for specific utilization requirements (Technological, Social, and Political-Economical experiments devised as part of HUD In-Cities program). They present universal evaluation and selection criteria from which it is possible to formulate a tailored set of selection criteria.

2-5 — 2-8
"Eggcrate environment" is the usual reaction to these examples of European housing systems projects. They include design variations of roof treatments (2-5), facade panels (2-6), massing (2-7), and textured end-walls (2-8).

2-5
NORTHERN IRELAND Sectra System. *Rathcoole Project, John Laing & Son, Ltd.*

2-6
FRANCE Tracoba System. *Project Meaux, Paris, France Tracoba*

2-7
CZECHOSLOVAKIA TO8B System. *Invalidovna Project, Stavebni Zavody Praha, Prague, Czechoslovakia*

2-8
GREAT BRITAIN Bison Wall Frame System. *South Kilburn Redevelopment, Brent, England Concrete, Ltd.*

2-5

2-6

2-7

2-8

DEVELOPING AN EVALUATION METHOD

The ECODESIGN Technology Search flow chart illustrated in Figure 9 shows the project effort and coordination with city/site field survey teams in the first HUD-sponsored general systems In-Cities study.

The Information Retrieval Sequence was initiated by contacting a recommended list of European and domestic sources by letter, telephone, library search, and interviews, both in foreign countries and at the Cambridge, Massachusetts project offices. An initial screening based on availability and feasibility provided the first-level selection narrowing over 360 systems and components contacted to approximately 200. The extensive *Systems Classification and Physical Performance questionnaire* which follows was used for screening, cataloguing, utilization purposes, and also for coordination with specific environmental inputs provided by city/site field survey teams which covered all geographic regions of the U. S. The data and information collected were to be reclassified for data storage and retrieval into a computer evaluation program known as "IBIDS"—Information Bank for Industrialized Design Systems.

2-9
ECODESIGN TECHNOLOGY TEAM FLOW CHART
ECODESIGN

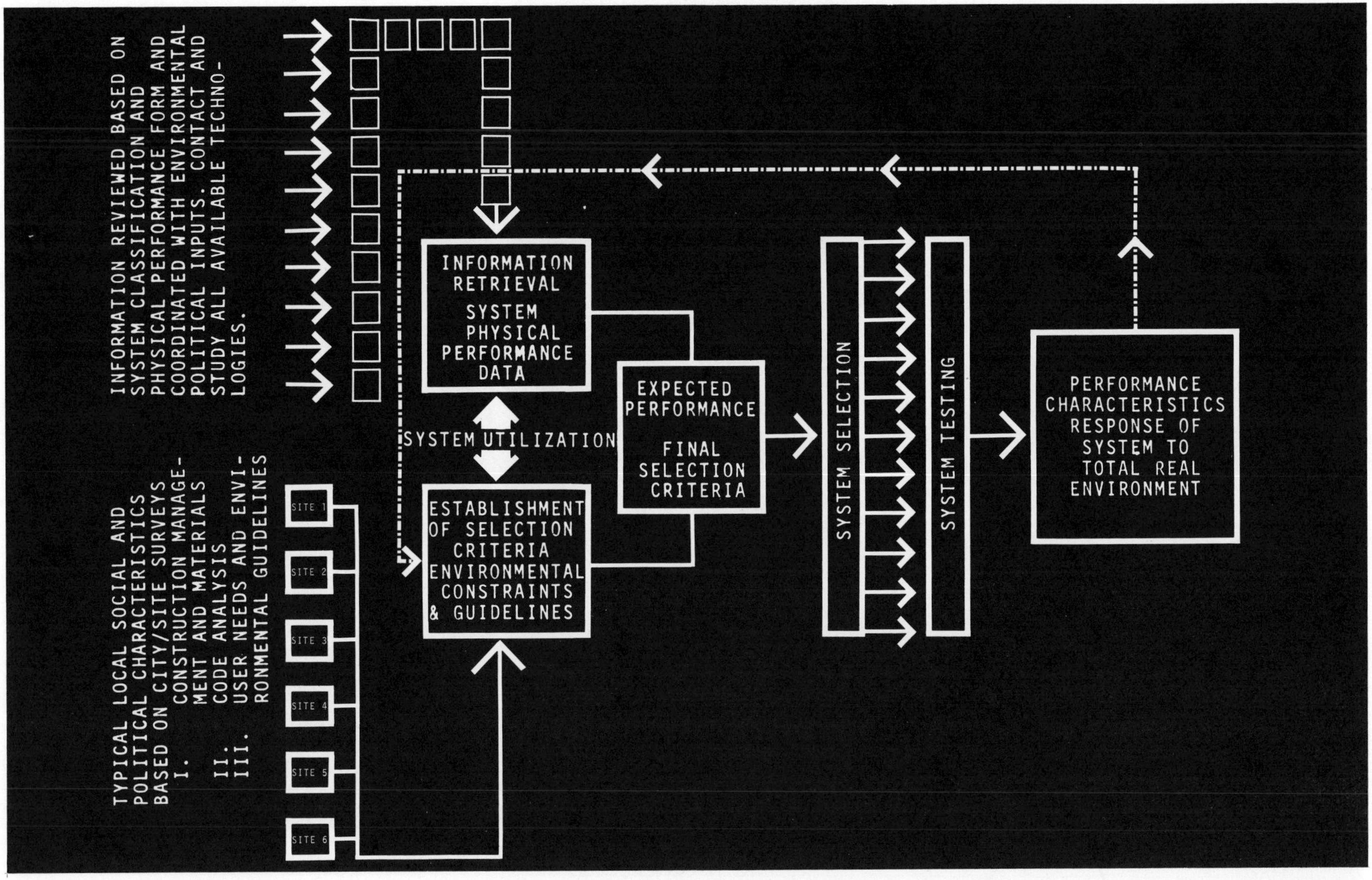
INFORMATION REVIEWED BASED ON SYSTEM CLASSIFICATION AND PHYSICAL PERFORMANCE FORM AND COORDINATED WITH ENVIRONMENTAL POLITICAL INPUTS. CONTACT AND STUDY ALL AVAILABLE TECHNO- LOGIES.
TYPICAL LOCAL SOCIAL AND POLITICAL CHARACTERISTICS BASED ON CITY/SITE SURVEYS
I. CONSTRUCTION MANAGE- MENT AND MATERIALS
II. CODE ANALYSIS
III. USER NEEDS AND ENVI- RONMENTAL GUIDELINES
SITE 1
SITE 2
SITE 3
SITE 4
SITE 5
SITE 6
INFORMATION RETRIEVAL SYSTEM PHYSICAL PERFORMANCE DATA
SYSTEM UTILIZATION
ESTABLISHMENT OF SELECTION CRITERIA ENVIRONMENTAL CONSTRAINTS & GUIDELINES
EXPECTED PERFORMANCE
FINAL SELECTION CRITERIA
SYSTEM SELECTION
SYSTEM TESTING
PERFORMANCE CHARACTERISTICS RESPONSE OF SYSTEM TO TOTAL REAL ENVIRONMENT

```
                                                         08/20/68  2:40    PAGE NO.  1

....*....1....*....2....*....3....*....4....*....5....*....6....*....7....*....8

                                            ***********************************
                                            *                                 *
                                            *  SELECTION CRITERIA  - ECODESIGN *
                                            *                                 *
                                            ***********************************
MANUFACTURER                               SYSTEM
  JOHN LAING CONST.LTD                       SECTRA
  MILL HILL
  LONDON, NW 7.ENGLAND
--------------------------------------------------------------------------------


1000     PERFORMANCE AND USER NEEDS

 1100 CLASSIFICATION                                   1300 PLAN FLEXIBILITY

  1. MONOLITHIC UNIT                                        *** 1 BDRM    *** EFF
  2. TOTAL SYSTEM  POURED IN PLACE SLABS + CROSSWALLS      *** 2 BDRM    *** BALCON
  3.  STRUCTURAL SYSTEM                                     *** 3 BDRM        YARD
      FRAME                                                 *** 4 BDRM        DUPLEX
      SHELL
      BEARING                                          1400 UTILIZATION
      OTHER
  4. NON-STRUCTURAL COMPONENT                               *** SAFETY
      EXT. PANEL                                            *** PRIVACY
      INT. PANEL                                                CHILDREN#S PLAY
      SERVICE CORE                                              ADAPT TO FAM. SIZE
      OTHER                                                     GROWTH
  5. SPECIAL CONST. TECHNIQUES  ***          (FACADE ONLY) *** DESIGN FLEXIBILITY
      SLIP FORM
      LIFT SLAB                                        1500 ACCESS TYPE
      TUNNEL FORM               ***
      ACCEL CURE                ***                             STAIR
  6. OTHER                                                  *** ELEVATOR
                                                                INDIV. UNIT ACCESS
 1200 DWELLING UNIT TYPE                                    *** SINGLE LOADED CORR.
                                                            *** DOUBLE LOADED CORR.
  SINGLE FAM. ATTACHED                                          TOWER ACCESS CORE
  SINGLE FAM. DETACHED
  WALK UP
  HIGH RISE                     ***
```

```
                                                         08/20/68  2:40    PAGE NO.  2

....*....1....*....2....*....3....*....4....*....5....*....6....*....7....*....8

2000     ADAPTABILITY

  COMPATABILITY WITH OTHER SYSTEMS AND COMPONENTS-MODULE      OPENED  *** CLOSED
  SITE AND SOIL- CONVENTIONAL FOUND.   WEIGHTS- DOES NOT APPLY
  CROSS VENTILATION- ALL UNITS         SUNLIGHT ORIENTATION- POSS. THROUGH UNITS
  CLIMATE- ANY TYPE OF FACADE PANEL MAY BE USED TO RESPOND TO CLIM. CONSTRAINTS
  SPAN SIZE- 18#-5## BETWEEN CROSSWALLS  OPENINGS- TO SPECIFICATION

3000     MATERIALS

  BASIC MATERIALS- POURED IN PLACE CONCRETE
  FIRE RESISTANT- TO REQUIREMENTS
  CODE CONFORMITY- SYSTEM CAN BE DESIGNED TO CONFORM
  FINISHING REQUIRED- SURFACES READY FOR PAINTING
  MAINTENANCE AND REPLACEMENT OF WORK PARTS- WORN PARTS CAN NOT BE REPLACED
  RESISTS- *** MOIST. *** STAIN + ODOR *** VERMIN TSP SOUND TRANS. TSP HEAT LOSS

4000     EFFICIENCY

  PRODUCTION-   ON SITE ***      IN FACTORY           WET ***      DRY
              SPECIAL EQUIP.- TOWER CRANE SCAFFOLD
  LABOR- 8 MEN + CRANE OP.                SEASONAL LIMITS- NONE
  TRANSPORT- NONE REQUIRED                COST-
  TIME- 2 APT#S/DAY                       SYSTEMS MGMT.- COMPUTER CRITICAL PATH

5000     APPLICATION

  AVAILABILITY-  CONTACT GUARANTEED      UNCERTAIN *** NOT AVAIL. BY OCT. 68
  LEGAL PROBLEMS-  FINANCIAL             FOREIGN ***
                   DOMESTIC              UNION PROBLEMS
  APPLICATION TO EXPERIMENT- INDUSTRIALIZED TRADITIONAL CONSTRUCTION METHOD
          ADAPTABLE TO MASS HOUSING REQUIRING QUICK COMPLETION
  MARKETABILITY OUTSIDE EXPERIMENT- YES-PLAN FLEXIBILITY IMPEDED BY CROSSWALLS
6000     GENERAL COMMENTS AND EVALUATION

EXAMINER- RNK                             DATE-
```

2-10
FINAL SELECTION CRITERIA These were compiled, and with the help of Design Systems, Inc., were expanded and resystematized into a computer program known as IBIDS (Information Bank for Industrialized Design Systems). *ECODESIGN*

A primary task involved cross-checking for compatibility the System Classification and Physical Performance data of each system with city/site field survey information. The feedback from these environmental and user needs and political and economic guidelines produced what became the ***Final Selection Criteria*** used for second- and final-level eliminations.

A series of questionnaires were devised within one city/site survey for use by field survey teams, in order to secure the local environmental input needed:

City/Site Survey—
- Part I: Construction Labor and Materials
- Part II: Building Code Analysis
- Part III: User Needs and Environmental Guidelines

The questions in these three forms were intended to determine exceptional characteristics of typical areas, cities, and sites being surveyed concerning materials availability, transport, unions, codes, user acceptance, and existing dwelling types. This information would effect the type and proportion of system types selected.

The second-level selection list presented systems evaluated by a ***System Utilization Chart*** to integrate technological, social, political, and management requirements.

The 90 techniques that rated best according to the selection criteria (immediate availability, technical performance, adaptability, utilization, cost, and application of city/sites) were reexamined in more detail. This was accomplished by an impartial review board made up of architects, engineers, planners, developers, economists, sociologists, etc. Approximately 40 industrialized techniques were selected and recommended to HUD as suitable for possible experiment.

COMMUNITY GROUP
COMMON SPACES
ADD-ON ROOMS
FINISHED UNITS
DAY CARE/MEETING SPACES
FEEDBACK TECHNIQUES
USER NEEDS SURVEY
MIXED LAND USE/COMMERCIAL
SOCIAL MIX
MIXED ACCOMMODATION TYPES
ALL-OUT COMMUNITY CENTER
CHANGING SPACE CONCEPT
MAINTENANCE SERVICE
VALUE MAINTENANCE
HOME OWNERSHIP TRADING
URBAN SUBURBAN EXCHANGE
NEIGHBORHOOD DISPERSAL
NEIGHBORHOOD RELOCATION
NEIGHBORHOOD TEMP RELOCATE
NEIGHBORHOOD REHAB
DO-IT-YOURSELF REHAB
DO-IT-YOURSELF HOME
HOME OWNERSHIP
COMMITTEE GROUP PROGRAM

GOVERNMENT MODULAR COORD.
CODE CONFRONTATION
UNION CONFRONTATION
UNION CONFRONT.(SOFT SELL)
SUBCONTRACTORS/SIM. COMP.
CONTR. CONSORT./BOND AGENT
RISK CAPITAL
HOME OWNER VALUE/MAINT.
NEGATIVE IMPROVEMENT TAX
REPLACEMENT COST SUBSIDY
SINGLE MGT/CITY COORD AGT
RECLMD HOME COMMITTEE MGT
CITY COMMITTEE MANAGEMENT
LAND RECLAMATION
AIR RIGHTS/OTHER ZONING
MULTIPURPOSE RIGHT OF WAY
INTERIM LAND USE
NEIGHBORHOOD DISPERSAL
URBAN SUBURBAN EXCHANGE
SINGLE FAM DETACHED FIN
SINGLE FAM DETACHED ACCOM
AUTOMATED ASSEMBLY
COMPUTER DRAWINGS & SPECS
MATERIALS HANDLING
UNION AGREEMENTS
NON-REGIONAL MARKET
WARRANTY SERVICES
ACT AS BUILDING MANAGER
ACT AS SELLER/RENTER
ACT AS MORTGAGEE
ACT AS EXPEDITER/PACKAGER
TRAIN LOCAL LABOR
ERECTION SPEED
TESTING

MONOLITHIC UNIT/BOX
TOTAL SYSTEM/PANEL
STRUCTURAL SYSTEM
ON-SITE TECHNIQUE
COMPONENTS & PERFORM. SPEC.
SPECIAL CONSTRUCTION TECH.
MATERIALS - CONCRETE
LIGHTWEIGHT CONCRETE
METAL: STEEL___ ALUM___
FIBERGLASS - PLASTIC
SPEC. MASONRY/MORTAR
CEMENT
WOOD: CONVEN__STRESSED__
MIXED COMPONENTS
MASS-PRODUCED COMPONENTS
PRODUCED COMPONENTS
CONVER. OF STRUCTURES
INSTANT REHAB
DWELLING UNIT MIXES
HIGHRISE, WALKUP, SINGLE
WALKUP, HIGHRISE
WALKUP, SINGLE STORY
SINGLE FAMILY DETACHED
MIXED COMMERCIAL/HOUSING
MOBILE HOMES
SINGLE
LOWRISE STACK
HIGHRISE STACK
LOW SOUND TRANSMISSION
UNIT ADDITION
ADD-ON ROOMS
INTERIOR WALL FLEXIBILITY
GROUP FORM SITE PLAN
LAND RECLAMATION
AIR RIGHTS TECHNOLOGY
MODULAR COORD AGENCY

2-11
SYSTEM UTILIZATION CHART This illustrates the integration of technological, social, political, and management requirements. *ECODESIGN*

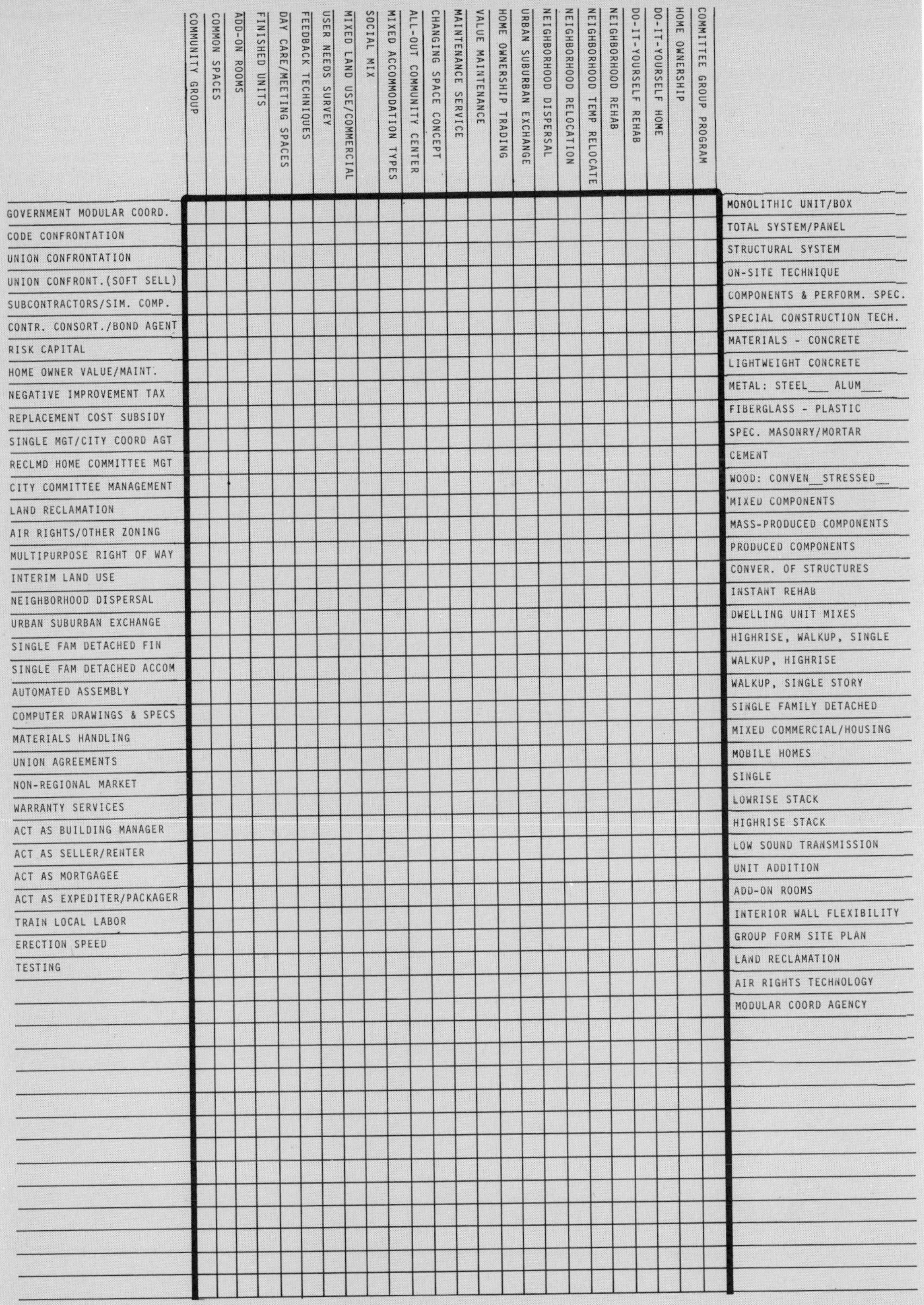

CLASSIFICATION AND PHYSICAL PERFORMANCE QUESTIONNAIRE

Name of Organization ____________________

Location ____________________

Firm _____ Franchise _____ Builder _____ Consortium _____ Developer _____

I. *Building Type Application*

Housing	Yes	No
Single Family	_____	_____
Multifamily:	_____	_____
Low-Rise (2–3 stories)	_____	_____
Medium-Rise (4–9 stories)	_____	_____
High-Rise (10 stories–up)	_____	_____
Dorms, Hotels	_____	_____
Hospitals	_____	_____
Offices	_____	_____
Commercial	_____	_____
Education/Schools	_____	_____
Garages	_____	_____
Other (Specify)	_____	_____

II. *Classification of System* (Check One)

_____ Monolithic Unit/Box ——— Heavy
_____ Light
_____ Total System ——— Structural Frame
_____ Structural Panels
_____ Infill Panels and Partitions
_____ Service Core and HVC
_____ Structural System ——— Frame
_____ Bearing
_____ Nonstructural Components ——— Infill Exterior
_____ Partitions Interior
_____ Service Core HVC
_____ Special On-Site Techniques ——— Tunnel Form
_____ Slip Form
_____ Jack Block
_____ Lift Slab
_____ Accelerated Cure
_____ Special Mortar
_____ Traditional Techniques

III. *Module and Compatibility*

Size of Module ____________________
Weight of Largest Component ____________________
Dimensions of Largest Component ____________________

IV. *Materials and Component Identification and Strength Parameters*

A. Foundation Size and Types Used:
Piles ____________________
Caissons ____________________
Footings ____________________
Foundation Walls ____________________
Slab on Grade ____________________

B. Structural Type:
Box ____________________
Frame ____________________
Bearing Wall Panel ____________________
Shell ____________________
Other ____________________
Span Sizes ____________________
Joist Size ____________________

C. Basic Material
PSI ____________________
Fire Rating ____________________
Decibels ____________________

D. Designed for Earthquake Zone 0_____ 1_____ 2_____ 3_____

E. Wind Bracing By ____________________

F. Maximum Grade ____________________

G. Joints: Wet _____ Dry ____________________

H. Components: ____________________

	Material	Finish	BTU/hr ft^2–f^0	Fire Rating
Load-Bearing Structure	______	______	______	______
Party Walls	______	______	______	______
Exterior Closure Walls	______	______	______	______
Floor	______	______	______	______
Ceiling	______	______	______	______
Interior Partition	______	______	______	______
Stairs	______	______	______	______
Roofing	______	______	______	______

I. Decibel Rating for Acoustic Separation Between Units:

Party Walls______________________________
Floors______________________________
Interior Partitions______________________________
Stairwells______________________________

J. Services Provided:

Integrated__________% Traditional__________%
Heating: Type__________ Central__________ Individual__________
Air Conditioning: Type______________________________
Plumbing: Rough__________ Finished______________________________
Electrical Material__________ Fixtures Supplied______________________________

K. Testing Methods and Standards Agencies Used:

____
____
____
____

V. *Production*

A. Years in Production Prototypes______________________________

B. Number of Completed Units:

Europe______________________________
U.S.______________________________
Other______________________________

C. Project Feedback:

Usage Record______________________________
Maintenance______________________________

D. Production Location:

Off-Site Factory______________________________
On-Site Factory______________________________
On-Site Special Construction______________________________
On-Site Traditional______________________________
Other______________________________

E. Minimum Total Order ____________________ Units/Sites

F. Maximum High Volume Production ____________________ Units/Year

G. Production Time/100 Units______________________________

H. Seasonal Limitations______________________________

I. Labor Skills in Plant______________________________

VI. *Erection Labor and Equipment*

A. Erection Time/100 Units (Excluding Foundations)____________

B. Special Equipment Utilized____________

C. Minimum Operable Lot Size____________
Minimum Number/Site____________

D. Labor: Skilled__________% Unskilled__________%

User/Sweat Equity____________
Local____________
On-Job Training____________

E. Use of Local Contractor____________

F. Transportation:

Maximum Economic Factory Radius______ Miles______ Hours
Widest Load____________
Transport By: Rail________ Highway________ Air__________
Site Factory____________

VII. *Management*

A. Which of the following services does your organization provide:

_____ Manufacturer
_____ Contractor
_____ Land Development
_____ Expediter/Packager
_____ Architectural and Planning Guidance
_____ Maintenance at Completion
_____ Warranties
_____ Act as Rental Agent
_____ Act as Manager
_____ Act as Mortgagee
_____ Marketing/Sales

B. What percentage of the construction work is:

Site Work____________
Utility Connections____________

C. Special Construction Management Techniques:

_____ Materials Handling
_____ Supervision
_____ Phasing
_____ Coding
_____ Teaching and Training Techniques
_____ Other (Specify)

VIII. *Codes and Unions*

A. Do you have agreements with:

National Unions____________
Local Unions____________

B. Does your system conform to the National Model Codes and Standards?

IX. *Marketing*

A. Existing Construction:

List location, number of units per project, and dates of construction.

____
____
____
____
____
____

B. Existing and Projected Market Areas (by state):

____
____
____
____
____

C. Production Capabilities (by region);

____
____
____

D. Minimum size project and time needed to establish a new area production organization:

____
____

E. Promotion and Advertising Plans:

____
____
____

X. *Costs*

A. Finished Construction—Total Square Foot Cost (excluding site-work)________________________________

B. Indirect Items (give percentages):

____ % Job Overhead
____ % General Overhead and Profit
____ % Architectural Fee
____ % License Fees or Royalties
____ % Bond Premium
____ % Other Items

C. Typical Unit Price (give unit size and configuration—assume 100-unit project)____________________________
Number of Stories/Building____________________________

D. Bid Procedures Accepted;____________________________

E. Optional Maintenance Price?____________________________

XI. *User Needs and Unit Types*

A. System Application to Unit Types:

Efficiency____________________________square feet
One Bedroom____________________________square feet
Two Bedroom____________________________square feet
Three Bedroom____________________________square feet
Four Bedroom____________________________square feet
Five Bedroom____________________________square feet

B. ____ Family Room
____ Separate Dining
____ Balconies and Decks
____ Utility
____ Duplex Units
____ Garage/Carports
____ Common Rooms and Commercial Space*
____ Pram and Bicycle Storage
____ Storage Standards

*Give maximum ground floor spans________________________

C. Cladding Materials Available________________________

D. Roof Line____________________________

E. Rate for Flexibility:
Unit Floor Plan__________
Adaptability to Dweller__________
Expansion Possibilities__________
Opening Locations__________
Cross-Ventilation__________
Interior Finishing Materials__________
Exterior Elevation__________

F. Floor to Ceiling Heights__________

G. Unit Identification__________

H. Configuration in Unit Groupings__________

I. Configuration in Building Groupings and Orientation:
High, Low, Medium-Rise Mix__________
Single-Loaded Corridor__________
Double-Loaded Corridor__________
Slab__________ Tower__________

J. Expansion: Horizontal__________ Vertical__________

K. Do you provide local user needs and preference surveys?__________

L. Are you willing to work with committee groups in preparing programs?__________

M. Resistance to:

_____ Moisture
_____ Stain and Odor
_____ Vermin
_____ Sound Transfer
_____ Heat Loss

CITY/SITE CONSTRUCTION AND MATERIALS SURVEY

I. *Construction Industry*

A. Identify the strong contractors and subcontractors. Obtain a resume of past projects and experience.

B. What are their attitudes towards entering into joint ventures with local counterparts?

C. Is there a strong local contractor, suppliers, or other concerns capable of organizing an entire construction process for the experimental project?

D. What construction techniques are utilized, primarily related to the various housing types (low, medium, high)?

E. Does the city have an organized method for analyzing existing housing in terms of rehabilitation; If so, what is an estimated quantity figure?

F. Are there any notable substructural or soils restraints related to the area as a whole? Also, if possible, give more specific soils information.

G. What is the utility availability in connection with each of the anticipated sites (gas, power, water, sewer)?

H. What new innovations and building experiments have occurred in the recent past? Have they failed or succeeded?

II. *Equipment*

Are heavy erection and assembly equipment and teams locally available? What types?

2-12
CITY/SITE CONSTRUCTION AND MATERIALS SURVEY
Equipment: Are heavy erection and assembly equipment and teams locally available? *The Wickes Corporation, Saginaw, Michigan*

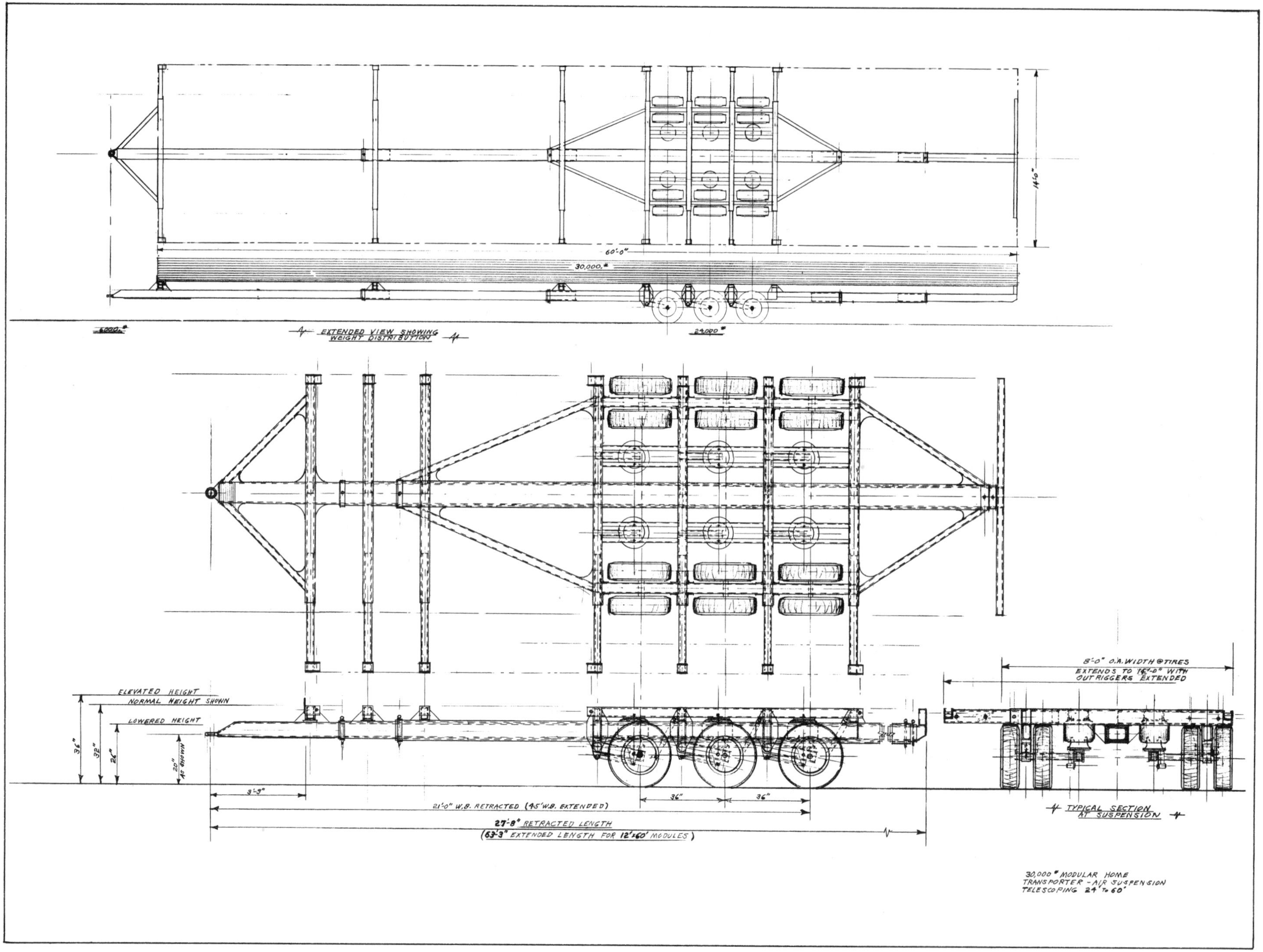
14'-0"
60'-0"
30,000#
6000#
EXTENDED VIEW SHOWING WEIGHT DISTRIBUTION
24,000#
ELEVATED HEIGHT
NORMAL HEIGHT SHOWN
LOWERED HEIGHT
36"
32"
24"
20" AS SHOWN
3'-3"
36"
36"
21'-0" W.B. RETRACTED (45' W.B. EXTENDED)
27'-8" RETRACTED LENGTH
(63'-3" EXTENDED LENGTH FOR 12'x60' MODULES)
8'-0" O.A. WIDTH @ TIRES
EXTENDS TO 16'-0" WITH OUTRIGGERS EXTENDED
TYPICAL SECTION AT SUSPENSION
30,000# MODULAR HOME
TRANSPORTER - AIR SUSPENSION
TELESCOPING 24' TO 60'

2-13
TRANSPORT FACILITIES Are there any specific problems related to the transportation of building materials or components? *General Trailer Company, Inc.*

III. *Building Materials*

A. What building materials are readily available?

B. Are there local reasons indicating that we cannot or should not consider any of the following building materials;

	Yes	No
1. Wood (lumber, plywood, prefab components, etc.)	_____	_____
2. Brick (local availability, etc.)	_____	_____
3. Other masonry (adobe, stone, concrete)	_____	_____
4. Steel (fabrication and erection)	_____	_____
5. Concrete (fabrication and erection) (lightweight, precast, prestressed, etc.)	_____	_____
6. Any other material (any reason)	_____	_____
Are there any building materials particularly well received or traditional in the area?	_____	_____

IV. *Transport Facilities*

A. What are local code requirements in regard to shipping (rail, highway) in terms of weight and size?

B. Are there any specific problems related to the transportation of building materials or components?

	Yes	No
1. Allowable highway loads	_____	_____
2. Particular materials	_____	_____
3. Large precast monolithic units	_____	_____
4. Moving existing houses	_____	_____

V. *Labor Situation*

A. Identify building trades union.

B. Are there organized apprentice training programs?

C. Retraining programs;

D. Membership statistics—skilled and unskilled.

E. Do union practices differ in this particular city? If so, in what ways?

F. Are there any special labor peculiarities *other than unions* which would militate against innovative experiments in any aspect of the construction process as it relates to low-cost housing?

	Yes	No
1. Local absence of certain trades	_____	_____
2. Local deficiencies in any trades	_____	_____

Are there tradesmen particularly proficient at their trade?

2-14
CITY/SITE BUILDING CODE ANALYSIS What are the requirements regarding structural characteristics? Wood, steel, concrete? *The Wickes Corporation*

CITY/SITE BUILDING CODE ANALYSIS FORMAT

A. What are the required fire ratings for:

Building Type	Ext. Walls	Party Walls	Egress Encs.	Floors	Roofs	Super Struc.
Single-Family Detached						
Low-Rise Attached						
Medium-Rise (4 Stories)						
High-Rise (Elevator)						

B. What are the requirements regarding sound insulation of:

Exterior Walls____________________
Party Walls____________________

C. What are the requirements regarding thermal insulation of:

Exterior Walls____________________
Roofs____________________
Grade Slabs____________________

D. What are the requirements regarding moistureproofing of:

Roofs____________________
Exterior Walls____________________
Grade Slabs____________________
Foundations____________________

2-15
PLUMBING DRAINAGE What are the restrictions relating to inspection of plumbing installations? Would factory inspections be acceptable? *Tyler Corporation*

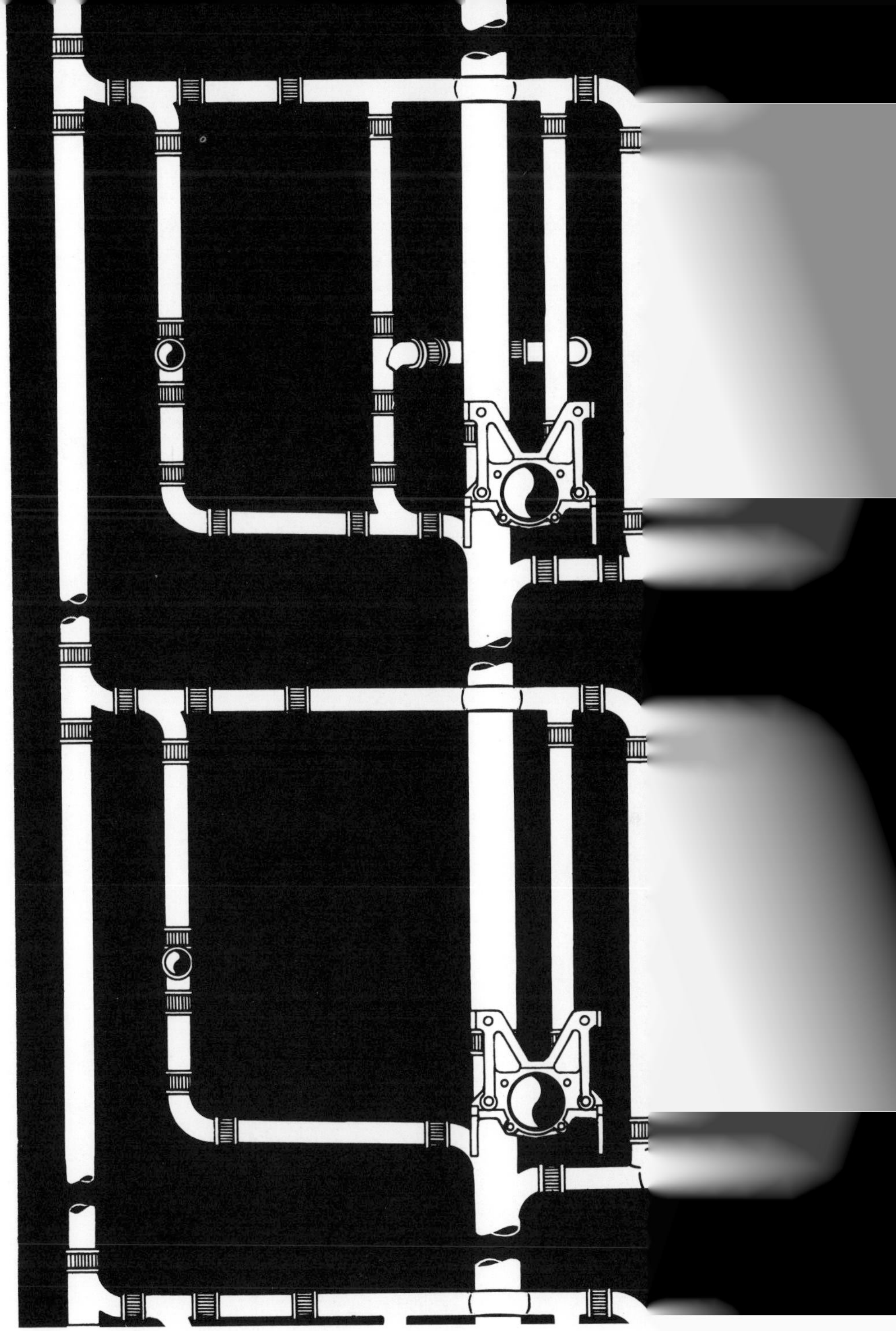

E. What are the requirements regarding vermin-proofing?

F. What is the attitude of the building code regarding experimental materials and testing?

G. What are the requirements regarding openings and ventilation?

H. Is the building code a performance code or a specification code?

I. What are the restrictions relating to inspection of plumbing and electrical installations? Would factory inspections be acceptable?

J. What is the attitude of the building department regarding construction undertaken by owner? Insurance requirements?

K. What are the requirements regarding structural characteristics?

	Wood	Steel	Concrete
P.S.I.			
Loads—Static			
Loads—Wind			
Modulus of:			
Fireproofing			

L. Zoning in project areas:

1. Land Use ______________________
2. Height limitations ______________________
3. Setbacks ______________________
4. Fire limits ______________________
5. Parking ______________________
6. Other ______________________

M. What are the requirements regarding room areas?

N. What are the requirements regarding means of egress?

O. What are the requirements regarding multiple-occupancy?

P. Are there any pertinent local requirements that differ significantly from the national building codes and/or the ACI and ACSI standards?

NOTE: Please make arrangements to obtain copies of pertinent documents.

CITY/SITE USER NEEDS AND ENVIRONMENTAL SURVEY

A. *User Preference/Housing Types* (See Chapter 4 also)

Do people in the Model Cities have decided preferences regarding:

1. Type of construction and materials (wood, brick, concrete)?
2. High-Rise Apartments ____________ %
   Low-Rise Apartments ____________ %
   Detached House ____________ %
   Ownership ____________ %
   Rental ____________ %
   Tenure ____________ %
3. Have cooperative and/or condominium concepts been introduced to housing market?
4. What is the availability and price range of housing types?
5. Are there identifiable neighborhoods? Type? Size? Social Mix? Density?
6. What is the visual character and fabric of the housing and the city in general? Scale? Roof line? Existing materials?
7. What are attitudes towards mixed usage, i.e., inclusion of commercial, office, educational, recreational facilities?
8. Common area requirements: day care, meeting rooms, laundry/utility?
9. What housing types are predominantly available (relate to income where possible)?

   a. Single family (detached)
   b. Low-rise attached (in townhouses, row houses, etc.)
   c. Medium-rise, walk-up apartments (4 stories)
   d. High-rise (elevator access)

2-16
CITY/SITE USER NEEDS AND ENVIRONMENTAL GUIDELINES *Messages from Perugia, Bruno Zevi, Industrie Buitoni Perugia, 1971*

10. Typical dwelling unit market:

    Square footage________________
    Number of bedrooms________________
    Number of baths________________
    Separate dining room________________
    Family room________________
    Garage/carport________________
    Balconies and decks________________
    Yard________________

B. User Needs

1. Safety: interior, exterior, equipment, circulation, pedestrian/vehicle
2. Open Space: heirarchy, private, commercial
    a. private—yard, court, garden, balcony
    b. small group—street, pedestrian way
    c. larger groups—common
    d. community—recreational lands and fields (conservation and buffer)
3. Children's Play
4. Daylight, ventilation, space around buildings
5. Sunlight—orientation
6. Micro-climate—controlled environment—space—shelter
7. Privacy—visual barriers and unit orientation
8. Noise—children—machines/autos—between units—between rooms
9. Opportunities for private ownership
10. Opportunities for self-expression
11. Landscaping—planting
12. Access—identity of units
13. Servicing and services
14. Emergency access
15. Public transit access
16. Removal—refuse collection
17. Deliveries
18. Organization of dwellings

C. Environmental Guidelines

1. Climatic description of area
2. Geographic description of area
3. Maximum wind velocity________ mph________________
4. Earthquake Zone 0________ 1________ 2________ 3________________
5. Winter snow: low________" average________" high________"
6. Flat site,________________ maximum grade________________
    Sloping site,________________ maximum grade________________

2-17
URBAN DESIGN FABRIC What is the visual character and fabric of the housing and the city in general? Does this Bison System project (high-rise tower buildings) work in this context? *Concrete, Ltd. copyright #A70-100-2*

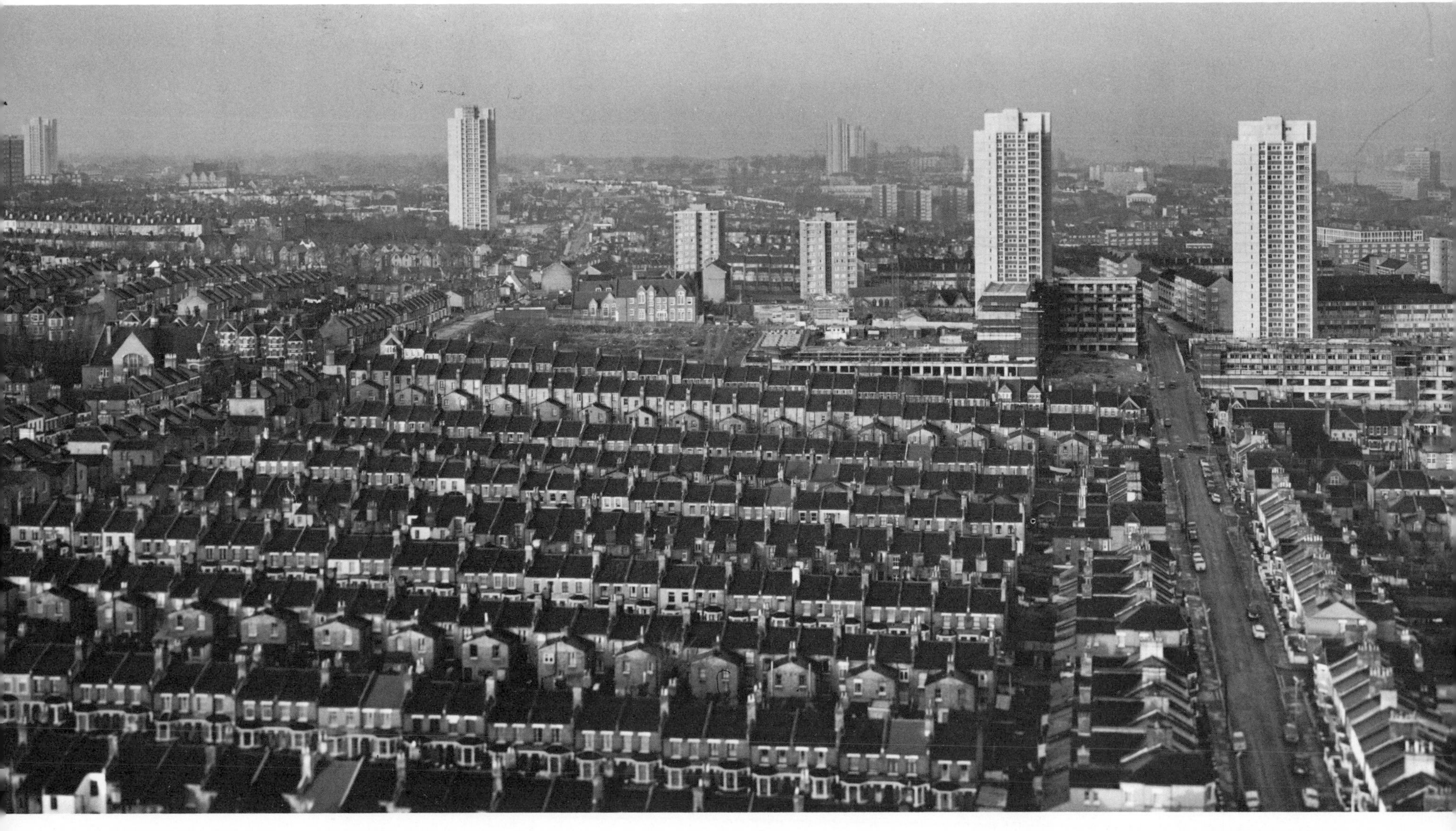

3 THE BUILDING SYSTEM AS A PROCESS

AN INTEGRATED SERIES OF STEPS

All those measures which are required in order to organize the building industry to work more like a mass production plant make up the industrialized building systems process. *This means not only the use of new materials and construction methods but also increased mechanization and control of on-site processes and improved management techniques.*

The management and corporate organization are the single most important ingredient in any successful building operation. It has frequently been observed that in Europe, for example, where circumstances have forced and fostered the development of many building systems, the most successful are those that are well-organized and well-managed, not the most technologically advanced or most industrialized. They are all, for the most part, technologically sound, but that alone has not been enough to insure their success. The same observation will undoubtedly apply to the U. S. and the means by which the major impact factors—labor cost, material cost, land cost, packaging and marketing, labor availability, construction activity, and transportation cost—are evaluated and utilized.

Everyone in the construction industry today states the commonplace phrase "the building process is a system of integrated parts," and it is worth restating here a diagram in which Professor Albert G. H. Dietz of M.I.T. has demonstrated these five interrelated steps (Figure 1):[1]

Need
Preliminary Planning
Design
Construction
Operation and Maintenance

Professor Dietz describes this integrated building process by illustrating how each step follows its predecessor in logical order, but each interacts with its predecessor, influencing it, and is in turn

1. Future Potential of Building Systems, Professor A. G. H. Dietz, presented to the ASCE, September 30, 1968.

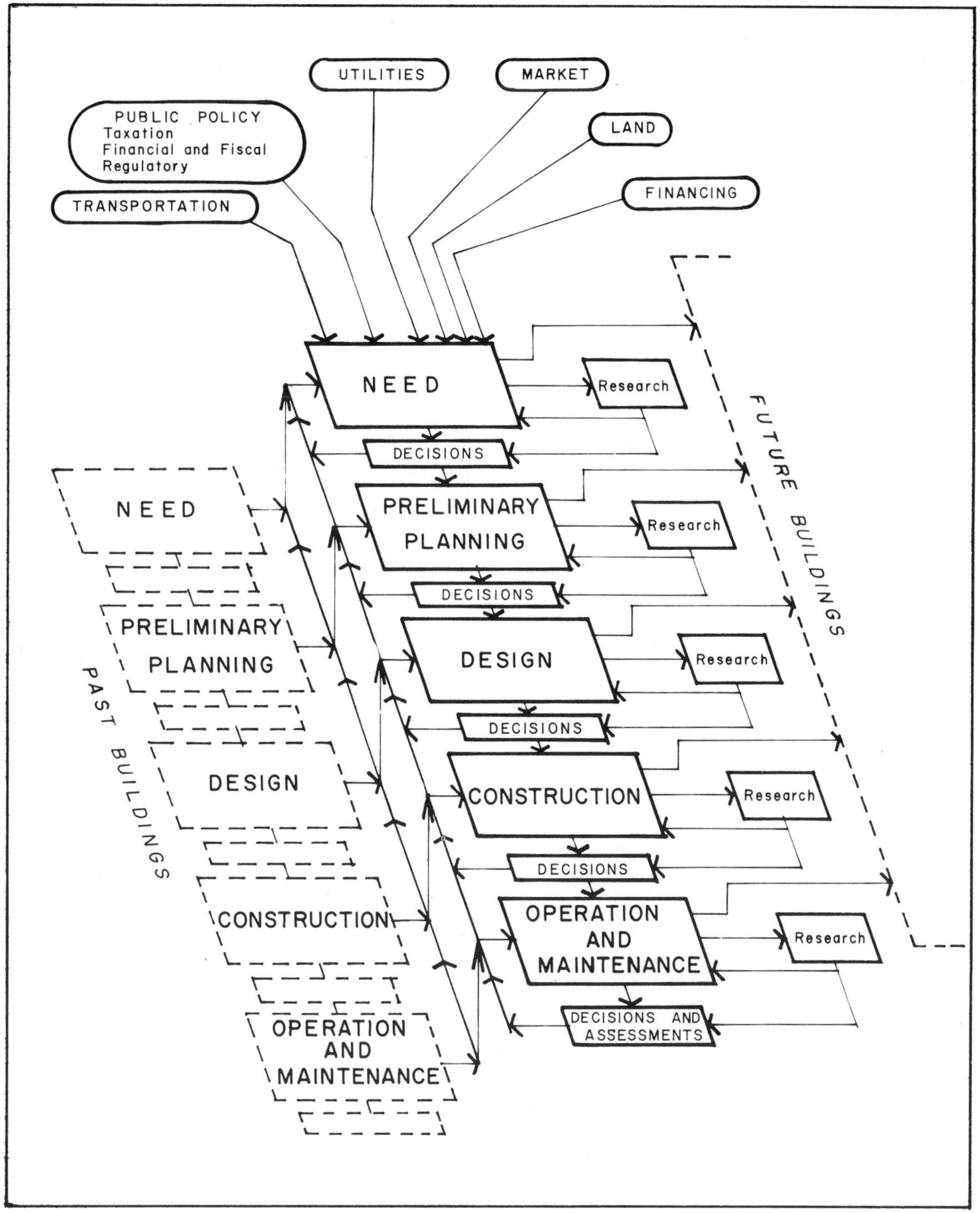

3-1
THE BUILDING PROCESS AS AN INTEGRATED SYSTEM OF STEPS RELATING TO PREVIOUS AND FUTURE BUILDING OPERATIONS, AND TO THE LARGER ECONOMY *Components in House Building (NAHB Components Study), ed. Albert G. H. Dietz, M.I.T., 1961*

12M Jespersen	Cycle of Operations

12M Jespersen	Cycle of Operations

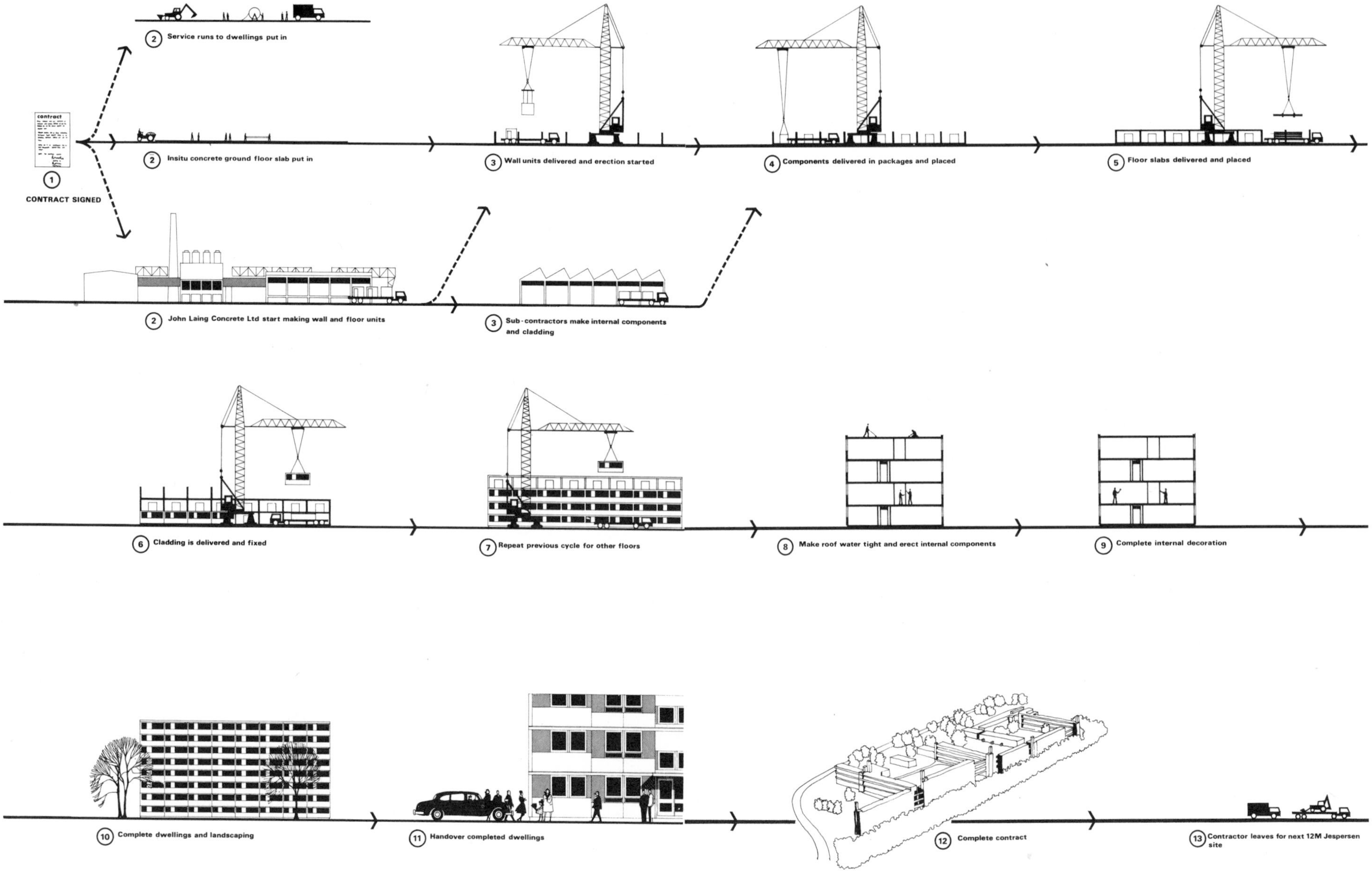

1 CONTRACT SIGNED
contract
2 Service runs to dwellings put in
2 Insitu concrete ground floor slab put in
2 John Laing Concrete Ltd start making wall and floor units
3 Wall units delivered and erection started
3 Sub-contractors make internal components and cladding
4 Components delivered in packages and placed
5 Floor slabs delivered and placed
6 Cladding is delivered and fixed
7 Repeat previous cycle for other floors
8 Make roof water tight and erect internal components
9 Complete internal decoration
10 Complete dwellings and landscaping
11 Handover completed dwellings
12 Complete contract
13 Contractor leaves for next 12M Jespersen site

3-2
INTEGRATION OF ACTIVITIES And the cycle of operations are of no small importance to factory systems. One delay can destroy the entire critical path and cause the systems operation to be more costly than more traditional construction. *John Laing & Son, Ltd.*

3-3
PRECAST CONCRETE PLANT FACILITY Design, firstly, to install the casting beds for single tee's and then, for the plant, using *only* two variations on the single tee, to "build itself." A low capital investment yet an efficient operating organism. *ECODESIGN*

influenced by the succeeding steps in the feedback process typical of systems operation.

Each step has a research loop indicating that problems must be solved before decisions can be made.

It is certainly recognizable from the process diagram that such integration of activities and personnel requires strong management and a clear organization plan.

As an extension of this notion of the building system as a building process, the total process an organization would go through in establishing an American building system—management activities, network analysis, and staffing—from the inception of the idea to full-scale mass production is outlined here in three phases:

- I Staffing and Design
- II Prototype Organization and Construction Plan
- III Full-Scale Production Plan

The management outline will be based primarily on the supposition of this text in that it will be oriented towards special on-site construction techniques (mobile plant) and/or plant facilities requiring a low initial capital expenditure.

The following is an index of the tasks and activities required during start-up. All activities have been grouped into three basic categories:

Direct Activities—related to creating the "products." to be offered—design and construction.

Service Activities—those elements whose function directly bear on the process of implementing the whole project—programming, planning and training.

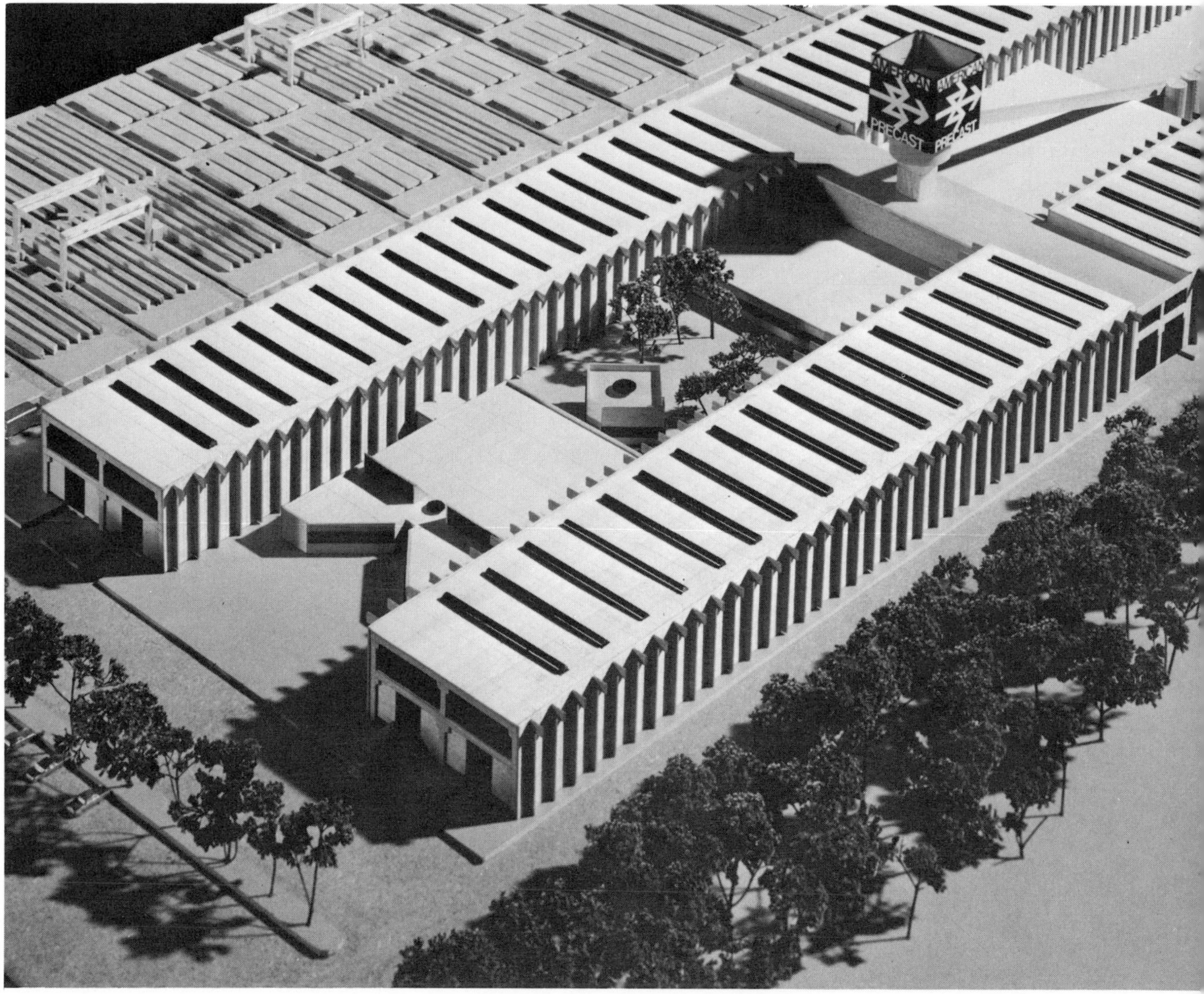

Management Activities—those activities which must be performed to assure that implementation is successfully achieved—marketing, negotiations, and financing.

The processes illustrated in the Network Schedule have been based on the following analysis and planning procedure.

1. Definition of current and future problems and the interrelationships among them.
2. Prediction of future conditions arising from identifiable problems.
3. Identification of parameters, boundary conditions, or constraints which determine the range of possible solutions to the total problem.
4. Determination of goals and objectives at varying levels: minimum, maximum, optimum; normative or utopian.
5. Formulation of alternatives.
6. Evaluation of qualitative and quantitative cost effectiveness.
7. Simulation of alternatives for performance standards.
8. Recommendation of alternatives.

Throughout the design and planning phase, this procedure will basically guide the tasks herein described and will be complemented with current techniques in systems analysis.

Index of Activities

1 *Direct Activities*

1.1 Architectural Design

- 1.11 Preliminaries
- 1.12 Basic drawings
- 1.13 Construction documents

1.2 Engineering

- 1.21 Systems and subsystems preliminary design
- 1.22 Basic drawings
- 1.23 Construction documents

1.3 Site Planning

- 1.31 Site reconnaissance
  - 1.31.1 Soils testing and flora evaluation
  - 1.31.2 Topographic survey
- 1.32 Preliminary design
- 1.33 Basic drawings
- 1.34 Construction drawings

1.4 Construction

- 1.41 Process reconnaissance
- 1.42 Adaptation of system to national and local construction standards and processes
- 1.43 Outline construction plan
- 1.44 Preliminary construction plan
- 1.45 Final construction plan
- 1.46 Prototype construction—execution
  - 1.46.1 Equipment set up
  - 1.46.2 Foundations
  - 1.46.3 Superstructure
  - 1.46.4 Prefabricated assemblies—subsystems
  - 1.46.5 Finishing
  - 1.46.6 Model suite

1.5 Cost Estimating

- 1.51 Budget goals
- 1.52 Preliminary cost estimate (comparisons to conventional construction)
- 1.53 Final cost estimate
- 1.54 Final construction cost
- 1.55 Adjustment to final construction cost

2 *Service Activities*

2.1 Facilities Program

- 2.11 Preliminary prototype program
- 2.12 Computer program model mass production
- 2.13 Test computer model

2.2 Standards and Controls

- 2.21 Local, state, and federal quality and performance analysis
- 2.22 Constraints for prototype construction
- 2.23 Construction testing plan
- 2.24 Mass housing quality and performance standards

2.3 Project Planning

- 2.31 Locational analysis—local, state, region
- 2.32 Municipal procedure
- 2.33 Location strategy for development
- 2.34 Master plan concepts

2.4 Training

- 2.41 Analysis of programs in training and education
- 2.42 Training program systems
- 2.43 Training program implementation
- 2.44 Preparation for production team operation

2.5 Community Participation

- 2.51 Analysis of selected community
- 2.52 Strata of group interaction
- 2.53 Identify leadership participants
- 2.54 Structure citizens participation program
- 2.55 Final citizens participation program

2.6 Management Activities

- 2.61 Program planning
- 2.62 Internal management
  - 2.62.1 Internal operation system
  - 2.62.2 Data handling system
  - 2.62.3 Marketing
  - 2.62.4 Communication and personnel
  - 2.62.5 Termination procedures
  - 2.62.6 Transfer of ownership and responsibility
- 2.63 Legal
  - 2.63.1 Contract negotiations
  - 2.63.2 Jurisdictions
  - 2.63.3 System patent procedures, royalties, and/or licensing rights
  - 2.63.4 Site ownership, title rights, etc.
  - 2.63.5 Permits and approvals
- 2.64 Finance
  - 2.64.1 Strategies and sources
  - 2.64.2 Vehicles for multiple participation
  - 2.64.3 Preliminary financial plan
  - 2.64.4 Test proposed financial plan alternatives
  - 2.64.5 Select alternative plans for use

Phasing is set within the network based on goals or milestones and time available and required. For example, the following first level milestones might define Phase I, *Staffing and Design:*

1. Decision to proceed or contract award
2. Program revision and beginning of operation teams
3. Environmental inputs for Phase I
4. Monthly progress reports
5. Preliminary program for prototype complete
6. Standards and controls determined as guidelines
7. Community analysis—for location of prototype
8. Subsystems preliminary proposal complete
9. Site planning preliminary proposal complete
10. Architectural design preliminary proposal complete
11. Budget goals determined
12. Final cost estimate complete
13. Final construction plan
14. Final construction costs
15. Begin construction of prototype
16. Begin documentation and final planning for Phase III—mass production

Similar milestones would define the following phases—Prototype Organization and Construction Plan (Phase II) and Full-Scale Production Plan (Phase III).

STAFFING AND DESIGN—PHASE I

Staffing

The staff put together for a total undertaking such as described in the index of activities is necessarily comprehensive and must be representative of all those affected by the housing market, including the users themselves, and it should operate as a team. The personalities and individual experience are decidedly more important than filling the precise "discipline slot" on our diagram, but for what it is worth, here is an optimum proposed team for Phase1—the manpower organization for industrialized building. The design stage of a construction project usually averages about six months from date of award of contract. Chart 2 also indicates a rough approximation of overlapping job influences and time of involvement.

Chart 1 diagrams the eventual type of corporate hierarchy and indicates the new teamwork necessary, especially between contractor and architects.

Chart 1

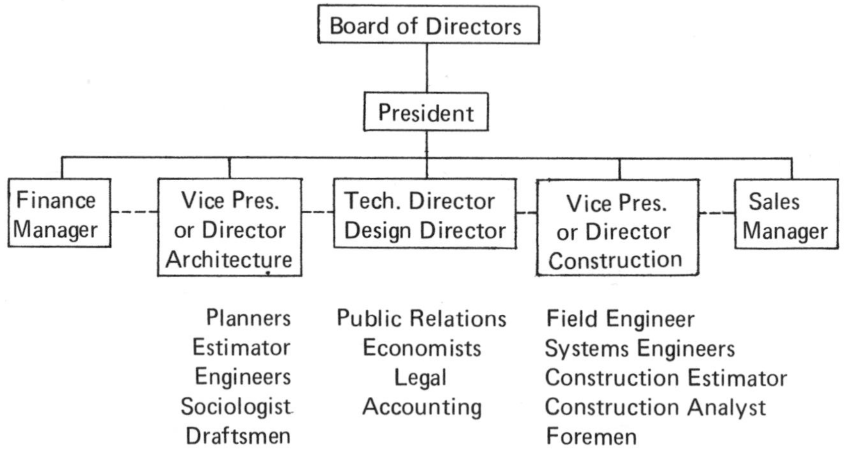

Design

The interrelationships required in the design and planning phase of such a system project deal with those major areas of work which represent crucial progress positions.

It is estimated that Phase 1 of a prototypical 150–200 unit project will take six months from decision to proceed. During this period, a coordinated system of events will take place, each varying in length of time, depending on their degree of complexity and longevity of influence.

The design and planning phase of this project will consist of the *Direct Activities* of investigation, *design,* and preparation of working drawings for the construction of a prototype housing development. The secondary activity is the formulation of a *construction plan* for the execution of this prototype. The third-level activity during this phase will be the preparation of a generalized *plan for the production* of housing on a mass scale.

Simultaneous to the prime activity leading toward the construction of prototype housing, a variety of service and management activities must take place. The effort made in these two areas will be directed toward the development of the prototype housing and a potential to be utilized on a national scale. The critical elements are a sensitivity to the market and programming.

Chart 2 Staffing Diagram

TITLE	NO.	MONTH 1	2	3	4	5	6
–Project Director	1	x	x	x	x	x	x
–Project Architects	2	x	x	x	x	x	x
Interior Designer	1	x	x	x			
Specifications	1	x	x	x	x	x	x
*Computer Graphics	1	x	x			x	x
Arch'l Estimator	1	x	x				
–Landscape Architect	1	x	x	x	x		
*Materials Specialist	1	x	x				x
Draftsmen	5	x	x	x	x	x	x
–Structural Engineer	1	x	x	x	x	x	x
*Mechanical Engineer	2	x	x	x	x	x	x
*Acoustical Engineer	1	x	x	x	x	x	x
*Soils Engineer	1	x	x	x	x	x	x
*Electrical Engineer	1	x	x	x	x	x	x
Arch'l Analyst	1	x	x	x	x	x	x
*Economist	1	x	x	x	x	x	x
*Sociologist	1	x					x
–Project Manager	1	x	x	x	x	x	x
Field Engineer	1	x	x	x	x	x	x
–Construction Estimator	2	x	x	x	x	x	x
Construction Analyst	1					x	x
–Form Engineer	2		x	x	x	x	x
–Systems Engineer	2	x	x	x	x	x	x
Analysis:							
*Legal		x	x	x	x	x	x
*Accounting		x	x	x	x	x	x
*Management		x	x	x	x	x	x
*Public Relations	1 PT	x	x	x	x	x	x
*City Planner (City)	1 PT	x	x	x	x	x	x
*Regional Planner (State)	1 PT	x	x	x	x	x	x

– = Senior Staff   * = Consultants

During this phase, physical design cannot progress without an understanding of the programmatic conditions. A program cannot be developed without the analysis of the community for which it is intended to serve. The advancements in technology cannot be utilized without the ability and know-how of professional and local participation. The total effort cannot run smoothly and lead to a successful end without a line of communications established through liaison and data distribution.

The architect, in his new role, will be asked to recommend a suitable system of industrialized building but he will also have to accept the design disciplines of the system chosen. He will find himself involved in relationships with the contractor for which he has probably had no precedent. He will have to evolve a new outlook of architectural aesthetics, new methods of preparing working drawings, new organization and management methods, and new administrative techniques. The new roles that this phase defines and requires for management architects, engineers, and contractors are discussed in Chapter 1, under "*The Nature of Industrialization*," Management and Design for Industrialization.

It will also be necessary to analyze the effects of these new procedures and techniques on the people involved and the work they produce both in the company's organization and in the architect's office. This has never been an easy analysis to make as the initiation of anything new may be subject to rejection without full consideration. The coordination of these disciplines and activities is, however, a tremendous challenge for the organizational capabilities of the company and for the architect, specifically, and it is of vital importance that this process is set in motion, as a team, from the beginning.

3-4
ON-SITE CONSTRUCTION TECHNIQUE Prototype construction illustrating the equipment setup, foundations (below), superstructure, subsystems, finish, and model dwelling units. The next sequence of photographs are through the courtesy of McKone Estates, Tallaght, Dublin, Ireland. *McKone Estates*

PROTOTYPE ORGANIZATION AND CONSTRUCTION PLAN–PHASE II

The following is an outline of the work elements necessary for constructing first units and establishing prototypical management and·construction methods.

1. *Staffing–Construction Teams*

   Construction teams and processes must be established including staff positions for management, personnel training, and labor schools. The majority of the trade skills required of the construction staff are traditional. The subsystem installation requires some degree of personnel training.

2. *Equipment Procurement*

   Prepare equipment for production. Final design, adjustments, dimensional, weight, and strength requirements, fabrication for purchase, preparation for shipping.

3. *Job Training*

   Training techniques: movies, models, color coding, etc.

4. *Tooling*

   Prepare special tools and devices for individuals working on and off site with forms or beds such as cranks, pins, spacers, jigs, etc.

5. *Testing and Assessments*

   Prepare for and execute tests to adjust equipment, teams, and methods.

6. *Schedule*

Program and confirm equipment cycle for specific project.

a. Size and complexity of the project
b. Completion target date
c. Optimum use of manpower
d. Full utilization of cranes and other equipment
e. Deliveries and materials handling
f. Subcontracted installations

7. *Maintenance/Quality Control*

Schedule regular inspection of equipment and forms during use periods and prepare for immediate repair of minor damages.

8. *Prepare for Erection*

Prepare facilities and equipment for assembling steel reinforcing.

9. *Construction*

a. Site preparation and crane set-up
b. Foundations
c. Superstructure (special technique of pre-fab)
d. Prefabricated assemblies and subsystems
e. Finishing
f. Project complete

10. *Management*

Financial proposals necessary for second-level funding would continue to be worked out. The full proposal would include:

3-5
Using a heavy-duty truck-mounted tower crane (capable of lifting one ton at 88½ feet) sophisticated steel tunnelforming is erected. The basic unit consists of two steel plates which are braced together at right angles in the form of an inverted L. The sections are connected together to form a tunnel section—in the form of an inverted U *McKone Estates*

3-6
Several sections can be joined together to form a tunnel section equal in length to the depth of the house. A number of inverted U sections are then placed parallel to each other but separated by the desired wall thickness. *McKone Estates*

3-7
Concrete is then poured between and over the tunnel sections to form walls and floors and the forms removed when the concrete has reached its specified strength. Accelerated curing of the concrete facilitates early stripping of the shutters and then infill facade panels are placed by a light crane. *McKone Estates*

3-6

3-7

a. *Introductory Paragraph*—indicating time and place of incorporation, corporation purpose and interest (to introduce and promote "X" building system, etc., and extent of development and involvement), general description of product or system, background and related experience of personnel, plan leading to full-scale production.

b. *Proposed Ownership Plan*—giving amount of capitalization required and ownership percentages, sources for further funding, attitudes on public issues, etc.

c. *Income Projection*—would be worked out over 1–10 year period. This projection would be based on incomes, sublicenses, royalties per dwelling unit, percentages of construction cost, per-square-foot fees, flat service or rental rates, or a combination of any of these sources. Additional income would be covered in a similar fashion for products or the full development packaging and Turnkey projects.

d. *Use of Capitalization*—would give full accounting for all expenditures and operation costs for initial year and following years. The set-up costs would include original licensing and/or patenting costs, plant, tooling or other machinery, initial promotion and system development expenses. Other annual expenses might include management, market analysis, promotion, design control and coordination, architectural supervision and liaison, promotion, construction supervision, materials control, market synthesis, contracts soliciting,

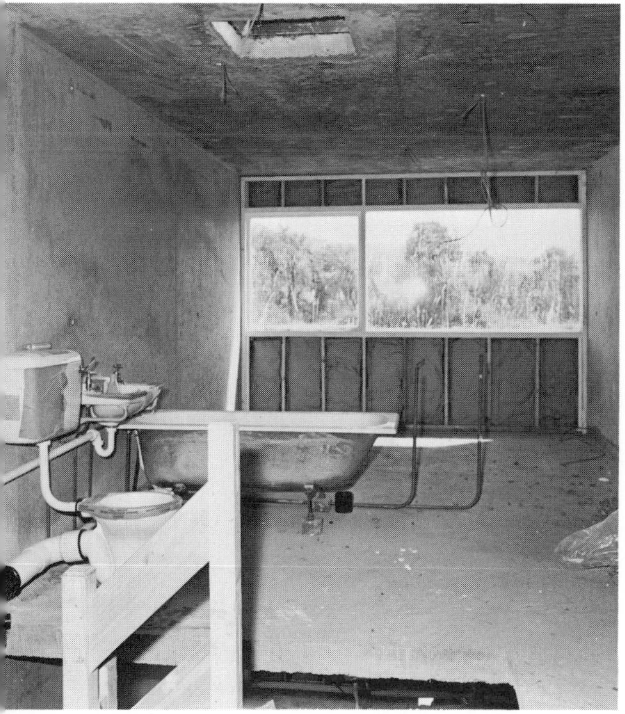

3-8
The stairs have been installed, the bathroom fixtures are in place, wiring is in its conduits, and the infill panel is fixed to the concrete bearing walls. The walls have a minimal amount of air holes due to the polished steel surface of the formwork. *McKone Estates*

3-9
A skin of concrete is built, forming a 1" cavity. Foam is injected into the cavity. The blockwork is plastered. U value of wall 0.17. *McKone Estates*

3-10
Factory-made nonloadbearing partitions are then erected. Door frames and all necessary hardware are supplied in a "package" for every unit. These sheetrock partitions do not require a skim coat of plaster. The panels are sealed and filled by painters and wallpapered direct. *McKone Estates*

3-11

promotion and public relations, legal fees, accounting, personnel control and business management, new products and systems development and adaptation, materials research, system continuity and construction supervision, equipment engineering and maintenance, training programs, cost estimating, clerical, salaries for full operating personnel, and general office overhead. Additional items would be included depending on whether actual construction is in-house or sublicensed and on extent of development role.

e. *Business Management*—section would cover general system description and promotional and sales attitudes, construction cost comparison (savings for contractor developer), marketing plan, as well as an organizational chart. This chart should indicate interaction between general contractor, architectural, and construction management and their methods of controlling project progress, accounting, and cost projection (include forms, schedules, reports, meeting techniques).

f. *Five–Ten Year Statement of Projected Profit and Loss*

3-11
The model dwelling units are completed! *McKone Estates*

3-12
FACTORY PRODUCTION Of large-scale heavyweight precast concrete panels incorporates a highly organized central organization and production runs reaching as much as 10,000 units per year. The next sequence of photographs are through the courtesy of Fram Higgs & Hill (Camus) Ltd. *Photos by John Mill, Liverpool, England*

FULL-SCALE PRODUCTION—PHASE III

Finalizing the organizational model will depend on whether the system intends to compete nationwide, operating out of a central organization office located in a major city, for either marketing or prestige purposes. Boston, Massachusetts, for instance, is often suggested as an appropriate location because of its worldwide reputation as a headquarters for technology and, specifically, building systems. Regional offices might then be located to cover specific regions of the country. A geographic breakdown might be North Atlantic, West Central, Great Lakes, Plateau, North Pacific, and Pacific, although these areas are more likely to be set by the para-political and marketing relationships of potential sublicensees than by the geographical regions themselves.

A central organization should provide planning, training, marketing, and costing activities for the regional offices. When a large-scale production plan is put into operation, reaching numbers upwards of 10,000 units per year, a communications center can facilitate maximum efficiency and maximum production. A communications center for the building system and its licensees, for instance, might be established for coordination of the use of the system and for training programs in construction and design.

a. Sales, promotion and market project packaging, cost estimating, marketing.

b. Architecture, engineering, and planning services for consultation on efficient use of the system in design.

c. Training program for construction teams and operations and construction management services for sublicensees and management personnel. Training films, full-scale models, color coding techniques and teams would be part of construction training.

d. Distribution of changes and/or improvements in the system developed by any user. There would be an intensive monitoring operation to insure quality control of construction, design, and planning and user needs.

e. Research, development, testing and feedback evaluation.

f. Community organization and user needs. In this way, large-scale production operations may act in direct response to a changing market demand.

3-13

3-14

3-15

3-13
Cement silos and aggregate storage bins provide a materials storehouse which can draw upon a particular customer's design requirements and needs without breaking a production run. *Higgs & Hill*

3-14
Casting beds are prepared for reinforcing bars, and electrical wiring may be installed in conduit at this point. *Higgs & Hill*

3-15
The concrete is poured, and in this plant, is hand-finished in a horizontal casting bed. Note the panel in the rear being cured. *Higgs & Hill*

3-16

3-17

3-16
Insulation panels of styrofoam may be set in the formwork to insure design specifications for U factors and acoustical privacy. *Higgs & Hill*

3-17
This particular panel has a smooth finish, but textures, aggregates, tiles, and even bricks may be inset to achieve particular "architectural effects." *Higgs & Hill*

3-18
Door and window openings are easily cast into the panel, and exposed stone aggregates are among the many finish options. The worker is washing excess concrete and oil from the stone textured surface. *Higgs & Hill*

3-18

3-19

3-20

3-19
Color-coded panels by project and in an erection sequence order are stored in large yards immediately adjacent to the casting beds. *Higgs & Hill*

3-20
When the proper moment arrives in the construction process, the prepared panel is trucked to the site and lifted into place. Note the holes in these panels to "key" to other panels, services, and design specifications for erection ease and efficient integration with subsystems. *Higgs & Hill*

3-21
MODEL FOR AN AMERICAN INDUSTRIALIZED SYSTEMS BUILDING ORGANIZATION demonstrates the full-scale model for "an American industrialized building organization" and the relationships and responsibilities to authorities and participants, such as committee groups, industry participants, codes, and labor unions.

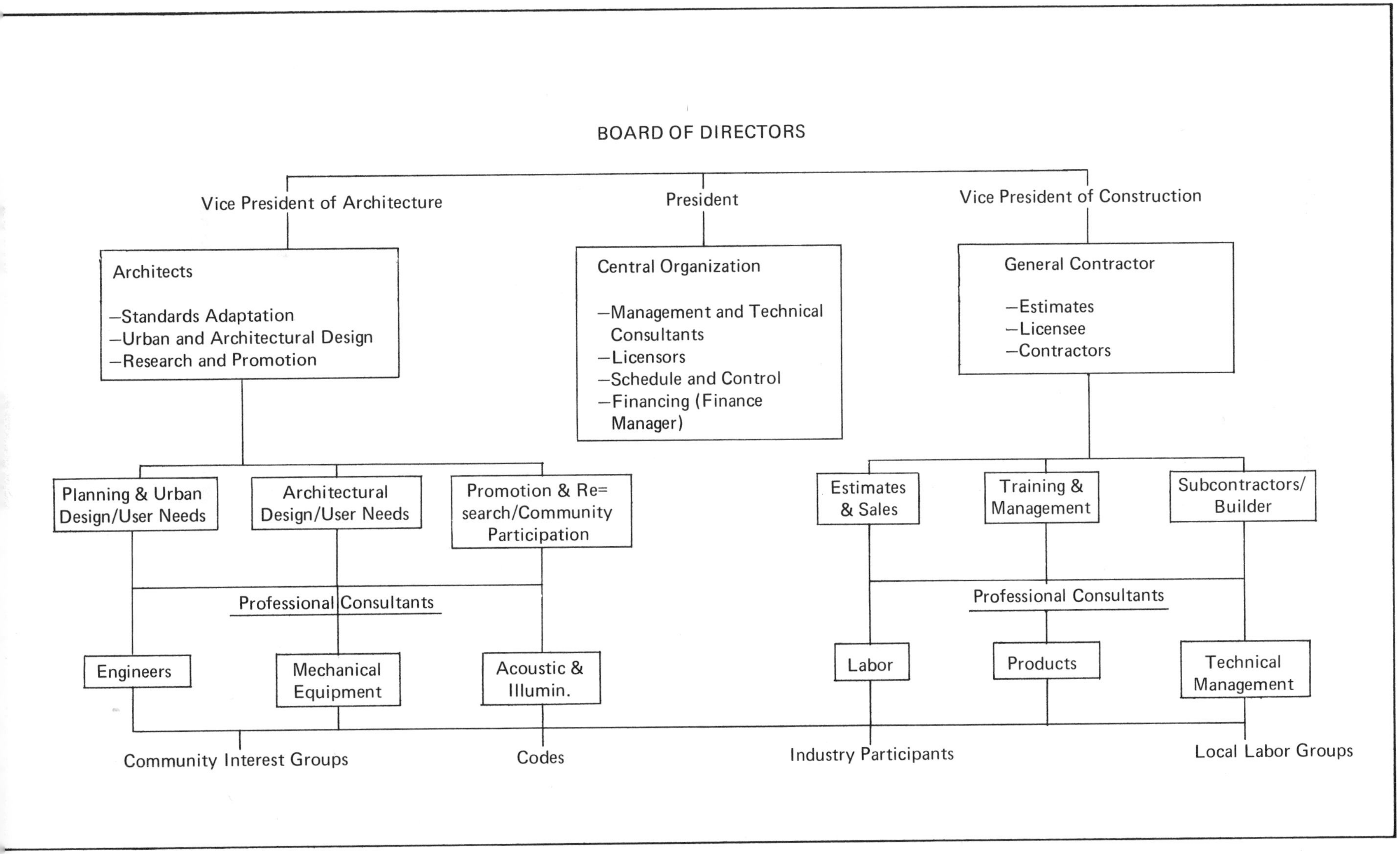

Community Organization and Management

Planning "with people" in place of "for people" is a concept which companies must act to promote. In the past, this nation's highly fragmented private housing industry has generally excluded racial minorities from the production process and access to a large majority of the supply.

Recently, William Morris, Director of Housing Programs for the NAACP, prepared a paper entitled "An Affirmative Action Program for Equal Opportunity in Systems Building and Housing Development Operations." In that paper, he states:

> "Black and other minorities must be made aware that mass production methods and new technology in housing can create pioneering opportunities for them to score a 'breakthrough' into the nation's economic mainstream of business and industry as producers of goods and services rather than the generally accepted role as consumers."

> "Housing corporations should develop a policy and program for equity participation on either a cash or contributing services basis by minority-group persons, a plan that will assure maximum opportunity and involvement in management and decision-making activities. One new innovation in systems building operations could include training (during the development and production stages) of local persons in property management, maintenance, and social service programs."

The Housing Market

This nation's stated policy and objective is to achieve "a decent home in a suitable environment for every American family." By 1985, our population is expected to increase to 260 million. This means that 20 million new households will be formed. The growing need for housing by new households will be further complicated by the accompanying deterioration of about one-quarter of our existing housing units. We must build, in less than 20 years, the equivalent of the entire housing supply of the United States in 1960. By the year 2000, the population increase is projected to be 69% or 338 million; by the year 2000, everything that has ever been built in the country from 1620 onwards will have to be built two and one-half times again. The modest goal of 26 million units set in the Housing Act of 1968 was to provide 26 million units in ten years. By mid-December 1971—three years had already passed—housing starts had still only reached 2 million.

It is also interesting to note that the 1970 census showed a 96.1% increase over 1960 for multi-unit dwellings in the suburbs and a 101.5% increase for mobile homes in the suburbs, with a 141.0% increase for mobile homes as a U. S. total.[2] The increase of black families owning their own homes soared to 42%. In absolute numbers, there were more than two new apartments in the decade for every new mobile home.

In 1942, during World War II, 8 million persons were housed within a four-year period by the National Housing Agency, in a magnificent organizational innovation. That innovation was the use of industrially manufactured building solutions virtually overnight for use as temporary housing quarters. Some of this so-called temporary housing is still in use today, but the point "where there's a will, there's a way" is well taken. It is an historic fact that the building industry has been motivated to change *only* when the needs of the client, the building owners and the users, have changed. These needs have slowly evolved.

As a result of the technological revolution in agriculture, employment opportunities for 3.2 million farm workers have disappeared over the last 15 years. Many are poor blacks, who have migrated to the city. Between 1940 and 1950, 1.6 million blacks left the South. Between 1950 and 1960, 1.5 million blacks migrated to northern cities and the migration has continued steadily at this rate between 1960 and 1970.

Twenty percent of families in our urban centers earn less than $5,000 a year for a family of four. In addition, a large segment of the urban population earns between $5,000 and $8,000 a year for a family of four who are ineligible for public housing, but who also cannot obtain adequate housing within their financial means in the privately financed housing market. To fully accommodate the housing needs described, rather sophisticated delivery systems must be employed.

Delivery System

When an order for a dwelling unit is received, a chain of events is set in motion in order to guarantee the proper handling of large-volume production. The scope of the entire project, its financial and its social parameters, are justified and the problem is attacked by the use of advanced statistical, analytical, and design techniques. The system

2. New York Times, December 9, 1971

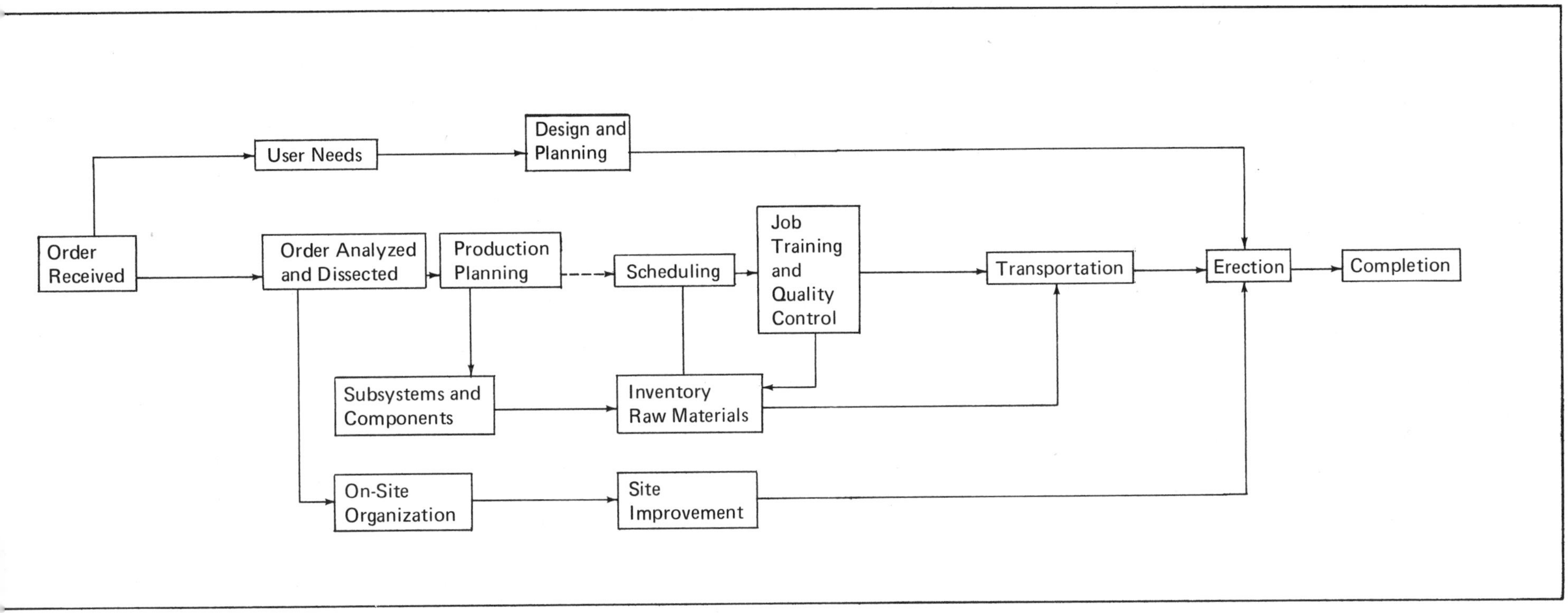

ROL SCHEDULE MODEL.

organization should have a staff with an exceptionally strong background in these methods.

Quantitative computer-assisted methodologies (design and costing systems) are significant to the entire process. It should be understood, however, that housing is a much more complex problem than any that systems analysis techniques have had to deal with hitherto. The unpredictable human element, the need for minority participation, the evolving government programs, and the reality of everyday living situations is difficult to model mathematically. This is why the informed judgment of experienced professionals with a wide variety of capabilities in all major social sciences and engineering disciplines is needed for the effective conduct of the management staff. The relevant constraints and hindrances to successful project completion are as follows:

1. Patterns of government institutions and policy.
2. Attitudes of labor organizations.
3. Lack of information about user needs and behavior.
4. Legal constraints.
5. Financial institutions.
6. The general disorganization of the construction industry.

A control schedule model is shown above.

On the individual project, the most important aspect of management is the smooth, coordinated flow of building components and labor to the site in the proper sequence and at the proper times for rapid, uninterrupted incorporation into the construction process. This involves careful scheduling and the optimum utilization of resources in manpower and equipment.[3]

Customarily, builders have employed the so-called bar charts for this purpose, and as of late, more progressive builders employ network diagramming techniques, also known as Critical Path Method (CPM) and Program Evaluation Review Technique (PERT), a variation of CPM which attempts to deal with uncertainty.

3. "Application of Statistical Decision Theory in the Building Industry," John N. Macrae and David C. Hamilton, under the supervision of Albert G. H. Dietz and Laurence S. Cutler, M.I.T. Urban Systems Laboratory, November 1969.

3-23
COMPARISON OF LABOR AND MATERIAL COST INDEXES 1950—1967/INDEXES OF SELECTED CONSTRUCTION UNION WAGE RATES 1950—1967
Labor and the unions are the well-worn scapegoats of the systems building movement, but it is a fact that wage rates have risen astronomically. *F. W. Dodge Report prepared for the President's Committee of Urban Housing*

CPM

It is important to note that basic CPM generally uses the "most likely" or model estimate for each activity duration. From these estimated durations, the "critical path" Is determined. Immediately, we are able to see the expected length of the sum of the mean durations of activities. Probability distributions for activity duration in construction are likely to be dropped, and this means that the sum of the modes is optimistic.

For more complex methods, such as those involved in a systems-built project, three basic problems arise. First, the variances and the shapes of the probability distribution associated with the activity times may cause the "real" total job time to be very different from the predicted. Second, in a moderately complex method, there will likely be a number of paths which are nearly as long as the "critical path." Thus, a number of paths may have some chance of becoming critical and the longest one will govern. Third, there may be degrees of covariance between distintly related activities which may either exaggerate or reduce the actual performance time.

PERT

PERT is an attempt to improve CPM by giving more consideration to the probability distributions associated with each activity duration. PERT calculates a mean duration time of each activity roughly; the most likely time is weighted 4/6, the earliest time is weighted 1/6, and the latest time is weighted 1/6. By adding the mean durations along each path, the critical path of the network is determined. This method makes the often incorrect assumption that there is only one possible critical path.

It is felt that probabilistic simulation offers the best opportunity to the systems management to show realistic expected times to project completion and realistic presentation of the variances associated with these expected times. It also offers a better opportunity to control the process by alerting the manager to those activities which actually might become critical as opposed to those which are near critical but unlikely to become critical because of small variance or large covariance.

When utilized to their full potential, statistical decision theories can be a powerful management tool, but few builders utilize them for more than scheduling. They can be used more dynamically for optimizing time-cost trade-offs to obtain minimum costs. Even more importantly, they can assist in smoothing out inefficient utilization of manpower and equipment. Time and motion studies are an additional tool to the understanding of production problems. However, continuity is critical, and fluctuation in volume leads to inefficiency and overall high cost.[3]

Labor Unions

Labor and the unions are the well-worn scapegoats of the systems building movement. Yet, little is really known of what the actual labor problems are and what sort of problems they will generate. One thing everyone is absolutely sure of is that, in the face of dramatic increases in demand, there can be only one effect: *drastic increases in cost.* Labor and management recognize this, and for the most part, they are jointly assessing what the future will bring. The three apparent areas of potential problems appear to be:

1. jurisdiction,
2. attitudes toward the individual trades and prefabrication, and
3. equal opportunity.

In Europe, labor problems were never a large issue, and in fact, in the United Kingdom at Liverpool (home of one of the oldest and most highly organized labor movements in the world), the Camus system has operated successfully for years without any serious incidents. In the U. S., national leaders of the AFL-CIO have emphasized their willingness to work with systems.

A recently formed total systems company has already secured a tri-trades agreement with the United Brotherhood of Carpenters and Joiners of America, United Association of Journeymen and Apprentices of the Plumbing and Pipefitting Industry of the U. S., and the International Brotherhood of Electrical Workers. In this agreement, the tri-trades unions will recognize that all shop employees will come under a shop rate which is approximately one-half the field rate. The shop union will also agree to furnish the licensed carpenter, plumber, and electrician for field operations, as well as a tri-trades union label for use on the products manufactured, plus an international agreement so that production will not be affected by local strikes and disagreements. It was also recommended by the unions that the system sign an agreement with the Teamsters to provide for the transportation of the building elements to the site.

Reese Hammond, Director of Research for the International Union of Operating Engineers, has

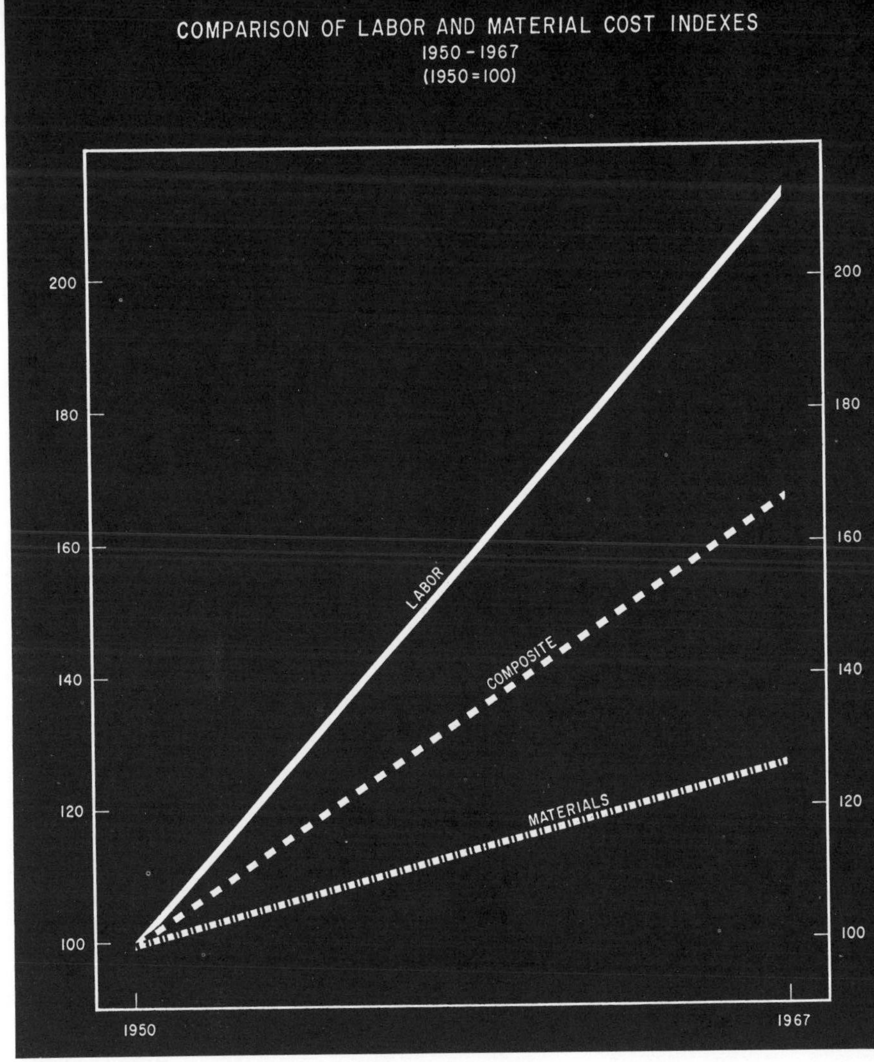

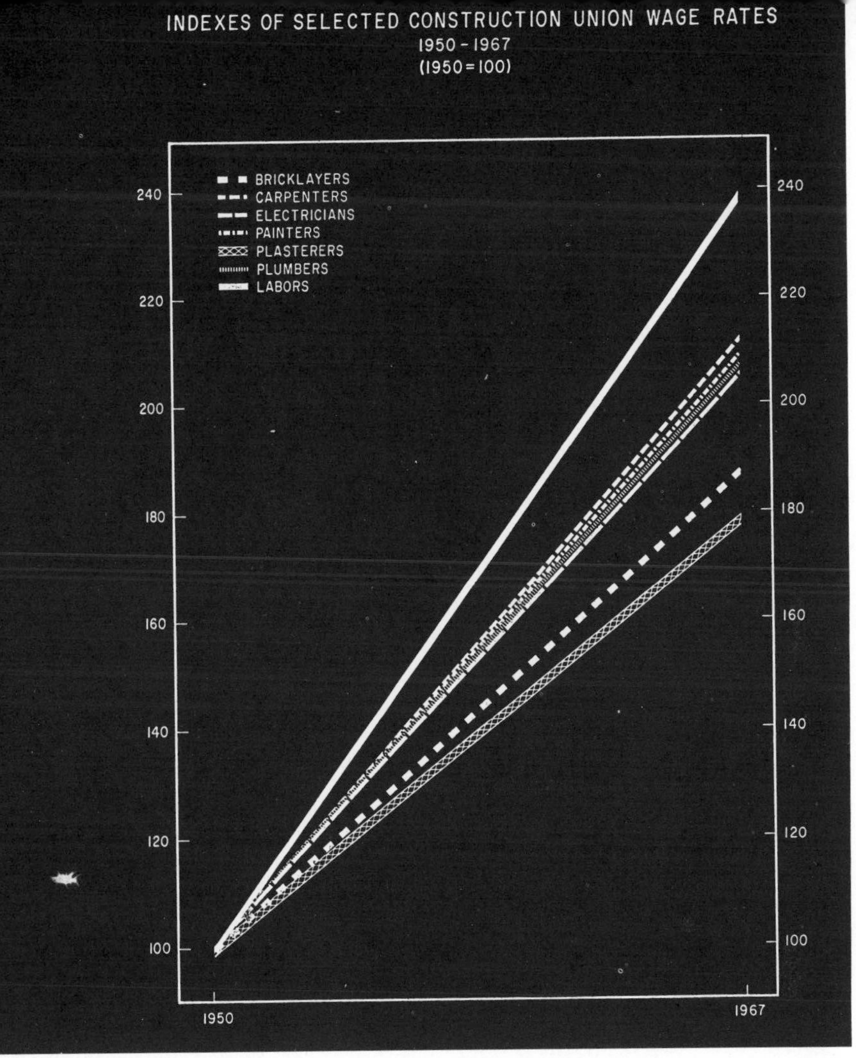

stated that there is a potential $40 billion residential building program, of which only some work is presently under union control; and the trade unions want to extend their control. Hammond further points out that with the projected substantial increases in subsidized building, the impact of industrialization will be felt more and more. He states that the unions were and are a response to industrialization, from the guild system to the industrial union to today's union, with the larger unions serving many industries, while the smaller ones are chiefly construction unions. There is a multiplicity of problems involved in building construction, which are outside the labor force, and for this reason, there is great sense in local negotiations, but the pension funds of the internationals and other union groups would even be interested in large capital investments in systems-built projects. So, it is apparent that the unions can clearly see the potential, the social significance, the increased membership, continuity of employment, and increased job opportunities.

Envisioning these opportunities, Walter Reuther, having taken the United Auto Workers out of the AFL-CIO, has formed the Alliance for Labor Action with the Teamsters. He has promised to become involved in low-cost, industrialized housing if the AFL-CIO building trades fail to respond. AFL-CIO leaders have begun to respond with a labor agreement between the Carpenters International union and Stirling Homex Corporation, (no longer in business) covering both in-plant and on-site work.

There will, of course, be a period of conflict and adjustment within the unions, while from without, they will be blamed for the delay of building systems industrialization in the U. S.

4 HUMAN ECOLOGY AND USER NEEDS

COMMUNITY PARTICIPATION

Any large-scale low-cost housing project or subsidized project, to be fully used and valued, must be an object of identification for its future inhabitants. No matter how good a plan is, if the people for whom it is made fail to feel that it belongs to them, it will not work successfully.

The systems developer should recognize resident participation as the critical element in the community's acceptance or non-acceptance of his housing system. This applies to lower and upper income housing as well, as evidenced by the desires which more and more are expressed in terms of special amenities—golf course, tennis courts, pools, security services, etc.—offered by upper-income projects. The success or failure of a development hinges on the degree of involvement of the local citizenry or anticipation of the needs of the tenants. Planning professionals, while they may be technically competent, are often not fully aware of the personal and group concerns of long-time residents of a community. Second, unless the residents themselves are directly involved in the planning process, they may be prone to undermine the planning and ultimate implementation of the housing system.

Resident participation satisfies the user's need for expression of his self-determination. It heightens his sense of self-worth and dignity, allowing him to know and understand that he has a voice in the decision-making process. The positive aspects of user participation provide the planner with the functional guidelines necessary for the most appropriate programming and implementation of his system, and the sense of personal investment made by the resident increases his concern for physical preservation of the housing development. Sociological studies have proven that most people take more pride in their housing *if* they have had some planning input. Resident participation of *both* direct and indirect users should be considered a function of the "total community."

The building system should be adaptable to

4-1
HUMAN ECOLOGY AND USER NEEDS Any large-scale low-cost housing project, to be fully used and valued, must involve the future residents in the planning process. *"Open Space for Human Needs," Marcou, O'Leary & Associates*

4-2
COMMUNITY PARTICIPATION The building system should be adaptable to the integration of behavior patterns and life styles of the user. *"Open Space for Human Needs," Marcou, O'Leary & Associates*

local conditions capable of integrating behavior patterns and life styles of the user. The programming for building in a particular community should endeavor to match this quality of response by establishing concensus within a democratic framework. In all communities there are either formal or informal groups (or both) which represent the peculiarities of its neighborhoods. In those communities where organizations are more formalized, the building system can work through these groups but not to the exclusion of the unrepresented individuals with no particular group affinity.

Recognition of the social forces of various subgroups within the community should be a key step during the survey and planning phase of any development. The focal point of concern is how members will come to be identified with and share social and even financial responsibility for the development of a community life which is alert to their particular and unique needs through these groups. In addition, there should be a concerted effort to link the smaller neighborhood unit (the family, the project) to the "total community."

Prototype Project Design

In order to fully investigate the economic and livability potentials of a building system for large-scale use, the systems builder should erect a prototype structure and undertake a phased analytic study into the viability of the system's particular effects, and once proven, into its ultimate use and feedback characteristics. Of course, what is outlined below is somewhat idealized as a testing pattern, and it may be altruistic for private firms to perform such an elaborate investigation. In fact, tests of this kind should be undertaken on a large

4-3
PROTOTYPE PROJECT DESIGN The system builder should erect a prototype structure and undertake a phased analytic study into the viability of the system.
ECODESIGN

4-4

4-4
PROTOTYPE NEW TOWN—ETAREA Satellite town to Prague with a proposed population of 135,000. A mass country retreat, excluding industry with the population living in a motley array of buildings, cars and pedestrians rigidly segregated, and an orgy of recreational activities, including 13 stadia. (Two views.) *Stavebni Zavody Praha*

4-5
PROTOTYPE NEW TOWN—ETAREA *Stavebni Zavody Praha*

scale by government agencies. The only similar kinds of studies to these suggested that we know of are in Czechoslovakia at the experimental housing estate known as Invalidovna.

Phase I will essentially be a planning effort for implementation of the prototype and preliminary planning for testing Phase II. Phase II will include implementing community involvement in the actual construction of the prototype and on-going community relations program of the prototype. In Phase III, one should evaluate the prototype from two separate points of view: the community and the professional. This third phase will aid in planning and construction on a mass housing scale:

Phase I—Planning

1. Communities are analyzed, including identification of ethnic groups, family composition and type (i.e., matriarchal vs. paternal, average size, age), life styles, economic level, leisure time activity, proximity to neighborhood services, religious affiliations, etc.

2. An inventory of local community formal and informal groups is prepared. This inventory will highlight such groups as civic and labor organizations, church groups, business organizations, youth groups, recognized and appointed leadership, and charismatic leadership. This effort is directly related to the team in building construction and finance, which will complement the inventory with a total survey of labor price and contracting capabilities in the community.

3. Professional and business services are listed, as well as community facilities, such as social

4-5

4-6
SUPERVISION Of the construction of the prototype by the community should be established with set procedures, *Magnolia Homes*

and medical services, transportation systems, laundries, grocery and drug stores, playgrounds and parks, schools, community centers, etc.

4. After collecting and assessing this information, a social science team will propose several action programs for community involvement. Through a collaborative effort, community residents and professionals will devise the program most complimentary to the community's behavior and life style. Community residents will be given staffing positions. Careful consideration will be given at this point to the basic system design as it will relate to the specific and projected needs and concerns of the community.

Phase II—Construction of Prototype and On-Going Planning

1. "Supervision" of the construction of the prototype will be the major function of this phase. Since those factors elicited from the community and other resources brought to bear in Phase I are strong determinants for planning the project, it is imperative a "community supervision procedure be established."

2. In terms of community involvement this will in fact be the most crucial phase. It is at this point that the community will actually witness and feel the labor of their Phase I efforts. The professional community worker will encourage discussion, feelings, and further inputs into the system so that members of the community will continue to feel that they had a contribution to make throughout.

3. The first step of construction as well as the final one will be celebrated by a community planned program to mark the beginning and the end of "their" community work and encourage more awareness on the part of the total neighborhood.

Phase III—Evaluation and Feedback

1. Following completion of prototype construction and after a specified period of time, the community will be asked to evaluate in detail the total process of their involvement and the completed structure. This should be thoroughly documented and used in the final housing development plans. At the same time as the community is drawing up its evaluation, the systems building team should be involved in a similar process.

2. Final evaluation of feedback on the completed project would analyze the completed housing at one year, five year, ten year intervals from a variety of viewpoints:

—realignment of desire lines from residents to jobs, shops, etc.
—friendship patterns
—preferred interior and exterior spaces
—walking distances
—adjustments and alterations to structure
—density form and social patterns
—rental/condominium
—continuing participation and management
This stage, usually not bothered with, can be the most rewarding in terms of meaningful information. The end result of both evaluations, however, should greatly minimize pitfalls and enhance the future effectiveness of planning and community schemes.

Levels of Participation

How can one elicit resident participation in four specific settings where one can visualize the greatest need for increased housing: (1) model cities neighborhoods, (2) urban renewal sites, (3) suburban/rural settings, (4) new communities?

Various methods of identifying patterns of behavior and life style can be used. All techniques available to the social science field should be analyzed to uncover those most appropriate for use on a systems project. Traditional techniques should also be employed; such as door-to-door surveys, written and verbal questionnaires and polls, fliers and leaflets mailed and hand-distributed, locations of site offices in communities which would be easily accessible; use of small groups of community residents for discussion with professional staff.

While very advanced and sophisticated techniques can be used, it should be a major aim to establish involvement on a *personal commitment basis*. This is the most effective vehicle for success. Rely heavily upon housing committees where they exist. Staffing can be made up of community residents and professional social scientists and planners. Group meetings at all stages will be encouraged with the prime goal of effecting the greatest amount of communication between the staff and the community.

Model Cities: The Model Cities Program provides a uniquely neat condition of an already established group of area representation. One of the basic precepts of Model Cities is the maximizing of participation among the inhabitants within a delimited geographic urban area specified by the city and the federal government. The common vehicle for channeling resident participation is through an elected board of residents and delegated representatives of existing community organizations and institutions. Initially, it is upon this formal group representation that the systems team can rely.

A group of social scientists can act as the "advance team" in making the initial contacts with the appropriate board representatives. As the project progresses, they endeavor to ferret out broad community needs, concerns, and interests, through in-depth studies and surveys of underlying cultural and subcultural factors. Based upon these findings and analysis, the program for construction and community participation can be established to suit the particular needs and goals.

Urban Renewal: It has been demonstrated in the past that physical improvements will not produce net gains unless accompanied by appropriate social planning. Urban renewal was conceived as a process by means of which cities could achieve the best environment for the greatest number. Historically, however, this process has tended to be one of demolition and reconstruction in blighted urban areas with little regard for the personal concerns of the citizens of the area. Although in recent years the need to extend social services and mobilization of community participation has become evident and put into practice, the bad taste of the term "urban renewal" lingers. A more apt and optimistic term which also expresses better planning theories in respect to preservation of "green areas" and avoiding deserted old city "wastelands" is the "New Town—In Town" concept.

Suburban/Rural: With the suburban/rural setting one is confronted with quite a different set of issues, including a wider diversity of socio-economic conditions, the potential for greater uses of amenities, requiring space and landscape.

A suburban/rural setting is generally assumed to be comprised of an upwardly mobile group of individuals. Value systems, standards, and life style on the whole are more consistently tied to a group identity. Its location away from the urban setting requires a more independent mode of transportation which may be a self-perpetuating fact behind a labor force and is most typically allied to the professional and businessman's sector. The average educational level is consistently higher than that of the urban center population, and likewise there is a heightened public interest in the community educational system. Small social and service groups play a big part in the life of the suburban/rural dweller, and there is generally a wider and more concerned interest in the civic affairs of the community. Generally, a low-rise capability in a system would probably be a prerequisite for entering suburban and rural areas.

4-7
URBAN RENEWAL "Physical improvements will not produce net gains unless accompanied by appropriate and adequate social planning. The pedestrian bridges in this scheme link housing areas to commercial and recreational areas with views to the water." *Scheme for Fall River, Massachusetts, 1968. ECODESIGN*

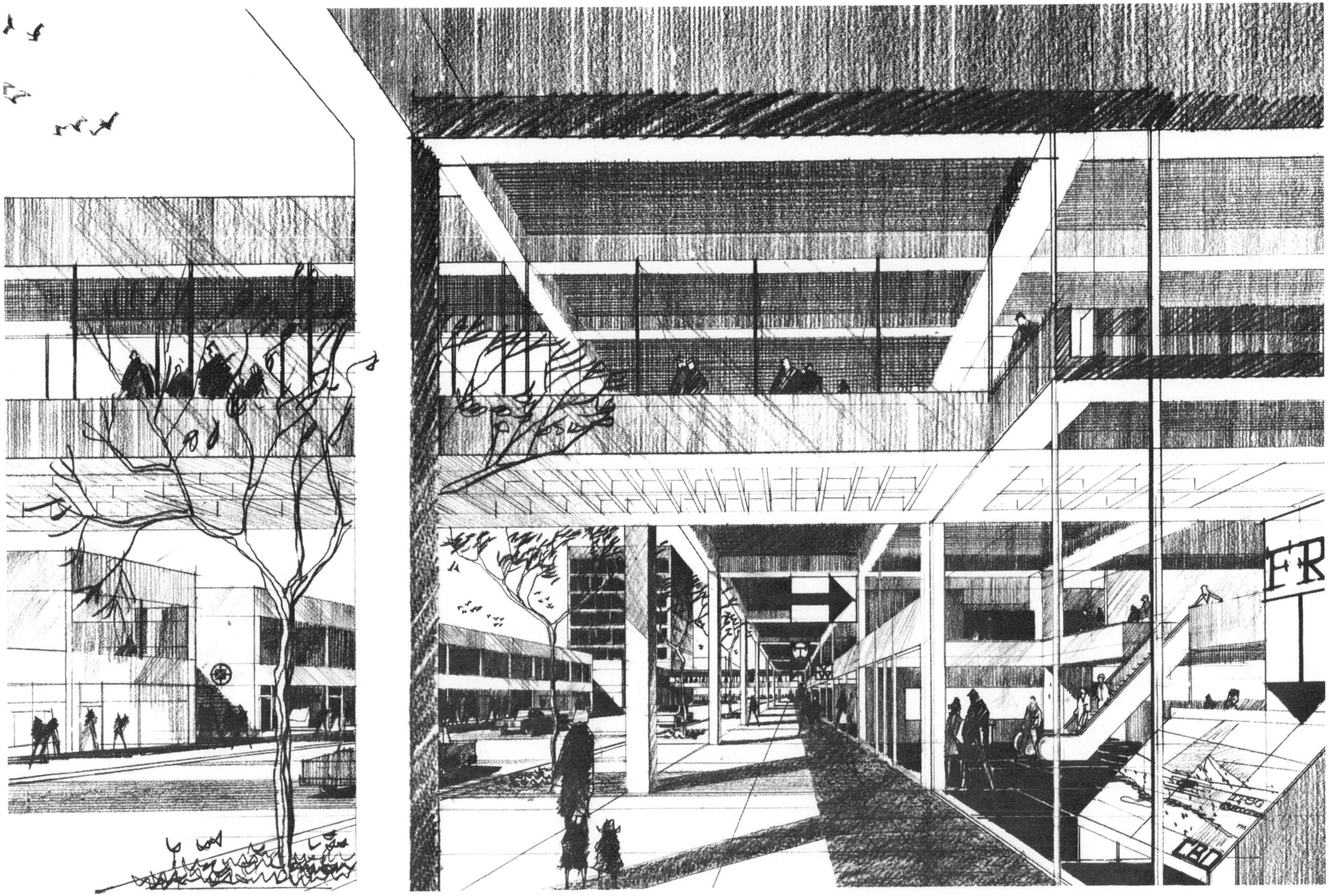
FR
CBD

4-8
Urban renewal has historically been a process of demolition and reconstruction in blighted urban areas with little regard for the personal concerns of the citizens of the area. This project in Rome, New York, retains many of the existing structures and links them with a space frame canopy.
ECODESIGN

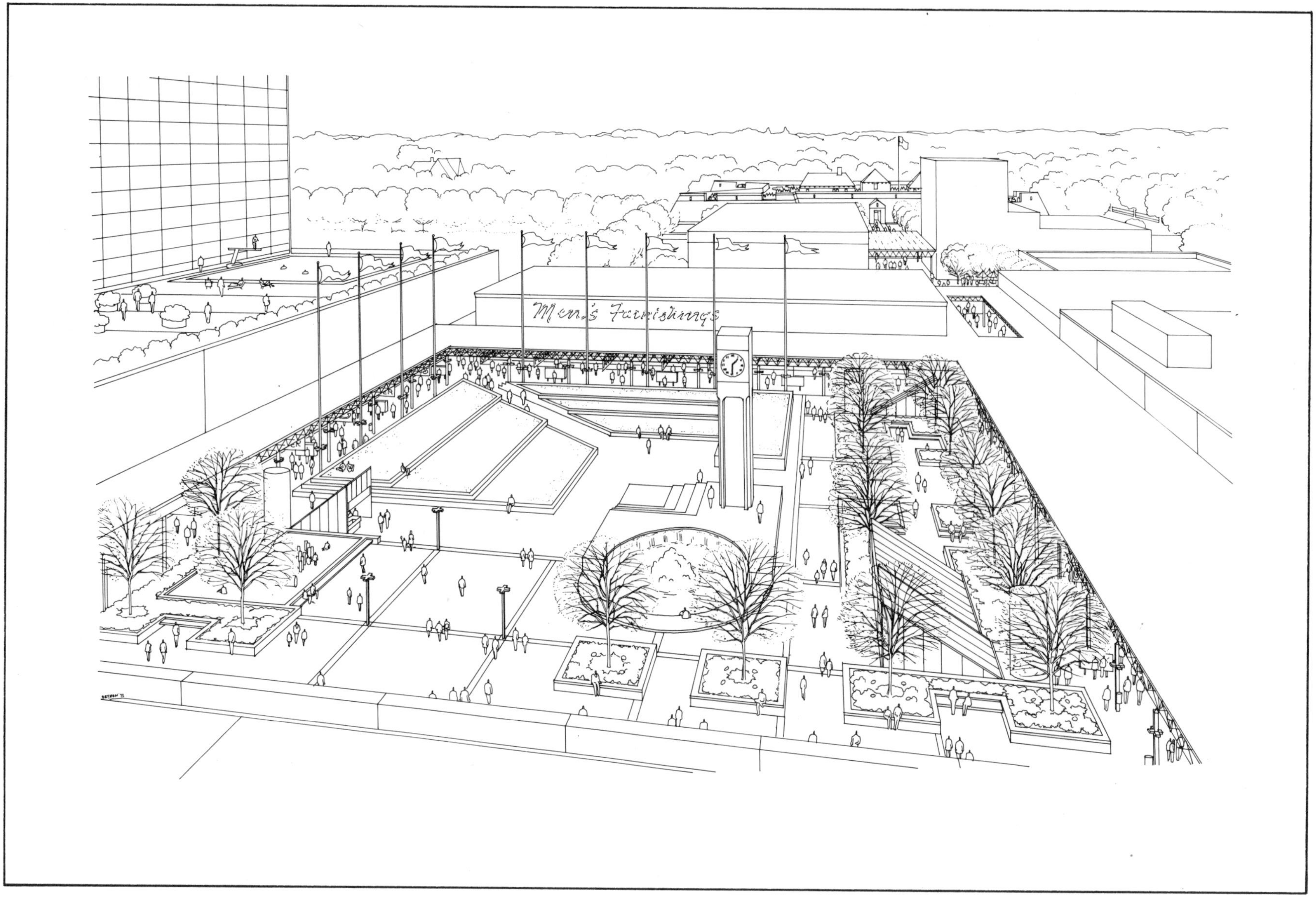

4-9
This view of Rome, New York's urban renewal area illustrates the thread, in the form of a space frame which joins together new projects and existing buildings to make a consistent environment for the community to enjoy the old and the new as part of a functional whole. *ECODESIGN*

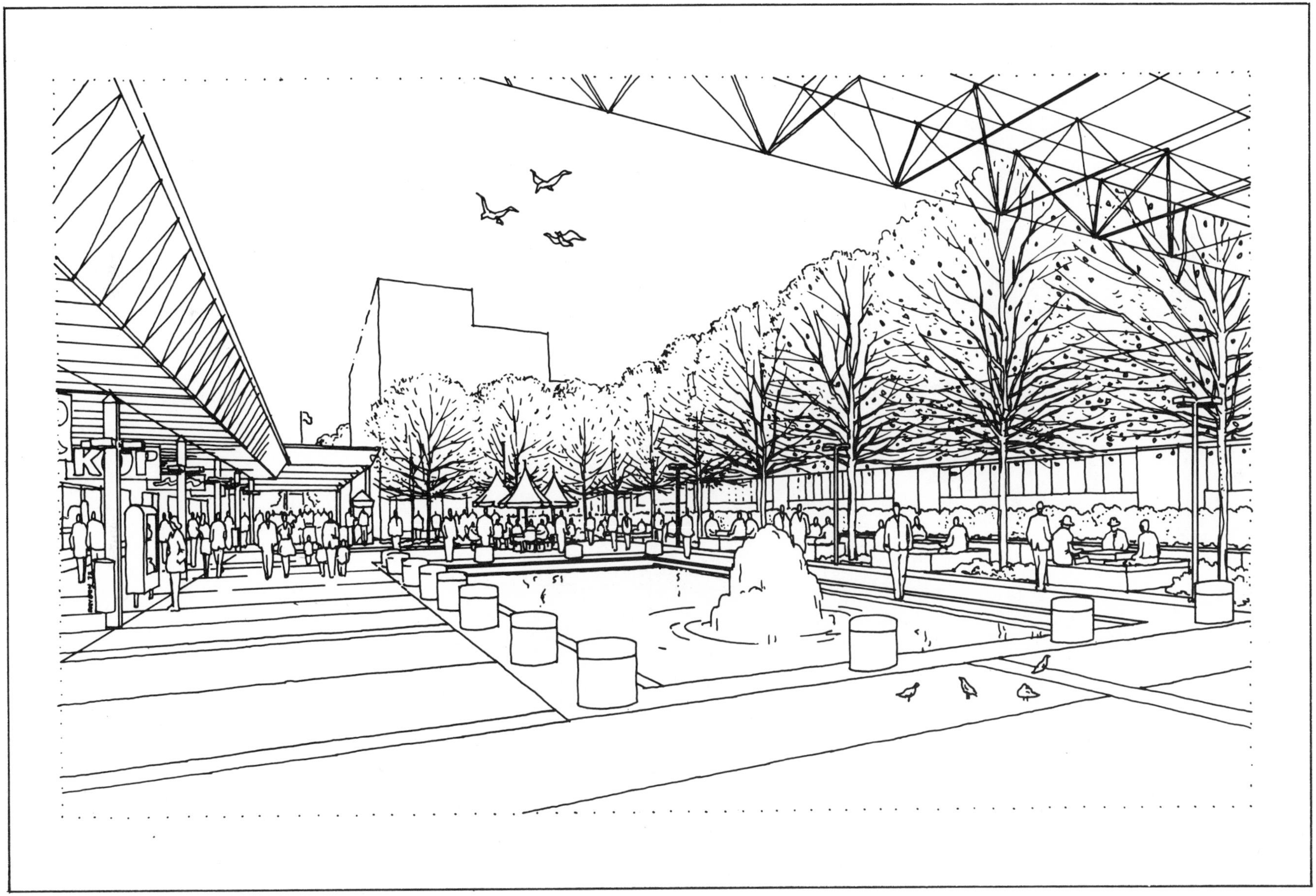

4-10

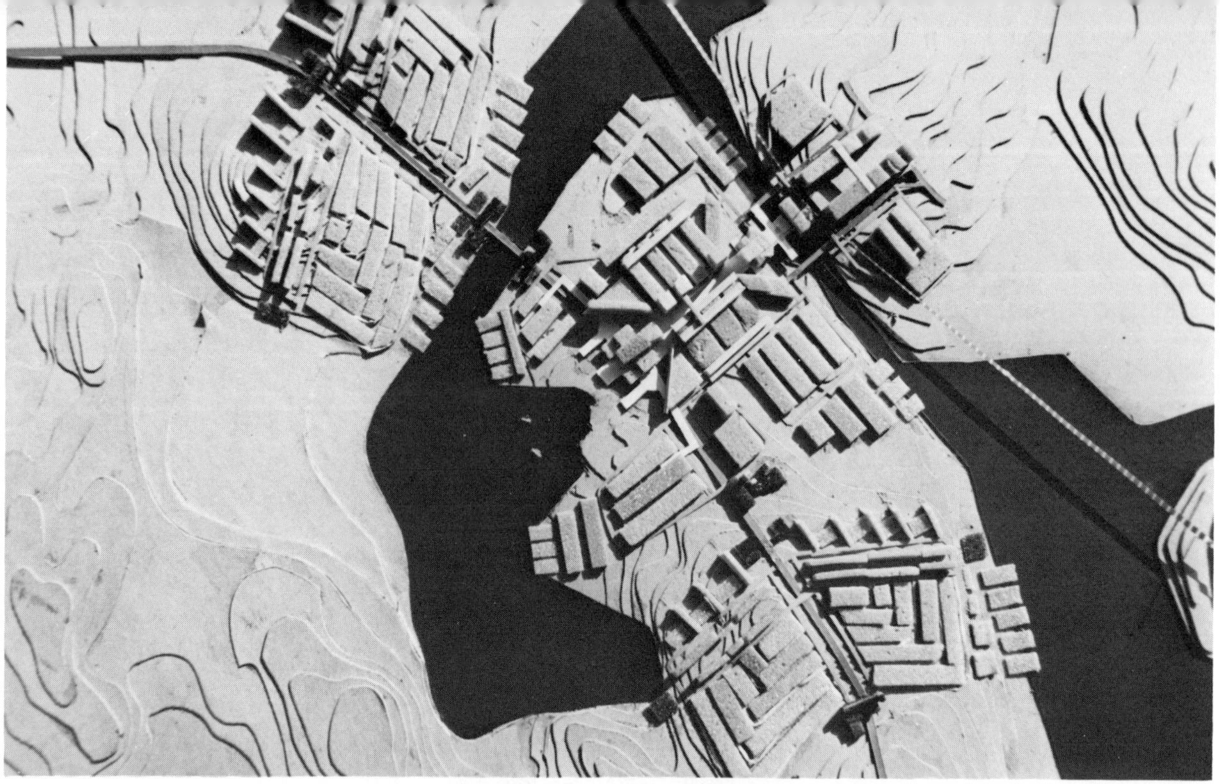

4-11

New Communities: The potential for innovative housing and community participation programs is clearly the greatest in the context of new communities. No preestablished biases exist, no ties to past methods and approaches must be overcome.

However, designing without any guidelines or instructions can be the most difficult. A building system should approach this potential market with these three areas in mind: (1) capability to build responsive projects to an evolving social and economic pattern, (2) potential for three-dimensional mixed land use with variety of building height and ground coverage, (3) capacity to keep pace with demand with no changes to the system and only minor production programming adjustment.

4-10
NEW COMMUNITIES No preestablished biases exist, no ties to past methods and approaches must be overcome.

4-11
NEW COMMUNITIES

4-10 and 4-11. *Baldwin, Cutler, Heder, Stephens: Masters Thesis, Harvard University, 1969*

Self-Help Completion

If the building system integrates mechanical into the structural shell, it may have special advantages in the area of local labor—in terms of use of local union and non-union contractors as well as the self-help tenant himself.

Once the superstructure is finished, it is complete with regard to the technical work which requires special trades and equipment, such as heavy cranes, electricians, and mechanical equipment contractors. The building could be sold at this point or turned over to local union and non-union contractors by government authorities or developers.

Local contractors, using local materials, could:

- clad the building either with traditional infill or with panels attached to fittings provided in the superstructure;
- place mechanical units on plugs provided in the slab or walls;
- install appliances;
- attach plumbing fixtures to wet walls;
- point walls and ceilings directly as formed or finish as system requires;
- lay flooring on power floated or screeded slab;
- install cabinets and partitions;
- apply roofing;
- do site work and utilities preparation prior to arrival of system;
- do landscaping and final site work.

With most systems, the completion of the superstructure and of the mechanical do not make a clear-cut stopoff point for this type of completion arrangement. With panel systems, the structural

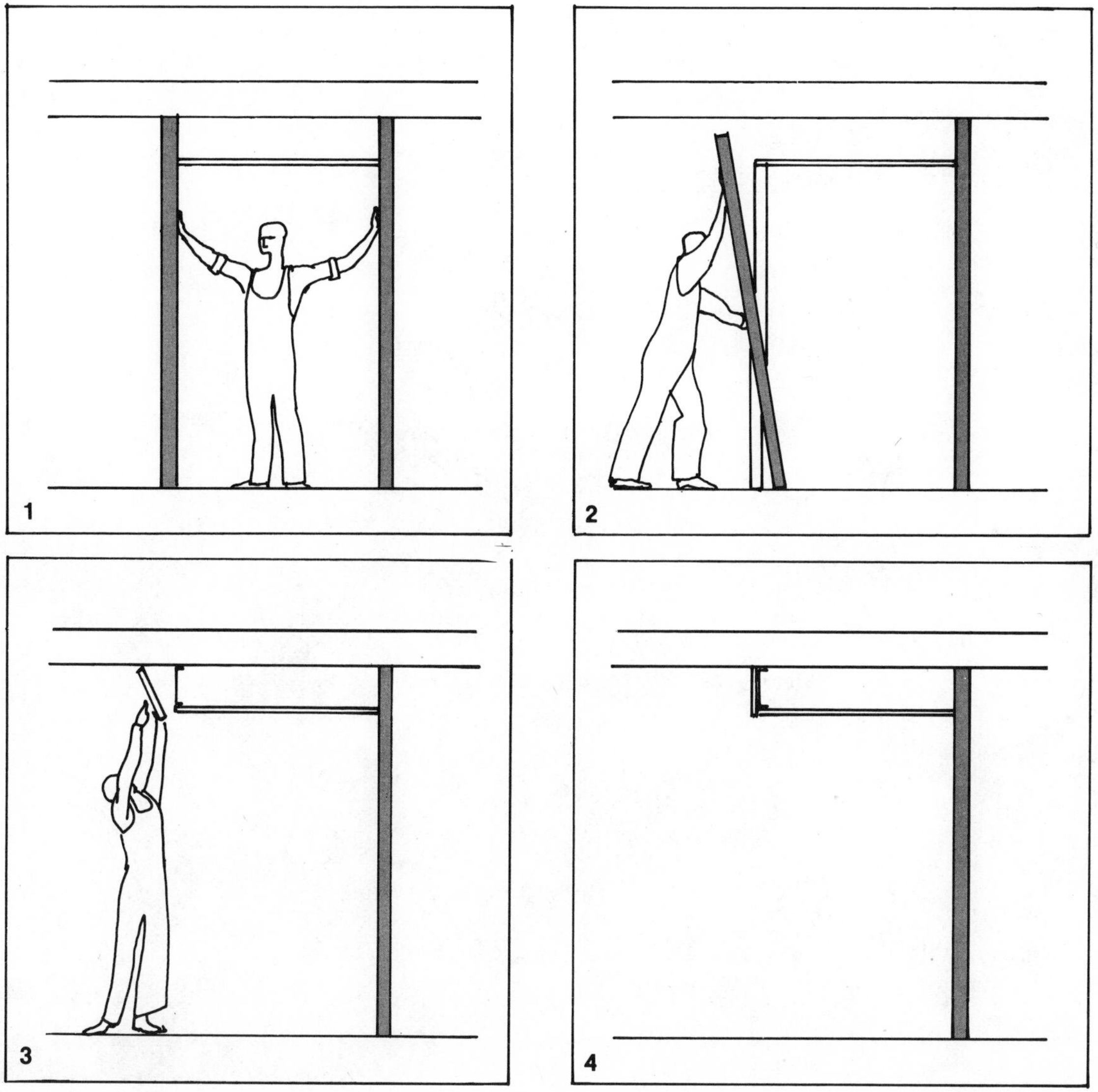

4-12
SELF-HELP COMPLETION If the building system integrates mechanical into the structural shell, it may have special advantages in the area of local labor. *URBS, Building Systems Development, Inc.*

4-13
NEW BUILDING TECHNOLOGY Technological change opens up new social possibilities in housing. Shown is a 5" wall casting of a monolithic box being removed from an on-site form by a gantry crane. *H. B. Zachry*
Photo by Tell-Pics

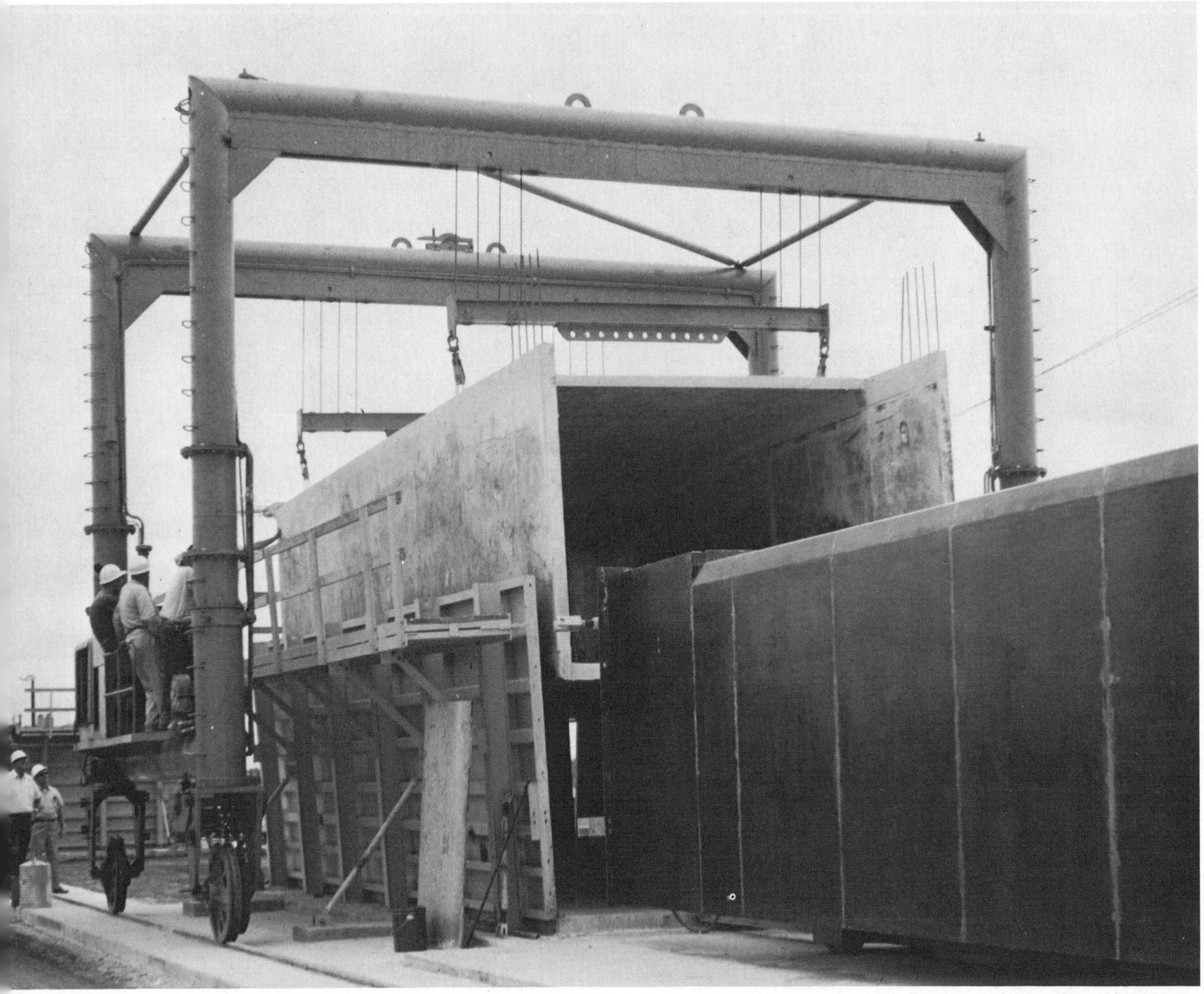

stability of the building is often tied in with the facade, or the method of attaching panels is so involved that the operation cannot be isolated.

Much of the work described above could be done by the occupants if self-help was a community-need requirement. They could install cabinets, partitions, closets, and some appliances, as well as apply flooring and wall finishes (paint or wallpaper). In some areas, it might even be feasible for the occupants to work with neighbors to clad the buildings.

This type of major completion work is obviously not often feasible because of special skills and investment in tools (carpentry, masonry) which might be necessary. In climatic areas where proper thermal seal is important, it is less feasible than in areas where closure is more a matter of privacy, sun, and rain protection. This is true with any building or system.

It is important to note, with respect to high-rise housing, that with an on-site concrete system the safety, mechanical, and code necessities are complete in the superstructure. Unlike frame systems, the units are divided from each other by concrete walls which provide fire and acoustic insulation to each dwelling. This cannot be easily assured when intra-unit partitioning is done in a self-help manner (i.e., frame infill).

NEW BUILDING TECHNOLOGY

Technological change should open new social possibilities in housing. Presently, low-income families are rarely able to obtain or finance housing on the open market and are forced to go through governmental agencies.

The social and institutional constraints operating against the construction of private low-income

4-14
DETERMINING USER NEEDS Normally, all users are anonymous to the architects, and they vary greatly according to their incomes, age, tastes, and habits. *"Open Space for Human Needs," Marcou, O'Leary & Associates*

housing at the moment outweigh the forces in favor of it, in terms of legislation, financing, methods, etc. Social change should and can make new demands on technology. Thinking of the future means more thinking about social opportunities and technological potential together and studying their likely interaction.

Until recently, the object has always been to provide "cheap" or "low-cost" housing without realizing the relative costs and benefits to be had by determining the user's preference and needs so as to satisfy him and decrease vandalsim and social maladies such as experienced in the infamous Pruitt-Igoe in St. Louis. Low-income families are not only interested in the housing as shelter but also in the services, safety, total environment, location, maintenance, and in ownership and alternative financing methods as well.

Understanding that, although technology offers the possibility of better, less expensive, and more accessible housing, it also offers: (1) large-scale use of machines, (2) large-scale use of factory-produced standardized building components, (3) large-scale projects constructed by repetitive processes, and (4) coordination of management leading to additional programming and control. In this light, it is easy to see that the user must participate in a process imbued with these prospects of inhuman solutions.

Determining User Needs

Normally, all users are anonymous to the architects or the developer and they vary greatly according to the composition of their families, their incomes, age, tastes, and habits. The important thing is to relate space and design criteria to these people's common and special activities in and

4-15

"SPACE IN THE HOME" An attempt by the British government to provide a specimen analysis of a dwelling unit plan. The British counterpart to our F.H.A. Minimum Property Standards are the Parker-Morris standards. The influence of this "Space in the Home" study has caused HUD to begin the use of new *Unit Design Criteria* to replace FHA and HAA design and construction standards, just as the Parker-Morris standards have grown obsolete. *"Space in the Home" Design Bulletin #6, 1968, Her Majesty's Stationery Office, London, England*

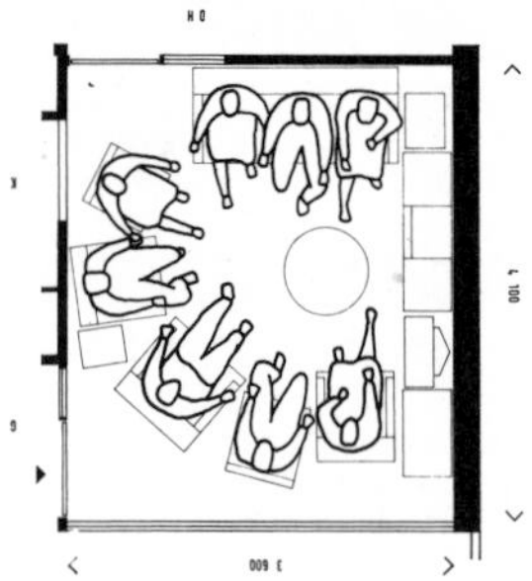

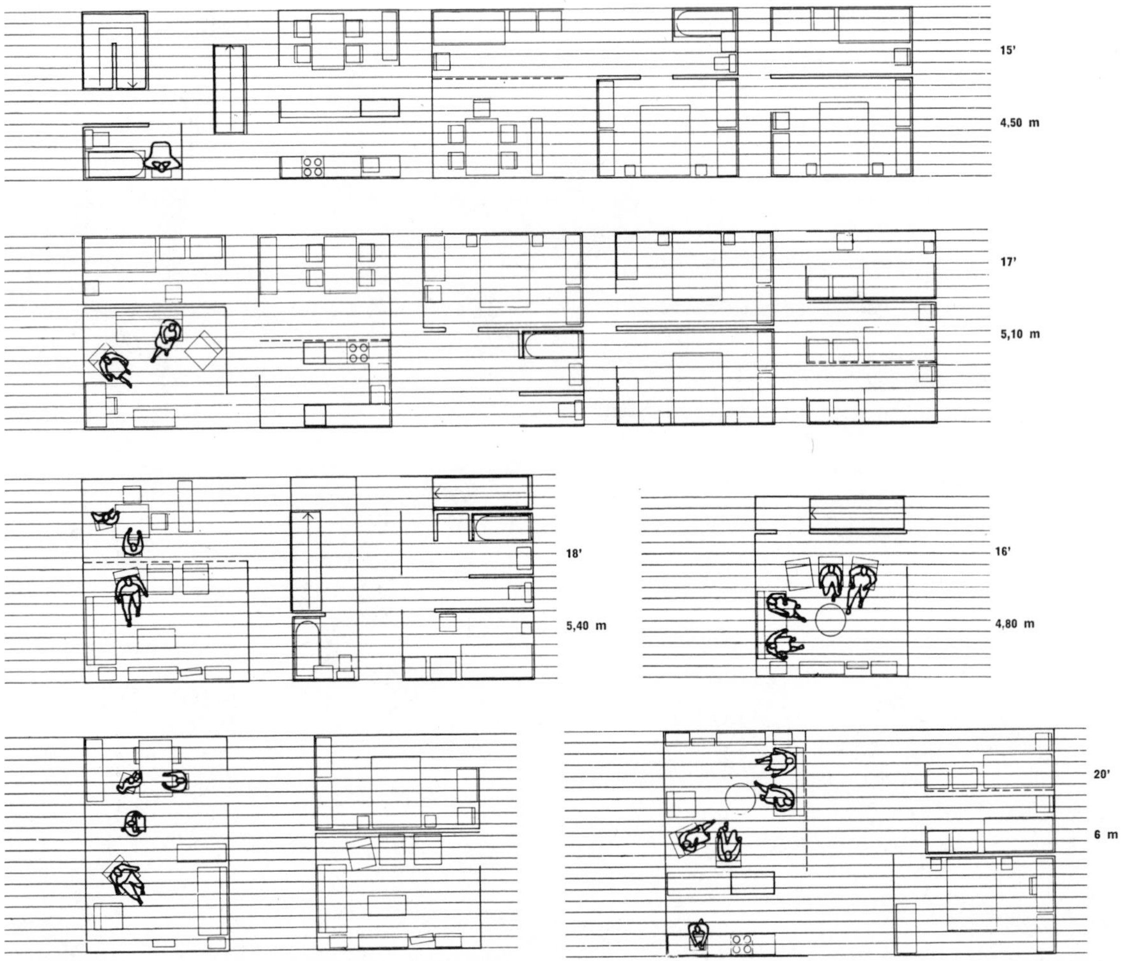

around the home and within the time constraints of the typical day or week and even lifetime.

The definition of user needs as a part of the design process is itself a product of the changing social environment. In fact, prior to the beginning of the modern movement (the early part of the twentieth century, which implicitly stated that buildings must function as well as provide shelter), architects researched styles or expressed the materials and only subconsciously expressed the social values of the period. In the most recent past, dimensional standards such as *Architectural Graphic Standards* (Ramsey & Sleeper) have been used and the architect's task was simplified through the use of such guidelines—measurements evolved through use—and his scale ruler. Today, the force of psychology plays a role in design, the molding of space, and the use of color and texture. The understanding of the kinesthetic experience—the movement of people through space require more than guidelines and standards—brings new tasks to the designer.

The new design tasks include:

Program development—determining the dimensions and configuration according to minimum space standards for daily functioning and family size, degree of privacy, openness versus closeness, and scope of the final building envolope massing, interior and exterior public space allocations—decisions on inclusion of recognized amenities such as garages, patios, extra bath, storage (based on economics, not cost/benefit).

Kinesthetic expression—the understanding of the movement of the users through the space, their activities and actions, the evolution of use and seasonality of interior and exterior spaces.

Urban context—the *relative* and *future* roles of the project are now considered as important as its initial *raison d'être.* The project infrastructure is architecturally influential on the ultimate morphology of the city. The flexibility of a city's buildings to adjust to new social patterns and uses can critically effect the longevity of entire sections of cities.

Circulation networks—transportation systems, meaning: pedestrian walkways, mass transit and vehicular traffic and parking, are prime considerations in project siting, scope, and form. The movement and flow of services relate to the densities, present and future, and the predicted or existing "desire lines." When the circulation networks and desire lines do not overlap, the problem must be dealt with before a project is sited. Only in cases of extreme size (e.g., Co-op City) can the developer assume that the size and location of his project will force a solution.

Building design schematics—planning the requirements for space and quality of environment, determining the overall cost yardsticks for the developers, allowances and percentages of open space, as well as the "suitable" and "saleable" appearance and scale of the project lie here, as does its physical adaptation to the existing environment. This is its adaptation in terms of massing, so as not to block sun, create wind tunnels or dead-ended spaces, and in terms of scale and materials, so as not to conflict with the existing "textures" of the area.

User needs—traditional standards for use involved accommodating the zoning and building codes and the requirements of unions, authorities. The new attitude towards determining and accommodating use has developed this aspect into a major task and a variety of methods have been devised for its determination.

User-needs gaming simulations have been utilized by social psychologists/architects to produce games which allow community participation that indicate a wide range of specific priorities of the residents. *The "gaming" approach has the advantage of introducing alternatives which are not a part of the actual experience of the user.* Without such an introduction of non-a priori alternatives, the only source of information is what the tenant thinks of his existing housing and how he would like to see it changed.

Relating the results of "User Needs" studies, evaluating them in terms of their "Cost Benefits" (e.g., would maintenance cost and vandalism be lower if garbage disposals were standard items?), and integrating these discussions into standard space/function programs of a "machine for living" involve questions which should be considered for each new situation.

The following examples of a survey of consumer needs and preferences 1–7 (From *Technology in Connecticut's Housing Delivery System*—Department of Community Affairs of the State of Connecticut) are typical of the new information which must be incorporated into the more traditional architectural programming tasks.

1. *Attitude towards construction techniques and durability*—Although general studies indicate that the technology issue per se was of little concern to users in relation to their interest in spatial and social wants, durability, general appearance, and function are key criteria for tenants. Although none of these is as yet fully prejudiced in the public's mind as being favorably or unfavorably related to systems building, the tendency is to associate the greater precision of advanced technology with better function and durability while general appearance is still unresolved.

2. *Social services and institutional operations*—Surrounding housing are more important to low-income households then is housing itself. The problems of garbage collection, snow removal, maintenance and repair, adequate schools, secure play spaces for parents with small children, and physical security on the streets of the neighborhood were more important than the problems of housing per se.

3. *The mixing of persons* having different socio-economic, household, and racial characteristics is crucial in multiple-family dwellings. Respondents felt that the clustering of a single category of low-income persons (mothers on welfare, young black couples with children, and so forth) reinforces the poverty cycle by depriving people of varied examples of ways to cope with and overcome poverty (for instance, by saving money to buy a house, or by undertaking one's own routine maintenance). Furthermore, such homogeneous units become derogatorily stereotyped, thus adding to the problems of self-respect and responsible behavior.

4. *Present public housing authority projects are perceived by occupants as being self-defeating* and detrimental to both the residents and the community. For the same reasons as set forth in 2 and 3 above, low-income projects have the following deleterious effects:

—They provide a negative incentive system in which the resident wishing to add to his income so as to save for private housing is penalized in the form of added rent whenever his income rises.

—They reward instability and disruption of the family via Aid to Dependent Children, reduction of rent for the unemployed, etc.

—They develop an unreal view of costs, responsibilities, and risks by providing tenants with a subsidized situation requiring little responsibility.

5. The exception to 4 is housing for the elderly poor, which seems to be well-suited to their needs; however, demand far exceeds supply.

6. *Opportunities for homeownership* and for the sharing of responsibilities for maintaining the environment appeal to a large proportion of low-income households, but:

—They need education in home care. An understanding of how to repair and maintain property is crucial to homeownership and was a key distinguishing factor between homeowners and renters in the black community.

—The housing must be small in scope and cooperatively owned rather than large in scope (hundreds of units) or condominium. Low-income persons, like middle- and high-income persons, want some degree of control over the type of neighbor they have and how his property is maintained.

—An insurance pool mechanism needs to be developed to minimize perceived and actual risk. When employment is uncertain or seasonal and savings are limited, to use savings for a down payment on real property involves a serious risk. The risk is perceived as being even greater than it actually is, and it is a deterrent to some families who would like to purchase homes. A simple insurance pool for full or partial payment of interest and accrued taxes during a limited period of unemployment could alleviate much of the perceived risk in such situations and encourage homeownership.

7. The households with the greatest needs are receiving the least attention or satisfaction.

—Welfare mothers need more services and have the hardest time finding suitable housing in the open market.

—Young couples with more than two children find it almost impossible to find landlords willing to accept them, and even project housing is insufficient in terms of space (bedroom and living areas).

Evaluation of Human Ecology and Feedback

The major source of information on ecology of a species is a study of existing habitats in detail. However, this is more often done for the sea otter, for example, than the human.

The English have begun this type of analysis of feedback on constructed projects, but even their *The Private Housing in London—People and the Environment in Three Wates Housing Schemes* by Shankland and Cox is unique.

This study was conducted by means of questionnaires and interviews and was made on three projects of varying size, age, location, building types, and densities and layouts. The areas of study were broken down as follows and results demonstrated with graphics and site and unit plans overlaid with comments.

People and movement

Previous dwelling
Reasons for buying
Turnover and potential movers
Reasons for moving
Conclusions and design implications

People and homes

Who lives where
Comments common to homes of all types
Comments on each type of home
Alterations and improvements
Conclusions and design implications

People and layout

District
Appearance and landscaping
Garaging and parking
Privacy and private open space
Children's play and common open space
Friendship
Maintenance and management

The information gathered was interpreted in the eight overlapping categories. The following summary of the more important conclusions shows the *definite* design implications that can be drawn from this type of conscientious preference study of as-built projects.

Common Open Space The policy of providing generous landscaped open space on all schemes

4-16
EVALUATION OF HUMAN ECOLOGY AND FEEDBACK
The remaining illustrations in this chapter are from "The Private Housing in London. *Shankland & Cox, London, England*

Likes and dislikes by dwelling type

NUMBER IN SAMPLE	29	10	21	13	28	4
	MAISONETTES	FLATS	2 STOREY STANDARD H'SE.	3 STOREY HOUSE	2 STOREY 4 BED. HOUSE	2 STOREY 3 BED. HOUSE
Convenient, easy to run	42	60	52	29	50	54
Space, size of rooms	38		14	43	75	46
Open plan	24	30	52	21	50	
Separate laundry room				40	25	
Garden		40	10		25	31
Being on one level		30				
Cupboards	38	20		46	75	46
Windows-light and bright	14	60	19	29		38
Central heating	14	20	48	32	25	31
Outlook	10		10	14		
Finishes and fittings	31			11		
Value for money	17					
First floor living				36		
3 Storey Arrangement				14		
Not enough space	17		29	57	25	38
Open plan		20	43		50	23
Want separate kitchen	31	30	10	18		46
Kitchen too small	21	60				
Want separate W.C.	24	30	29			15
Not enough space in hall	10		10			
No private open space	35					
Not enough storage		20		14		
Stairs				18		
Garden			19			
Heating-hot water	65	50	33	18	50	38
Detail design points	48	50	48	40	100	46
Planning faults	21	20	24			
Dustbin cupboard smells	28	20		14		
Windows - size and shape	28		14	14	25	31
Kitchen fittings	24		10	11		
Internal sound insulation	17		10	11		31
Outside or party insulation				14	25	15
Condensation problems	28	30				23
Layout faults				14	25	15
Flat roof						23

* BARS REPRESENT PERCENTAGES

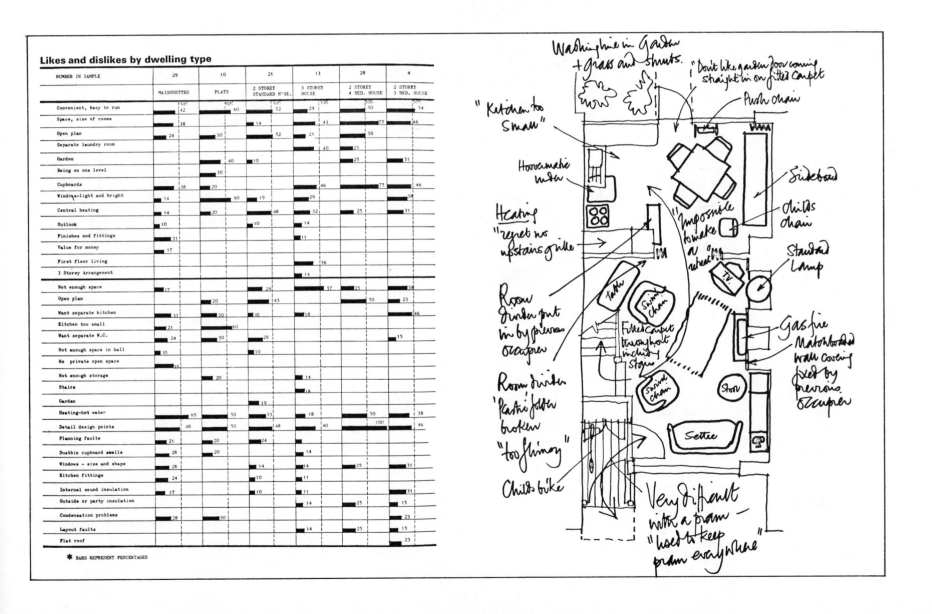

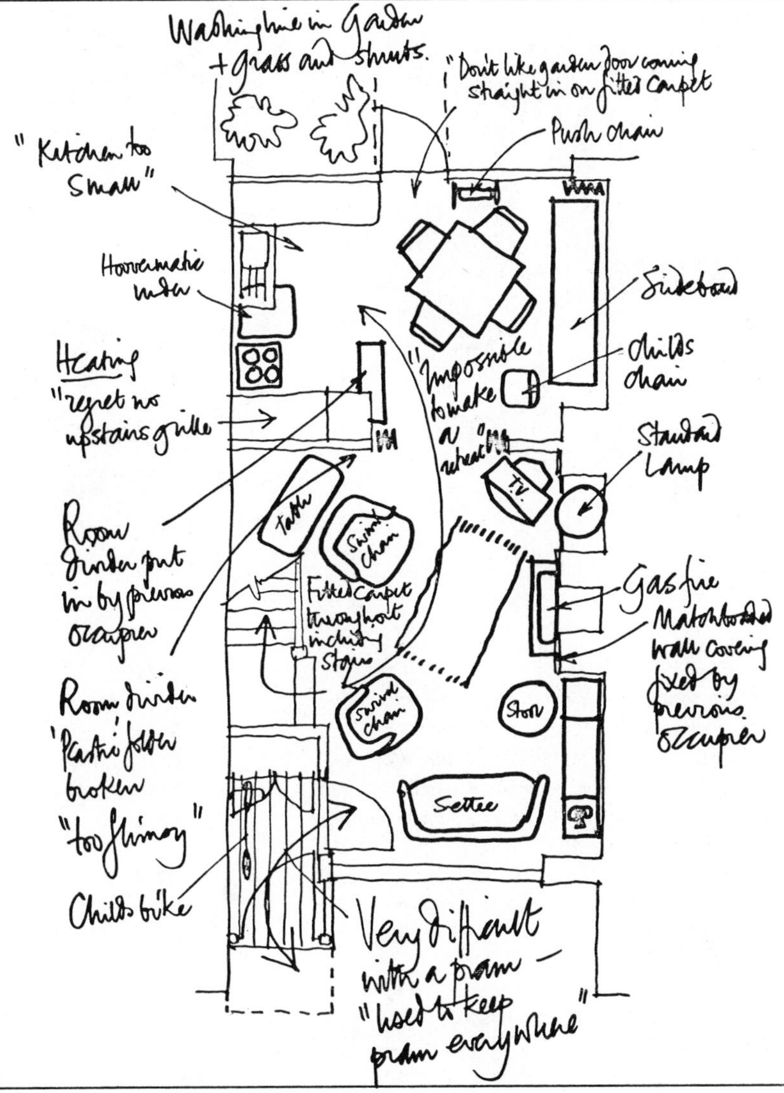

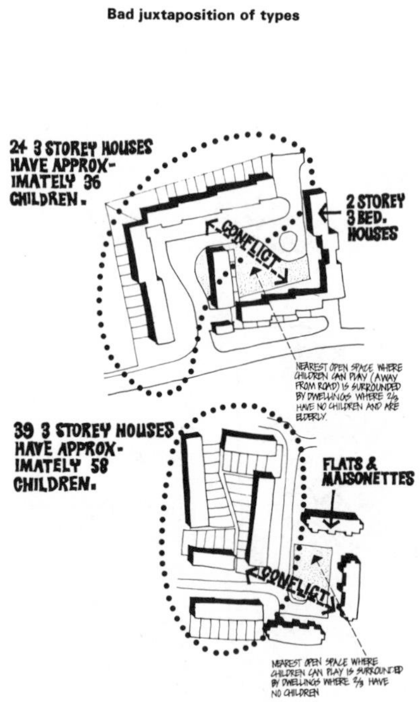

4-17

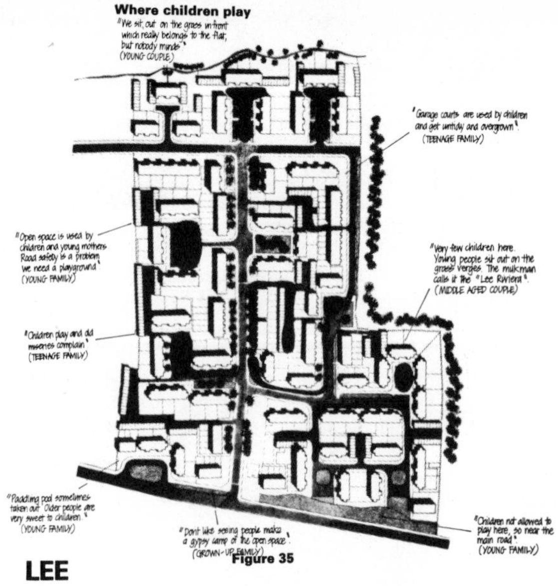

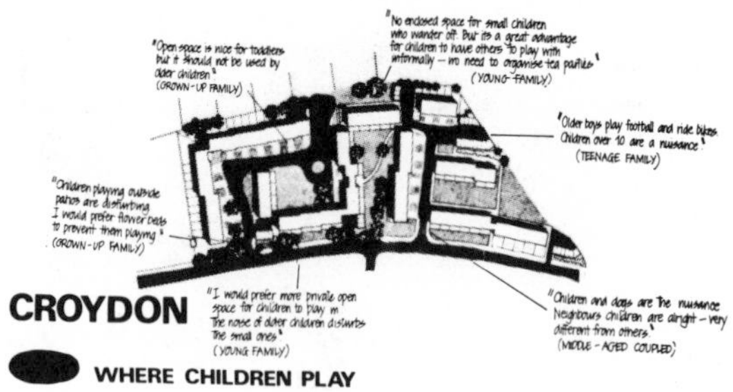

4-17
CHILDREN AND OPEN SPACE
Shankland & Cox, London, England

4-18
PRIVATE OPEN SPACE *Shankland & Cox, London, England*

was a primary source of satisfaction to most people. Many expressed pleasure in the appearance of those grassed and planted areas which were well landscaped and maintained. Maintenance was generally best in those schemes where a management company operated. Collective responsibility for management and the more informal layout associated with it appeared to people's satisfaction and noticeably increased the number of local friendships.

* Each scheme must have a reasonable landscape investment or a good inheritance (i.e. mature trees).

* Each scheme requires good-quality landscaping, the more so if it lacks mature trees. Where money is tight, the quality must be preserved and effort concentrated on fewer areas.

* It is a mistake to mix management methods on the same scheme. Layouts of management schemes should be so designed that all contributing members have a sight of the amenities for which they are paying.

* Further experiment is needed to discover how big management companies should be and what relationship they should have with the developer.

Children and Common Open Space In common with schemes at much higher densities, children's play was the major source of conflict and difficulty. Some people wanted open space for their children to play on—others as something to look out on.

* At least two distinctive types of landscaped area are required: those suitable for children to play on (the majority) and those from which children are excluded.

* All hard surfaces (including garage courts) must be designed for children's play, and made as safe and interesting for them as possible.

* Concentrations of dwellings likely to produce a high local child density must be avoided (e.g., a cluster composed only of three-story narrow-frontage houses).

Private Open Space Most households strongly demanded some genuinely private open space. Most people did not want big gardens, but those with children considered the "patio" inadequate.

* Every dwelling, including those on the upper floors, should have some private open space, at least part of which is well screened.

* A few larger dwellings should have gardens of over 750 sq. ft.

Garaging and Parking The current level of car ownership on these schemes (1.1 cars per house-

Conclusions and Design Implications

Densities of around 15 dwellings per acre, in part a reflection of high land costs, inevitably produce conflicts between economy and privacy. It is significant however that on these schemes the general balance between private and public open space was considered about right. Most people wanted some genuinely private open space, however, and the patio was considered inadequate by families with children mainly because it was too public.

Overlooking was not a serious problem, being mainly confined to closely spaced parallel terraces where living spaces and bedrooms were exposed to view from upper living levels, and particularly where the kitchen required people to be near windows for a long time.

* Every dwelling, including those on upper floors, should have some private open space, at least a part of which is well screened.
* A few larger family dwellings should have gardens of over 750 square feet.
* Open fronts require special landscape protection from dogs and children, and from tradesmen taking short cuts.

Average garden size

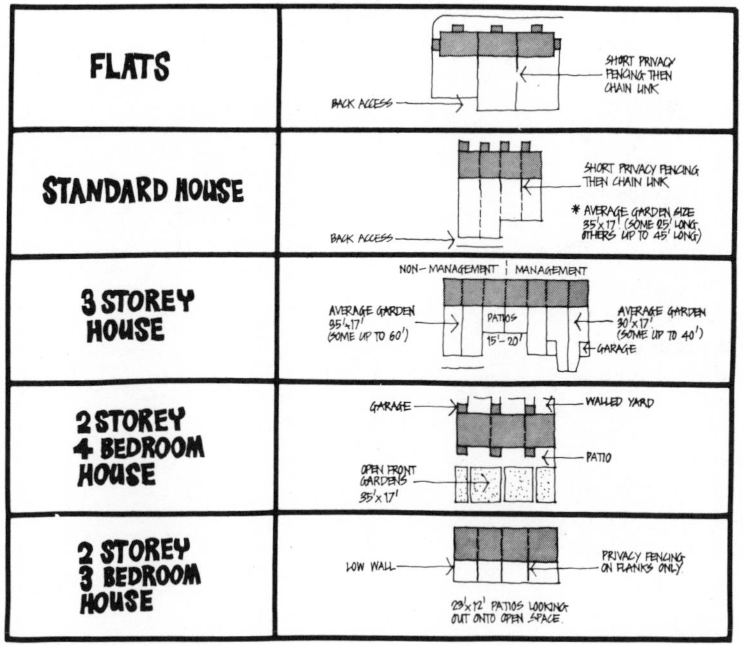

Appearance and landscaping

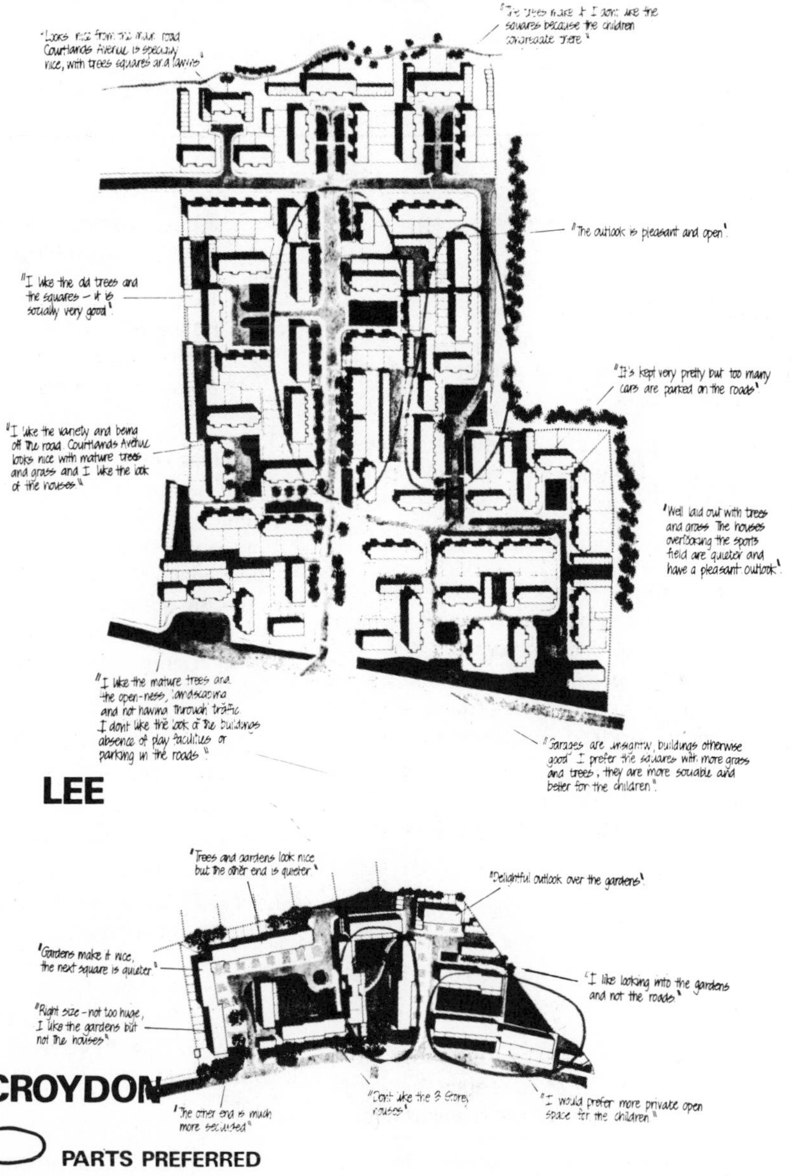

hold) is beginning to cause difficulties in parking. (1 garage and 0.3 visitors parking spaces per household).

* Schemes should be designed for 2 car spaces per household (1 garage and 1 visitors parking space) and allow for the addition of further open parking later.

* Garages must be big enough to take storage or workbenches in addition to cars and be equipped with water supply and electric power.

Families and Dwellings There is a consistent relationship on all schemes between type of dwelling and type of household. This means that it is possible to predict the range of households likely to be attracted by current designs. The present policy of providing some low-priced dwellings in the mix meets a real need for those predominantly young families purchasing for the first time to whom low cost is of paramount importance. This policy should be continued.

* The developer should indicate to the local education authority at the initial planning stage of future schemes the likely nature and growth of the child population.

* Layouts should be designed with the likely occupants in mind so as to reduce the danger of conflict and maximize the range of choice.

* There should, however, be a greater degree of separation in some parts of the scheme. In particular, dwellings likely to attract families with children should be clustered together around common open space, on the "private" side of the dwellings, suitable for play.

* In other clusters, dwellings for older childless households should predominate. These should enclose amenity open space to which children do not have access.

Families and Friendship Families with young children had more local friends than any other household type. They made friends quickly, supported one another, and particularly appreciated a layout that enabled their children to play safely together near the home without supervision.

Households generally appeared to make most local friends amongst households of a similar type to their own. Where dwellings were grouped with patios around a common open space, people had more local friends—this was true not only of young families but of older people as well.

Each housing scheme needs a variety of dwelling types so as to offer freedom of choice and to counter the danger of physical and social uniformity. This argues for a mixture of dwelling types. On the other hand, there may be tensions between residents if, for instance, the dwellings likely to appeal to older people are placed in close proximity to common open spaces where children are likely to play. Layouts should be varied to cater to both those who like "mixing" and those who do not.

* There should be some mixture of dwelling types both in the scheme as a whole and in its various parts.

4-19
FAMILIES AND FRIENDSHIP
Shankland & Cox, London, England

The proportion of different kinds of family in each dwelling type

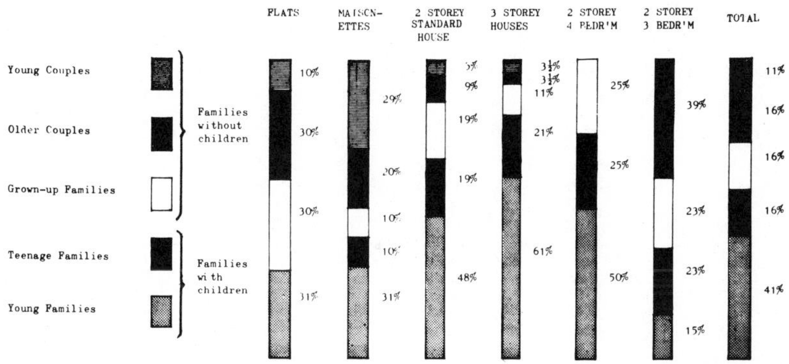

THE NUMBER OF PEOPLE PER DWELLING

	FLATS	MAISON-ETTES	2 STOREY STANDARD HOUSE	3 STOREY HOUSES	2 STOREY 4 BEDR'M	2 STOREY 3 BEDR'M	TOTAL
	2.8	2.5	3.2	3.6	4.75	3.2	3.2
NO. OF BED SPACES PER DWELLING	3	4	5	5	6	5	

The number and ages of children in each dwelling type

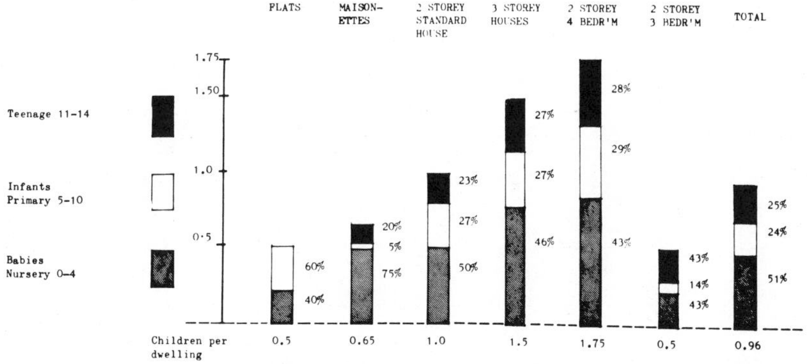

FRIENDSHIP PATTERNS

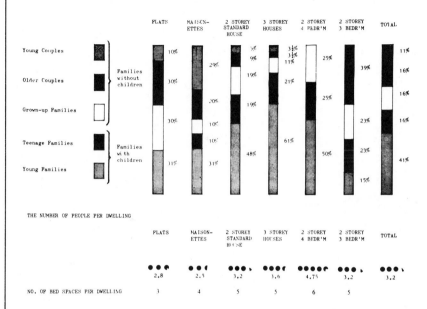

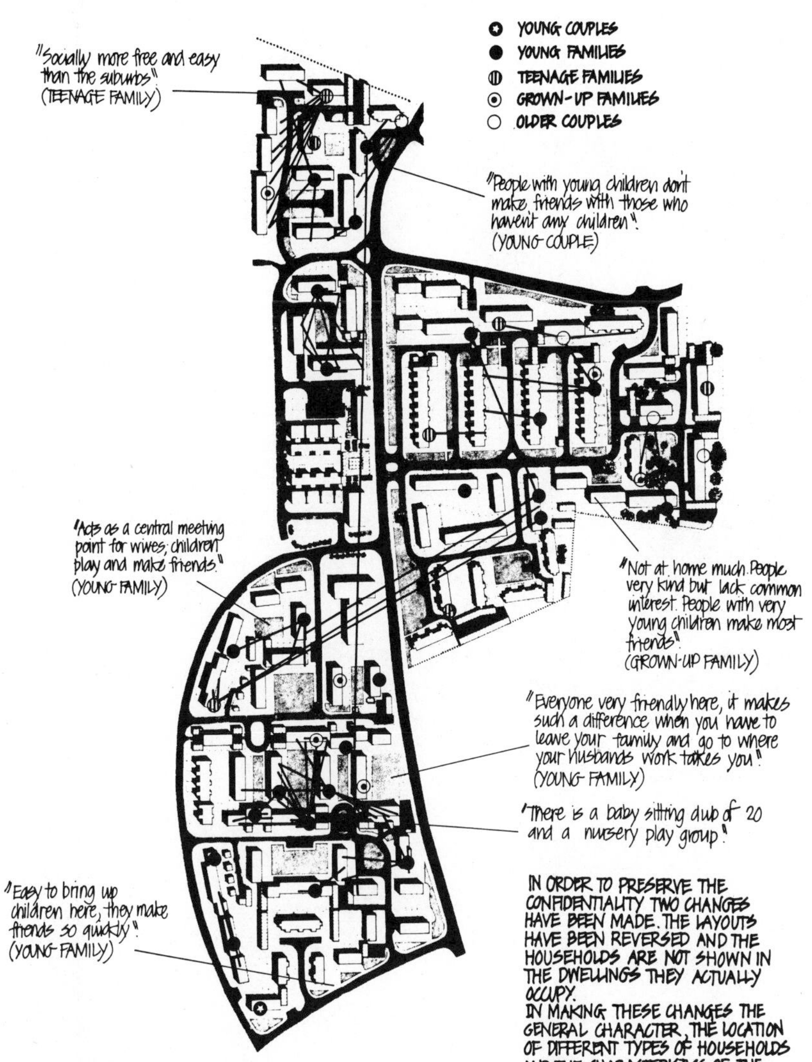

HURST PARK

IN ORDER TO PRESERVE THE CONFIDENTIALITY TWO CHANGES HAVE BEEN MADE. THE LAYOUTS HAVE BEEN REVERSED AND THE HOUSEHOLDS ARE NOT SHOWN IN THE DWELLINGS THEY ACTUALLY OCCUPY.
IN MAKING THESE CHANGES THE GENERAL CHARACTER, THE LOCATION OF DIFFERENT TYPES OF HOUSEHOLDS AND THE CHARACTERISTICS OF THE FRIENDSHIP PATTERNS HAVE NOT BEEN DISTORTED.

Range of Dwellings and Movement The range of types attracted an unbalanced population with a high proportion of families with young children, a few with older children, and very few old people. Turnover appeared to be higher than the national average. This was due in part to the inclusion of less expensive dwellings particularly aimed at first purchasers whose mobility rate was understandably high and in part to a deficiency of four-bedroom dwellings suitable for older and larger families.

* A higher proportion of four-bedroom dwellings should be provided if possible at a relatively low local density. They are likely to be occupied by families with older, noisier children.

* Further experiment is needed into dwellings designed to be extended or adapted internally to meet changing family needs (to reduce movement caused by the family outgrowing the dwelling).

Opinions about the district

	LEE	HURST PARK	CROYDON
Number in sample	43	42	20
ADVANTAGES	%	%	%
Shopping good	53	29	65
Convenient to town, country, work	60	33	70
Near beauty spots (Greenwich, Hampton Court, River)	14	69	-
Open, quiet, fresh air, near parks.	23	36	60
Transport good	14	2	5
'Good' neighbourhood Friends and relatives near	14	7	-
Schools near, good for children	7	-	15
Near Croydon Town Centre	-	-	35
None	7	2	-
DISADVANTAGES			
Transport problems	37	64	-
Crossing main roads (Eltham Rd.-Lee, Hurst Rd. Hurst Park)	35	14	-
Distance to shops	26	38	30
Distance to State primary schools or shortage	12	33	-
Lack of character Dislike of new development	7	16	20
None	14	7	35

4-20
RANGE OF DWELLINGS AND MOVEMENT
Shankland & Cox, London, England

Potential moves, reason for moving, district desired and kind of home sought

	LEE	HURST PARK	CROYDON	TOTAL
Total in sample	43	42	20	105
Total potential movers	31	27	9	67
REASONS FOR MOVING Totals exceed 100% because some people gave more than one reason.	% OF POTENTIAL MOVERS	% OF POTENTIAL MOVERS	% OF POTENTIAL MOVERS	% OF POTENTIAL MOVERS
Growing family.actual or anticipated	55	37	33	45
Job	3	26	-	12
Schools inadequate	-	19	-	7½
Progress to something better	3	11	11	7½
'Way of Life'.	29	11	33	22
Other	9	15	33	16
DISTRICT DESIRED				
Within 5 mile radius	32	44	78	43
Elsewhere in London or home counties	10	4	-	6
Other	32	37	11	31
Don't know.	26	15	11	20
KIND OF HOME				
Detached house and garden	16	18	22	18
House and garden	32	18	22	25
Detached bungalow	20	8	-	12
Other	6	-	-	3
Don't know	26	56	56	42
AGE OF NEW HOME				
New home	32	52	56	43
Second-hand	23	22	11	31
Don't know or don't care	45	26	33	26
NUMBER OF BEDROOMS WANTED				
2	-	11	-	4
2/3, 3	42	18	33	31
3/4, 4	45	41	56	45
4/5, 5	6	15	11	10
Dont know	7	15	-	10

Dwelling Design The appearance of the individual dwellings mattered less to people than the layout and landscaping of the scheme as a whole.

* Elevational treatment should relate more to the whole than the individual unit and should be more fully integrated with the planting scheme which should include climbing plants on walls.

"Open plans" were liked by small households but disliked by larger ones. Dwellings generally lack definite spaces suitable for children's play.

* There should be a mixture of types in any scheme, some with open plans and others in which rooms likely to contain noisy or messy activities could be easily separated from the main living space.

Kitchens were too small for the equipment people used.

* They should be enlarged and where possible clothes washing provided for elsewhere.

Combined living/dining/kitchen space was not popular with larger families.

* Kitchen, dining space, and living spaces should be designed as separate units which could be opened into one another rather than continuous spaces flimsily subdivided. Kitchens should have separate access from the hall.

Access to the garden through the living room was objected to.

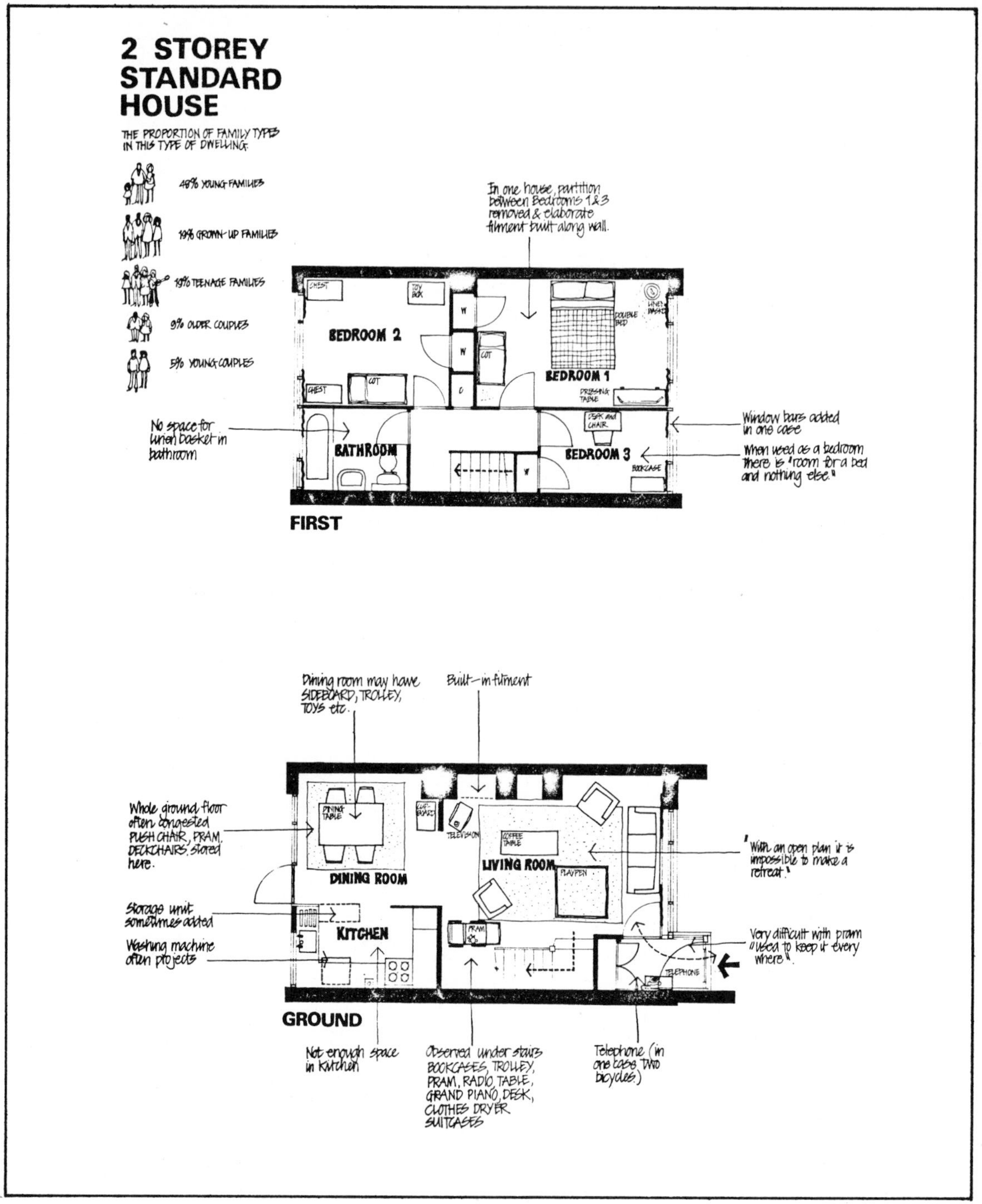

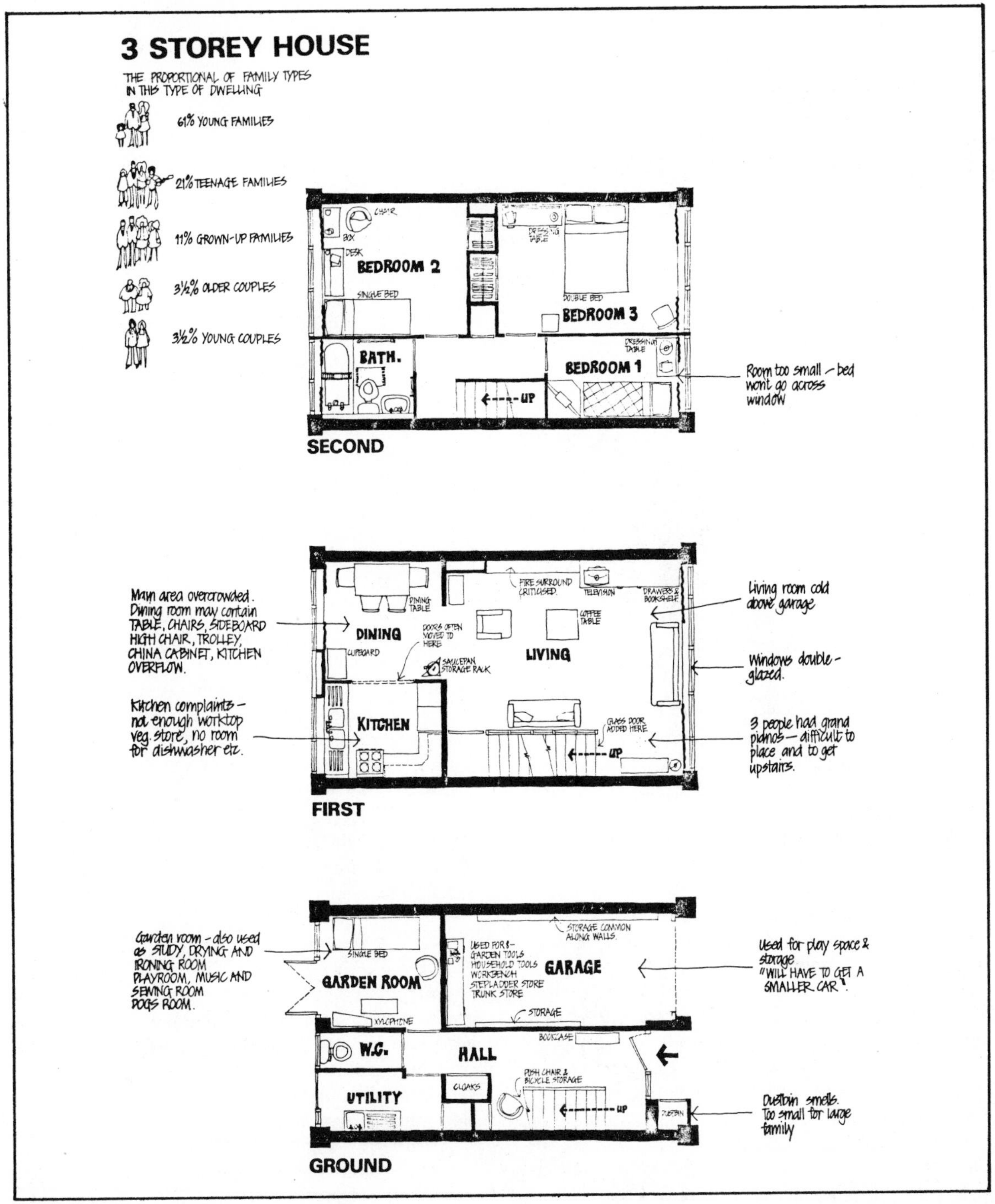

3 STOREY HOUSE
THE PROPORTIONAL OF FAMILY TYPES IN THIS TYPE OF DWELLING
61% YOUNG FAMILIES
21% TEENAGE FAMILIES
11% GROWN-UP FAMILIES
3½% OLDER COUPLES
3½% YOUNG COUPLES
BEDROOM 2
BEDROOM 3
BATH.
BEDROOM 1
UP
Room too small – bed won't go across window
SECOND
Main area overcrowded. Dining room may contain TABLE, CHAIRS, SIDEBOARD HIGH CHAIR, TROLLEY, CHINA CABINET, KITCHEN OVERFLOW.
Kitchen complaints – not enough worktop veg. store, no room for dishwasher etc.
DINING
LIVING
KITCHEN
UP
Living room cold above garage
Windows double-glazed.
3 people had grand pianos – difficult to place and to get upstairs.
FIRST
Garden room – also used as STUDY, DRYING AND IRONING ROOM PLAYROOM, MUSIC AND SEWING ROOM DOGS ROOM.
GARDEN ROOM
GARAGE
W.C.
HALL
UTILITY
UP
Used for play space & storage "WILL HAVE TO GET A SMALLER CAR".
Dustbin smells. Too small for large family
GROUND

* An access way must be provided through the hall/utility room or kitchen to the area of private open space, and a place provided at the point of entry for coats and boots.

Clothes hanging space and general storage space on the ground floor was inadequate.

* A length of 4 ft. per person is required. General storage space to Parker-Morris standard (See Glossary) is needed with some on the ground floor.

4-21
DWELLING DESIGN *Shankland & Cox, London, England*

5 THE ECOLOGIC SYSTEM: I-WOOD, II-STEEL, III-CONCRETE

THE STRUCTURAL SYSTEM

Structure: Open vs. Closed Systems

One danger indicated in dealing with system construction, either open or closed, is that the building industry will become dominated by a relatively small number of mutually exclusive systems which could smother its progress. Also, the lack of incentive for improvement might so standardize the built environment that it will become intolerable for human habit.

Already the general direction of the "small number of nationally available systems" has been towards large concrete panel systems economically capable of exclusively high-rise structures and lightweight modular boxes for exclusively small low-rise projects, while social-needs studies have been advising that a greater compatibility and mix between low-and high-rise structures should occur.

Already the general direction of low-rise systems building has been towards volumetric factory-made modules, while the location of market needs and project sizes are not always compatible with factory transport radii and minimum unit number size.

For most systems available today—volumetric or panels—the single most common deterrent to usage is their *"minimum economical project size" requirement.* The number of units or safe size of the average housing project anticipated by the "typical small developer" (who still, as a category, builds 85% of our housing in the U. S.) more often than not fall short of the minimum project size required for economical factory fabrication. For the small developer, the size of his project is usually determined by financing, immediate marketing capabilities, and site parcel size. Parcel size is particularly a factor in cities where, lacking public authorities' powers of eminent domain, only limited-size sites are often available. Even public authorities find that public opinion often limits the parcel they are able to put together, and this, in turn, limits their project size. Infill housing has a more acceptable end application than urban renewal.

Market location is the second most common deterrent. With the exodus from the cities continuing and the vacation or second home in the country becoming even more common with the shorter work weeks, the tendency is towards even more decentralization of projects to tax minimum economical transportation radii and market aggregation requirements.

European systems were developed for European geography and postwar political/economic framework—large, concentrated housing projects heavily

5-1
PROJECT SIZES Are not always compatible with factory transport radii and minimum unit number sizes. *IBIS System, Richard Thomas and Baldwins Ltd. and the Pressed Steel Co., Ltd.*

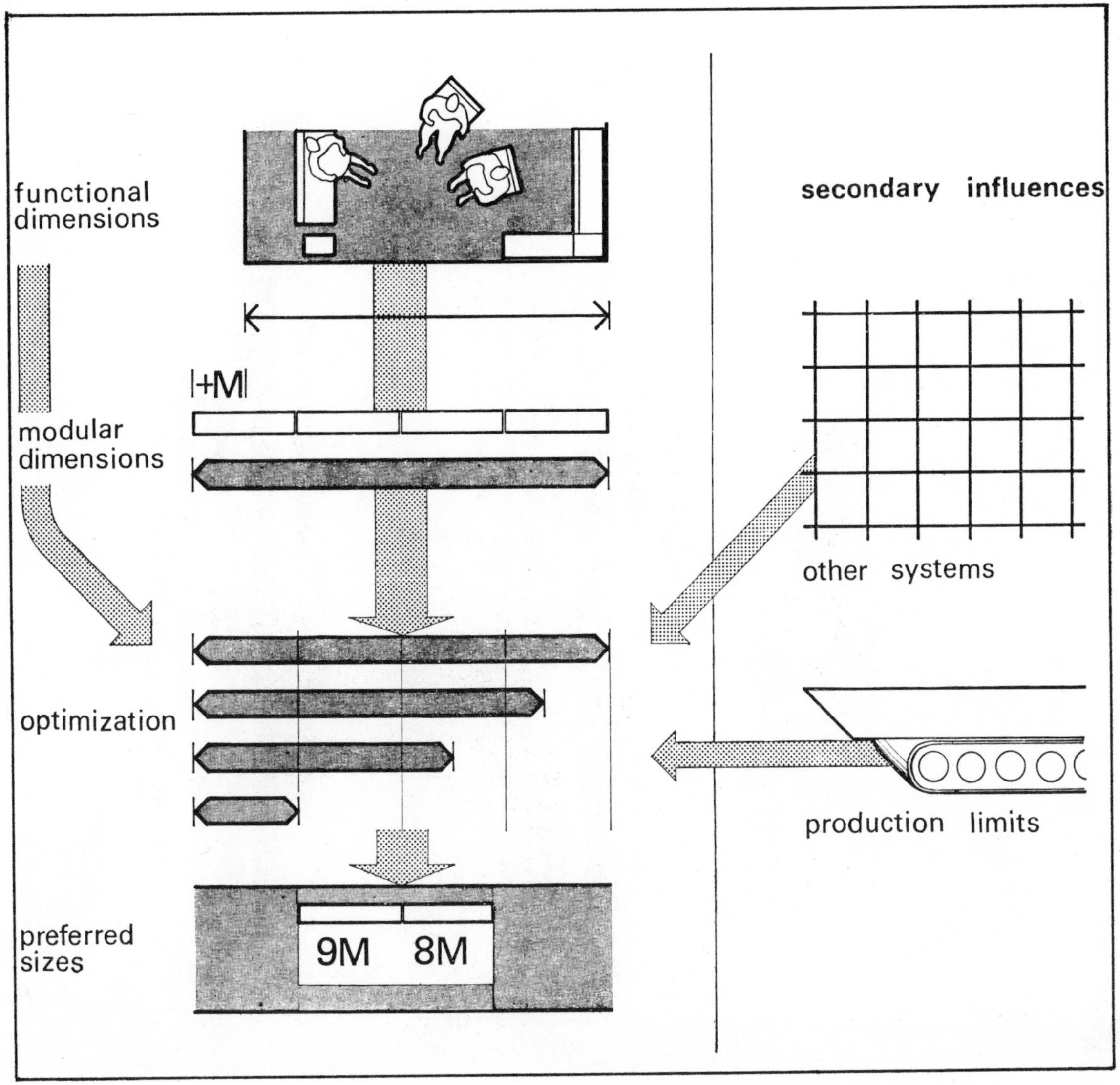

5-2
SMALL DEVELOPERS Accounted for 85% of our housing in the U.S., most of their projects are 75 dwelling units or less. *Boston Sunday Globe, 1 August 1971*

nday Globe August 1, 1971 A—51

LAND FOR SALE

One of the East's
Foremost Building Companies . . .

WANTS TO BUY LAND

with the following requirements

1. 15 to 20 acres zoned for town houses.
2. Sewer and water immediately available to the site.
3. Minimum of 75 units

subsidized by governments. If the majority of U.S. housing does continue to be built as relatively small projects (25–100 units) in decentralized areas of a vast country, the European type of systems approach will continue to offer no solution. No immediately apparent "American" systems approach seems in the offing.

The indigenous American—the small scale developer/contractor who wants to rationalize his approach without taking the enormous risks involved in paving the way with a totally new concept for in the U.S.—is the element to whom this book is directed and for whom we have outlined the transitional building system—ECOLOGIC.

The Borderline Case

For this element of the housing industry, the norm is the borderline case—a project not quite large enough or not quite close enough or with not quite enough flexibility in plan or materials to allow for factory fabrication.

What the borderline case indicates is that the truly "open system is the one that is capable of many degrees of both factory and on-site fabrication and one capable of savings for the small project as well as the large. The materials must not be precluded by codes. local climate, and user preference, and so inevitably a variety of construction materials must be offered. Again, the borderline case syndrome shows that no *one* material offers the key to all situations—there is still *a place for wood, a place for steel, and a place for concrete.*

If industrialized housing is the result of a balanced approach using "Modern Methods and Specialized Equipment harnessed and built into the Design of the Housing Units," then to be a truly "open" and universally available and applicable system, the material, the equipment, and the designs must be infinitely flexible—even to the point that one's choice of systems manufacturing should not commit him irrevocably to either factory or on-site production.

Basically, there are three methods of manufacture. Namely, on-site factory, permanent factory, or a combination of the two. The final choice is affected by many conditions, some of which are organization, number of units required, site conditions, and continuity of program within a defined area. Whatever determines the final choice, what is common to all methods is the requirements of the equipment and the material.

The three-part ECOLOGIC system is designed to be capable of this kind of flexibility. The three parts— I, wood; II, steel; III, concrete structures—are each able to adjust in varying degrees to either on-site fabrication, permanent factory, or a combination of the two. The equipment in all the cases is lightweight, mobile, and requiring only the very minimum capital expenditure. It therefore imposes only the minimum in terms of market aggregation and economical project size.

The *ECOLOGIC SYSTEM—I, Wood; II, Steel; III, Concrete* is, by its very concept, capable of the full range of use with respect to material, but also with respect to design. The Kit of Parts (Chapter 6) and the typical Generic Plans (Chapter 8) are, in effect, a superimposed *design system* that overlays the *structural system;* and, through this modularization of the structural, closure, and finish components, it makes enables ECOLOGIC I, II, and III to work together and to offer the most economical low-, medium-, or high-rise solution—or combination—to a given situation. Without sacrificing material or design flexibility. The result is a full "stable" from which it is possible to select just the right horse to ride.

5-3
EUROPEAN SYSTEMS Developed for European geography and post-war political/economic framework, these large, concentrated housing projects are heavily subsidized by their governments. *Tracoba System, 840 dwelling unit project at Vigneux, France*

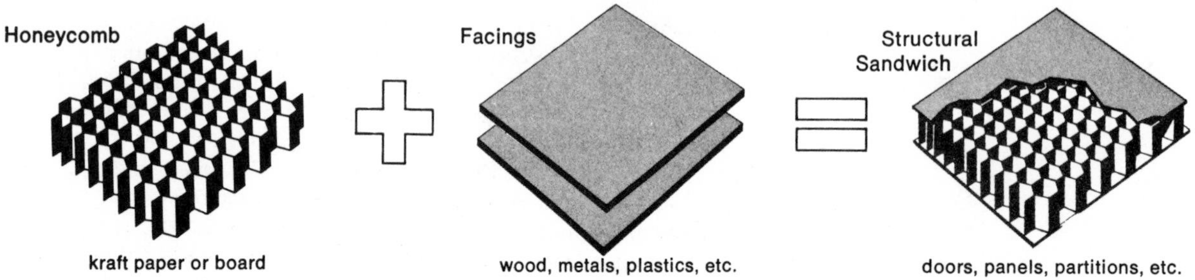

5-4
Union-Camp Corporation, Wayne, New Jersey

ECOLOGIC I–WOOD/HONEYCOMB

Wood/Honeycomb Panel Structure

Essential for a complete "stable" applicable to all major housing areas is a unit which is a potentially competitive alternative to the mobile home. One alternative which is advanced in concept, lightweight in structure, and basically provides insulation and finish materials in one building element, is the unit we developed using wood on honeycomb in structural sandwich panels.

Honeycomb structural core material, such as that designed by Union-Camp for use in the fabrication of stressed skin and sandwich panels, is manufactured of high-strength Kraft paper to form an open-celled structure of low density with high compressive and shear strengths.

Using the principle of form rather than mass to achieve its performance characteristics, this unique, celled material provides a low-cost, lightweight, high-performance structural core. Honeycomb if manufactured of high-strength Kraft paper, either in its natural state or impregnated with phenolic resin for additional strength and durability. Special fire-retardant grades are available, and when high thermal insulation is required, the honeycomb cells can be supplied partially filled with polyurethane foam. Once combined with the desired facings, honeycomb gives efficient structural sandwich panels.

Chapter 6 shows the developed wood/honeycomb unit with wall panels of 3" honeycomb expanded with ¾" urethane and exterior faced with 3/8" Texture 1-11, finished on the interior surface with ½" wall board. Roof and floor panels are of 4" urethane-infused honeycomb faced with ½" plywood. The steel frame indicated is necessary only for multistory structures, as panels are structural.

An order-bid blank for a prototype unit, based on a single presently operating panel fabricating facility:

ITEM NO	QUANTITY	DESCRIPTION		PRICE
		S.S. PLYWOOD PANELS		
		FLOOR		
1	9	4'-0" X 12'-0"	1/2" CD Plywood Skin 1/2" Underlayment ext. Plywood Skin 4" Honeycomb/Urethane—Union Camp Waterproof Glue, Pressure-Glued 5" Panel Thickness	
		ROOF		
2	9	4'-0" X 12'-0"	1/4" AC Plywood Both Skins 4" Honeycomb etc.—Same as Item #1. Panel Thickness 4 1/2" Waterproof Glue	
		EXTERIOR WALLS		
3	15	4'-0" X 8'-0"	3/8" Ivy League 1/2" Gypsum Board 3" Honeycomb etc.—Same as Item #1. Waterglue Glue 3 7/8" Panel Thickness	
		INTERIOR WALLS		
4	9	4'-0" X 8'-0"	1/2" Gypsum Board—one sie only 3" Honeycomb etc.—Same as Item # 1.	
		Notes: 1. Panels have Wood Framing. 2. Steel Angles should be pre-punched for nails. 3. Use skirt detail to eliminate 9' Plywood.		
		Alternate: Price for 10 Basic Units produced and and delivered same time.		
		Base Bid	1 Unit	
		Alternates	10 Units	

If a sufficient market can be provided to a manufacturer for this type of panel, the costs can be reduced to the point where plywood-faced honeycomb panels can be provided in a volume of business where costs can be more dramatically less than traditional stick construction.

The most important reason for the inclusion of the wood/honeycomb panel in the ECOLOGIC stable is its capacity for "industrializing" the very small stick builder. For an extremely low capital investment—$3,000 or less—this small contractor can purchase a cold press and facilities to enable him to produce these panels himself. The transport of unexpanded honeycomb to remote sites for expansion and framing at a mobile facility offers another intriguing capability. Even the higher-capacity and more efficient hot press at approximately $25,000 is still an industrialization expenditure which is feasible for the small home builder.

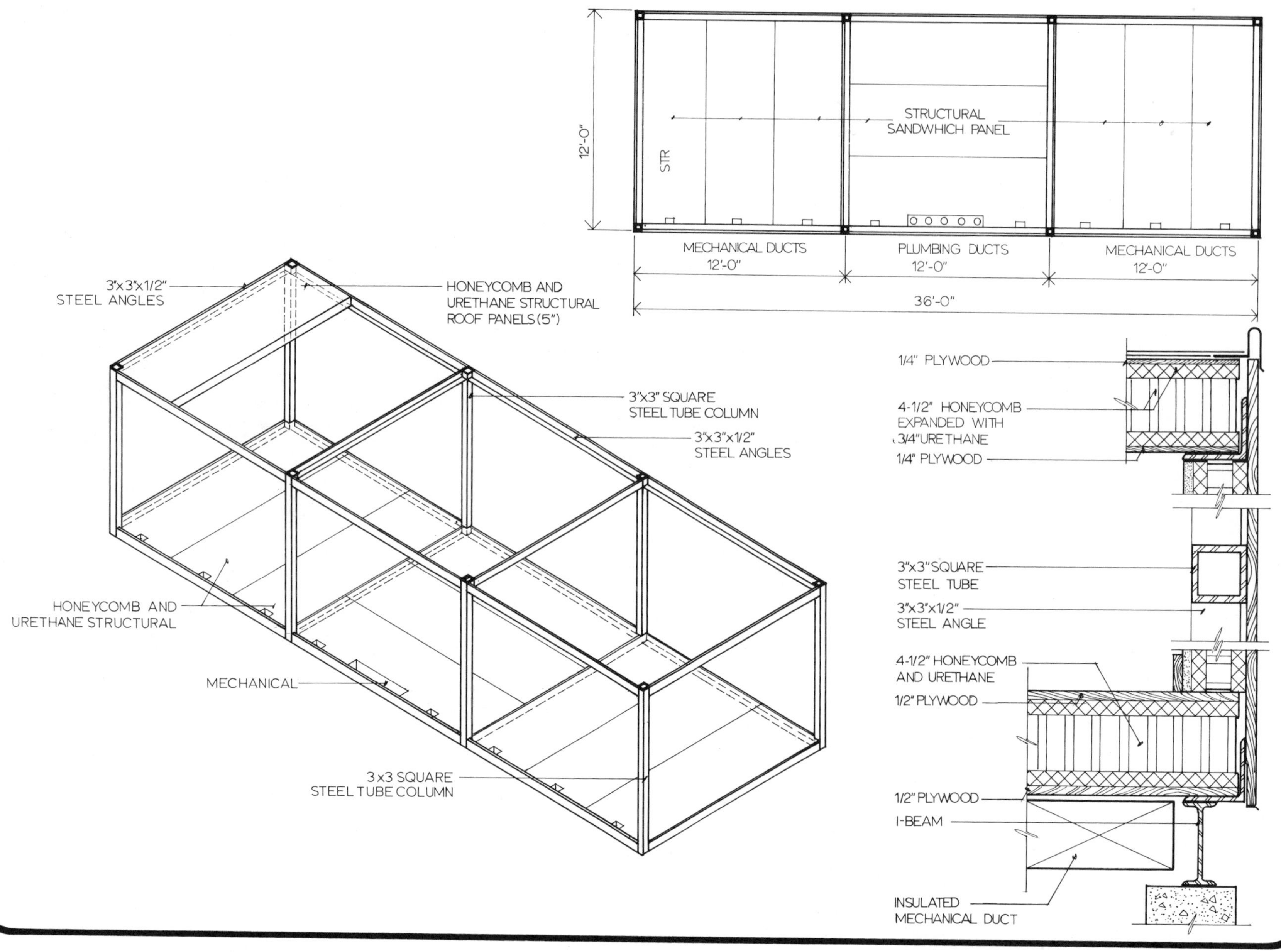
12'-0"
STR
STRUCTURAL
SANDWHICH PANEL
MECHANICAL DUCTS
12'-0"
PLUMBING DUCTS
12'-0"
MECHANICAL DUCTS
12'-0"
36'-0"
3"x3"x1/2"
STEEL ANGLES
HONEYCOMB AND
URETHANE STRUCTURAL
ROOF PANELS (5")
3"x3" SQUARE
STEEL TUBE COLUMN
3"x3"x1/2"
STEEL ANGLES
HONEYCOMB AND
URETHANE STRUCTURAL
MECHANICAL
3 x 3 SQUARE
STEEL TUBE COLUMN
1/4" PLYWOOD
4-1/2" HONEYCOMB
EXPANDED WITH
3/4" URETHANE
1/4" PLYWOOD
3"x3" SQUARE
STEEL TUBE
3"x3"x1/2"
STEEL ANGLE
4-1/2" HONEYCOMB
AND URETHANE
1/2" PLYWOOD
1/2" PLYWOOD
I-BEAM
INSULATED
MECHANICAL DUCT

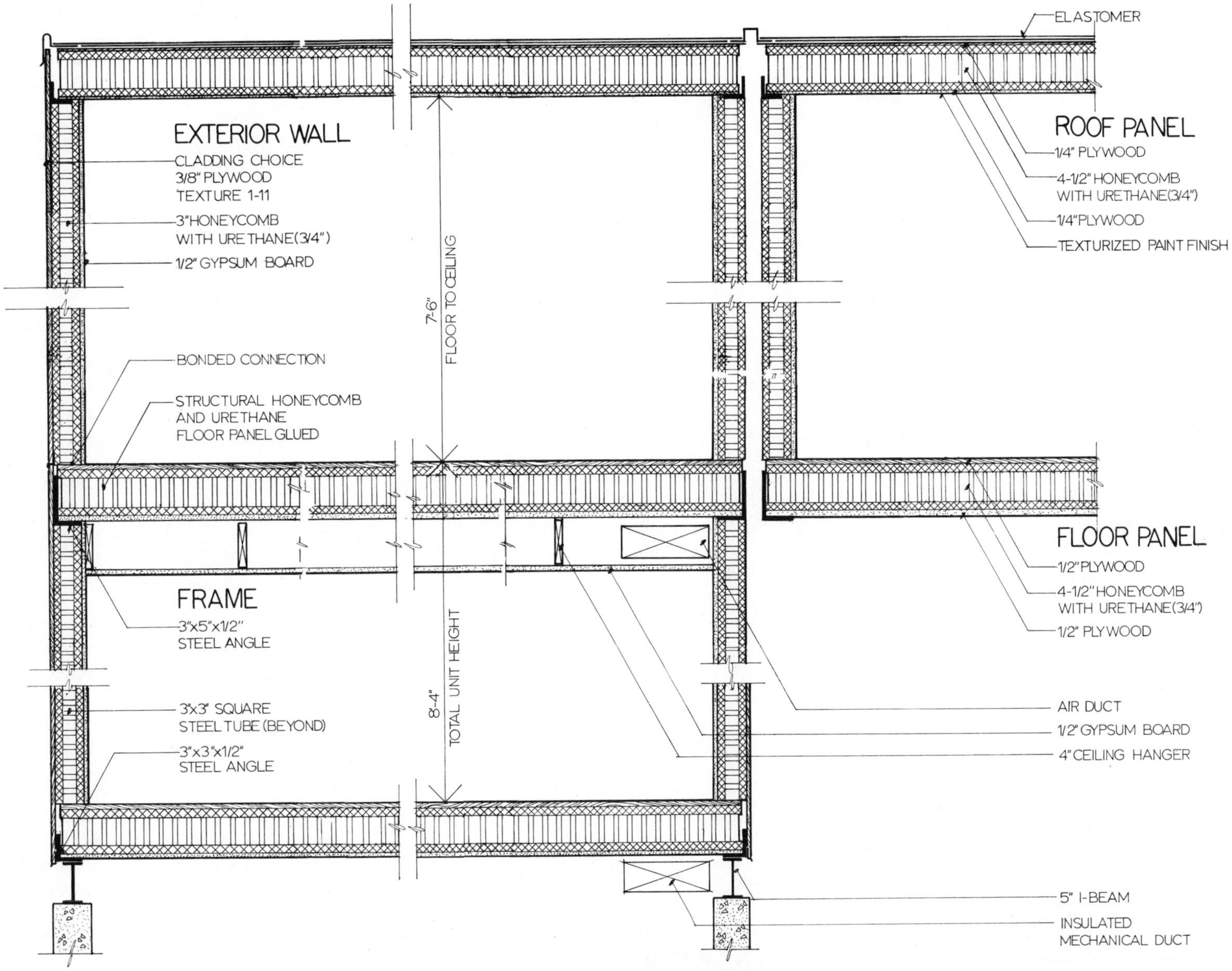
ELASTOMER
ROOF PANEL
1/4" PLYWOOD
4-1/2" HONEYCOMB
WITH URETHANE(3/4")
1/4" PLYWOOD
TEXTURIZED PAINT FINISH
EXTERIOR WALL
CLADDING CHOICE
3/8" PLYWOOD
TEXTURE 1-11
3" HONEYCOMB
WITH URETHANE(3/4")
1/2" GYPSUM BOARD
7'-6"
FLOOR TO CEILING
BONDED CONNECTION
STRUCTURAL HONEYCOMB
AND URETHANE
FLOOR PANEL GLUED
FLOOR PANEL
1/2" PLYWOOD
4-1/2" HONEYCOMB
WITH URETHANE(3/4")
1/2" PLYWOOD
FRAME
3"x5"x1/2"
STEEL ANGLE
8'-4"
TOTAL UNIT HEIGHT
3'x3' SQUARE
STEEL TUBE (BEYOND)
3"x3"x1/2"
STEEL ANGLE
AIR DUCT
1/2" GYPSUM BOARD
4" CEILING HANGER
5" I-BEAM
INSULATED
MECHANICAL DUCT

ECOLOGIC I
WOOD

5-7
Union-Camp Corporation

Honeycomb can be obtained in six basic forms:

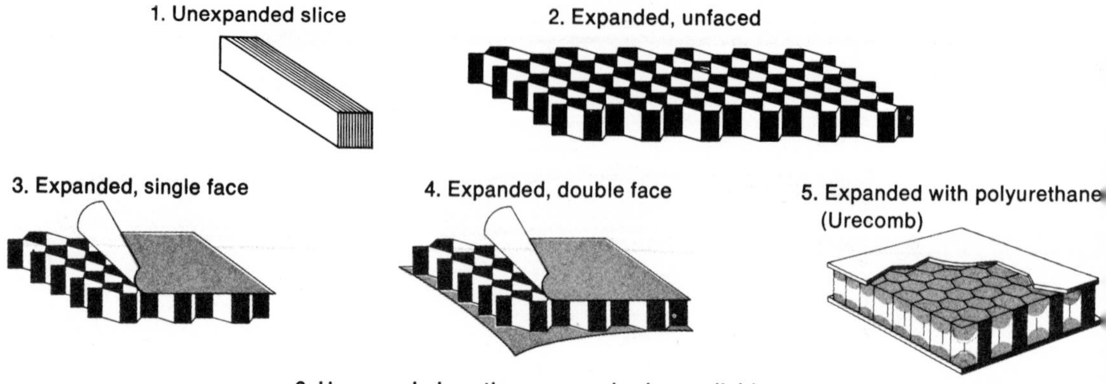

6. Unexpanded continuous core is also available.

Material Properties

Honeycomb is made in three standard Kraft weights—70 pound, 80 pound, and 99 pound—and three standard cell sizes—½″, ¾″, and 1″. Size limitations are a maximum expanded length of 12′. Honeycomb can be obtained in six basic forms:

Structural characteristics, fire retardancy, sound and thermal insulation, and other properties are outlined in Fig. 5-7

Panel Fabrication

The following comments deal with some common questions involving panel fabrication. For the most part, panels are assembled by one of the following methods.

Hot press—Single or multiple opening hot press, in which only one panel is pressed in each opening.

Cold press or clamps—Baling press or screw clamps, in which pressure is applied to a stack of panels while the adhesive is curing at room temperature.

Pressure rolls—Used in combination with special synthetic rubber adhesives.

Each of these methods offers certain advantages. The one that is ultimately adopted will largely depend on the user's past experience, equipment available, and capital available for the purchase of new equipment.

The adhesive selected will also have a bearing on the bonding procedure. For example, certain adhesives must be cured at high temperatures, in which case a hot press is the only economical answer. In many structural applications involving metal skins or facings, suitable adhesives must be cured at *approximately* 300° F if the desired properties are to be obtained. Other adhesives are specifically formulated for cold pressing.

Core Thickness

One of the first questions which generally arises deals with core thickness. The answer is best given by emphasizing the fact that intimate contact must be obtained between the faces and the core during the pressing operation. Where the honeycomb is the only material between the faces, there is no problem, and the core can be cut to the nominal thickness.

Where the panel is surrounded by a frame, or contains inserts of high-density material, the selection of the proper core thickness becomes very important. It is obvious that the core must be equal to or thicker than the frame or insert, if contact with the skins is to be assured. Knowing the permissible thickness tolerances of the component parts, it is possible to arrive at a core thickness, over and above the nominal, which will guarantee contact.

In a hot press, where only one panel is pressed per opening, the core is crushed down to match the mating parts.

In a stack of unseparated panels being cold pressed or bonded, it is sometimes difficult to control the crushing of each layer of core, and unequal deformation results in distorted panels. This is often the case when the cell size is greater than one inch. However, the problem is sometimes overcome by passing the individual core and frame assemblies between sizing rolls before pressing.

Press Pressures

Again, where honeycomb is the only material be-

5-8
SYSTEM ECOLOGIC A case in point, illustrating a dwelling unit which does not subject the panels to substantial loads.
ECODESIGN

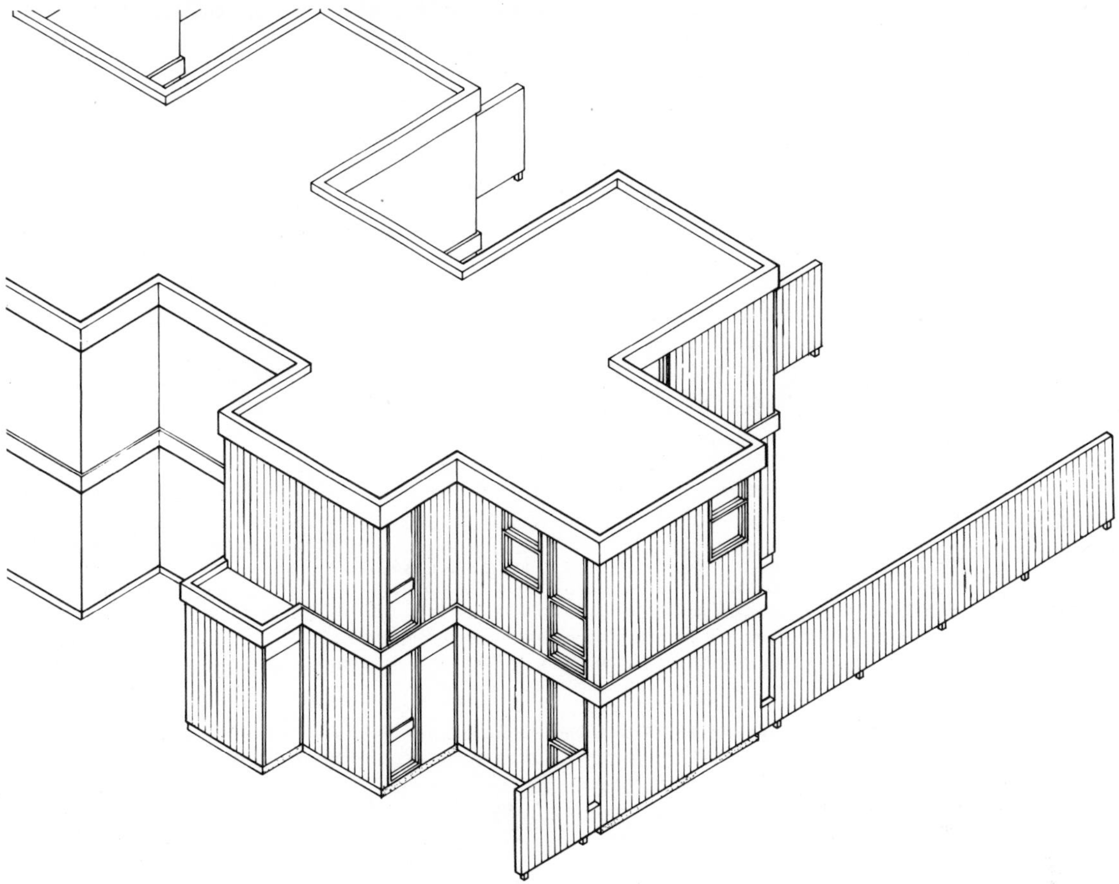

tween the faces, the problem is relatively simple. All that is required is good contact, which can be achieved with pressures in the neighborhood of 5–20 pounds per square inch of panel area.

Where frames and inserts are included, however, sufficient pressure must be applied to slightly crush the core and still supply the desired pressure to the other components. Since compression strengths vary considerably for available honeycomb grades, the most suitable press pressure should be determined for each individual case.

Adhesive Application

Conventional coating equipment such as glue spreaders, hand roller coaters, and spray guns all have been used successfully. For primary structural panels, both core and faces are coated. In cases of less critical applications, the adhesive is often applied only to the faces of the core. When the core alone is coated, the "sizing" operation mentioned previously (see "Core Thickness") will be useful as a means of creating a folded over edge of the honeycomb cell ends, thus making it possible to retain an adequate amount of adhesive on the surface of the core edge to insure proper bonding of the faces.

Continuity of Cores

Another fundamental point worth noting deals with the continuity of the core in a given panel. In structural panels, it is highly desirable to have one-piece core. If, however, it is necessary to use two or more pieces to fill the required space, the adjacent pieces should be well nested and bonded together if possible. Where panels are not to be subjected to substantial loads, structural continuity is not so important, and the core can even be constructed from random scrap pieces nested together.

Production Problems

In the early stages of product development, difficulties are often encountered which are frequently attributed to the honeycomb core. In reality, the basic cause can generally be traced back to improper design, faulty assembly technique, the influence of the components other than the core, etc. "Show through" of frames and inserts sometimes causes a problem which may be solved by reducing the pressure during panel fabrication. Warping is another common problem and it is well to keep in mind that honeycomb cores do not cause this difficulty. The inherent stability and flatness of honeycomb panels is recognized, but the core itself may not prevent other stronger components from inducing warpage.

Structural Characteristics

Standard Grades (S),
Selected Custom Grades (C).

Special high-strength cores for very high performance applications are available—data sheets on request.

	Kraft weight	Impregnation	Cell size	Wt. per cu. ft.	Comprehensive strength	Shear strength L	W
C	80#	(18%)	½"	2.25#	170 p.s.i.	59 p.s.i.	40 p.s.i.
C	99	(18)	¾	1.87	105	42	27
S	70	(11)	½				
S	99	(11)	¾	1.73	94	37	23
S	99	(11)	1	1.30	53	25	16
S	70	(0)	½				
S	99	(0)	¾	1.56	48		
S	90	(0)	1	1.15	33		
S	80	(0)	1	.94	20		
		grade identification					

L=Length W=Width

Fire retardant grades of Honeycomb

Grade	Pounds per cubic	Compressive strength	Shear strength L	W	Flame spread rating (ASTM E-84)
99# F (0%) 1"	1.15	33 p.s.i.	25 p.s.i.	16 p.s.i.	15*
99# F (11) 1"	1.30	53			44

Standard grades of honeycomb provide a significant degree of fire retardance by eliminating flue lines within panel or partition construction. Fuel and smoke ratings are very low.

Sound insulation

The resistance of honeycomb cored structures to sound transmission will depend upon the nature of the facing material. Sound transmission for typical panel structures are shown below.

For unusual applications there are types of honeycomb construction more highly resistant to sound transmission. Please contact our representative or home office for information on these constructions.

Average transmission loss (TL)—Various constructions

Panel Facings Honeycomb Core 1" Cell (not filled)	Panel Thickness 1¾"	3"
Steel—16 Gage	34.5	30.3
Steel—20 Gage	29.6	25.9
Hardboard—1/4"	25.4	24.6
Gypsum Board—3/8" thick	26.6	25.9

Thermal insulation

Standard honeycomb contributes to the insulative qualities of a structural panel. When more thermal insulation is required, a unique urethane honeycomb combination is sold under the tradename URECOMB.

Insulation properties of typical HC panels
Shown in terms of "U" value

	Panel Thicknesses ¼" plywood facing 1"	2"	3"	Metal facing 1"	2"	3"
Open Honeycomb	.39	.30	.21	.45	.33	.23
Urecomb ¾" foam	.19	.15	.13	.21	.16	.13
Urecomb 1¼" foam		.12	.10		.12	.10

Durability

Untreated honeycomb has exceptional durability, but impregnated grades are recommended for construction applications and other situations where the panel will be exposed to weather, temperature extremes, moisture and other uncontrolled conditions. Impregnated honeycomb is highly resistant to moisture, decay, fungus rot and insects.

Panels incorporating kraft honeycomb, tested after more than 20 years of severe exposure to the elements, have retained their original structural integrity. The Forest Products Laboratory, U.S. Department of Agriculture, which conducted these and other tests, have numerous publications attesting to the durability and performance of kraft honeycomb as a building panel component.

Building construction

Numerous tests have demonstrated that honeycomb sandwich panels meet the strength and rigidity requirements of housing construction.

Some typical examples of panel thicknesses required with various facing materials are shown below:

Wall assemblies

Requirements: 20 lbs per square foot load with deflection L/360

Skin / Span	Panel Thicknesses 4'	8'	12'	16'
1/4" Fir Ply	1.1"	2.7"	4.7"	
3/8" Fir Ply	1.1"	2.5"	4.2"	
.025" Steel	5"	1.3"	2.4"	3.7"
.050" Aluminum	.6"	1.6"	2.9"	4.6"

Floor assemblies

Typical building panels and honeycomb core assemblies capable of supporting 40 PSF with L/360 deflection:

Skin / Span	Panel Thicknesses 4'	6'	8'	1C'	12'
1/4" Plywood	1.25"	2.3"	3.5"	4.8"	6.0"
3/8" Plywood	1.4"	2.3"	3.3"	4.4"	5.8"
1/4" Hardwood	1.4"	2.4"	3.5"	4.8"	6.1"

ECOLOGIC II—STEEL

Lightweight Steel Framing

Steel's strength, durability, and ease of handling make it a "must" for a "stable" of building systems. It is applicable to every type of construction—long-span facilities, parking, commercial, and now, with the advent of lightweight framing members, for short-span housing applications. Research on lighter-gage design, fastening techniques, and engineering know-how has enabled the steel industry to incorporate economical steel wall stud systems into building systems.

A product that is representative of this versatility and design freedom is Speed-Steel, a lightweight, easy-to-handle, structural framing technique developed by Keene. Fabricated from structural grade strip steel by cold forming, these nailable joists and double studs are equipped with nailing grooves which speed and simplify the attachment of collateral materials, such as wallboard. A built-in feature of all double studs and joists, it is obtained by welding or crimping two elements together in such a way that a nail, driven into the space between the two, is not only held by friction, but is also deformed to provide maximum holding power. Tests show the nailing grooves to have a holding power of three times that of wood. Automatic screw guns and self-dulling, self-tapping fasteners are used with "screw studs" and "cee studs."

The nailable joists are manufactured in depths of 6", 8", 9", 10", and 12" with punched webs. Screw studs are made in 16 and 18 gage, painted or galvanized, in five depths: 2-½", 3-¼", 3-5/8", 4", and 6" with 1-3/8" knurled flanges, channel studs, screw cee studs, as well as unpunched studs. This full range of sections, each designed for maximum economy, strength-to-weight ratio, contribute to the elimination of waste and high costs due to overdesign. All are available and have been tested and meet the requirements of ASTM specification. The load-bearing studs meet all FHA and local building codes.

Solid bridging is required as follows:

Up to 10'—one row near center
11—15'—two rows at approximately 1/3 span
15—20'—three rows at approximately 1/4 span
20—25'—four rows at approximately 1/5 span

Rafters consist of joist, stud, or channel members as determined by the live load and dead load requirements. The heel of each rafter should be fastened at the wall to the top track by a rafter plate by means of welding, or sheet metal screws or bolts connecting each flange to the rafter plate. Rafters will be flanged to the rafter plate. Rafters should be secured at the ridge by butt welding, by cutting away adjacent flanges and bolting through webs, or by welding or bolting to a continuous ridge plate.

This type of structural framing is designed specifically for panelization and is also applicable to prefabrication of boxes. Track sections are sized to fit over the flanges of joists, studs, and channels. Bridging fits within the flanges of studs and channels.

All sections are made of uniform-quality structural-grade steel with precisely engineered and tested strength characteristics. The use of a standard safety factor of 1.85 in design assures a safe, sturdy frame.

Wall and ceiling panels, framing and roof trusses, can be fabricated on the site as panels or in factory as panels or boxes. On-site jigs can be set up for cutting, assembling, and welding the roof trusses and larger panels. Most smaller fabrications can be handled under cover during periods of inclement weather.

After fabrication, roof trusses are raised into position by a mobile crane and securely welded in place in a matter of minutes. Smaller panels were light enough to be carried and erected without the use of handling equipment.

The Speed-Steel type of structural system provides an incombustible frame that is lightweight for transport economics and for elimination of heavy handling equipment in positioning panels. The nailability and light gage make the cutting and nailing versatility akin to the on-site adaptability of wood and eliminate many problems which arise on-site from damaged or unalterable concrete panels. Wall panels can be more quickly and accurately applied than with wood and with no warping or nail popping to correct later. As the sections are prepunched for electrical and plumbing access, no skilled work time is lost drilling holes, and the conduits and piping are quickly and efficiently run through openings in the studs.

Lightweight steel sections are incombustible (reducing insurance costs) and impervious to warp, shrinkage, and termites. They are well used in earthquake design, as they reduce dead weight without reducing structural strength. With this wall design, it is a simple matter to install diagonal tension members to form "shear panels" for resisting lateral forces.

Hollow wall design permits improved acoustic and thermal insulation and maximum flexibility and for future changes in location of electrical

5-9
KEENE-SPEED STEEL *Keene Corporation, Parkersburg, West Virginia*

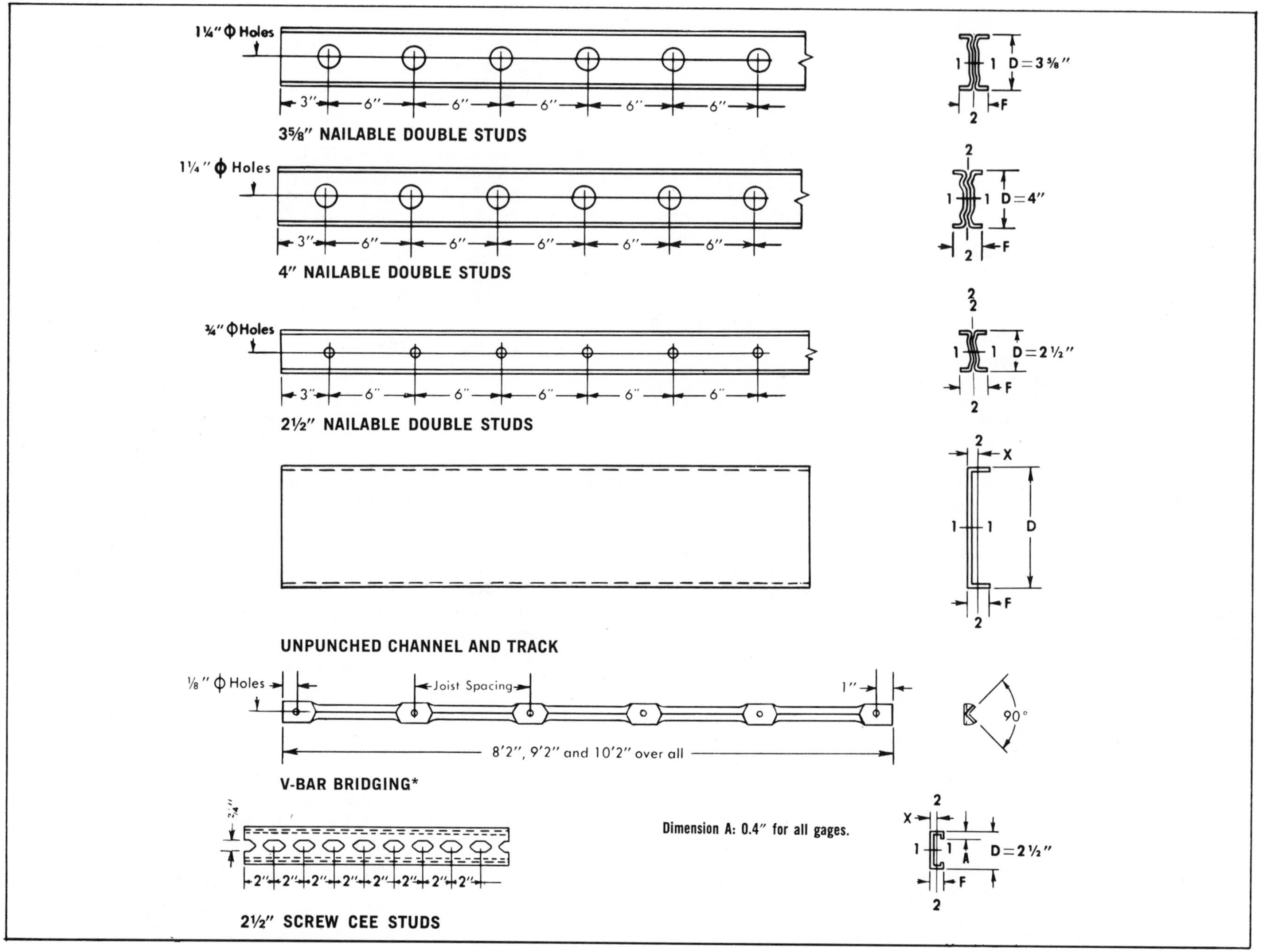

NAILABLE JOISTS, NAILABLE STUDS, UNPUNCHED CHANNEL

Dimensional Data

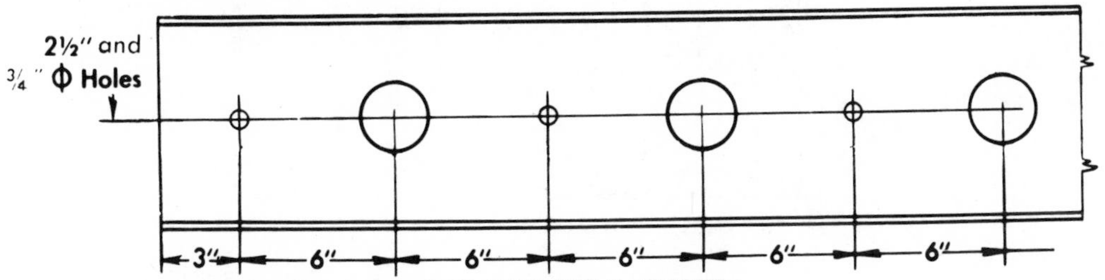

12", 10", 9", 8" and 6" NAILABLE JOISTS

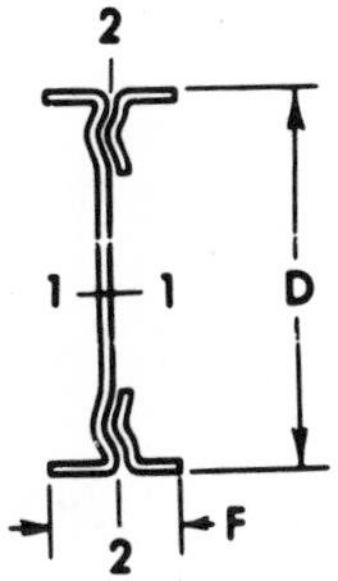

CHANNEL STUDS SCREW STUDS,

Dimensional Data

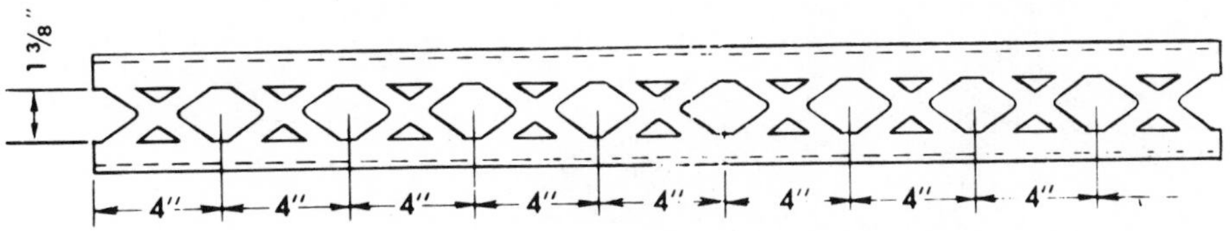

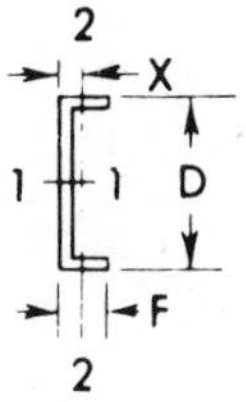

3⅝" PUNCHED SCREW STUDS AND 3¼", 3⅝" AND 4" CHANNEL STUDS

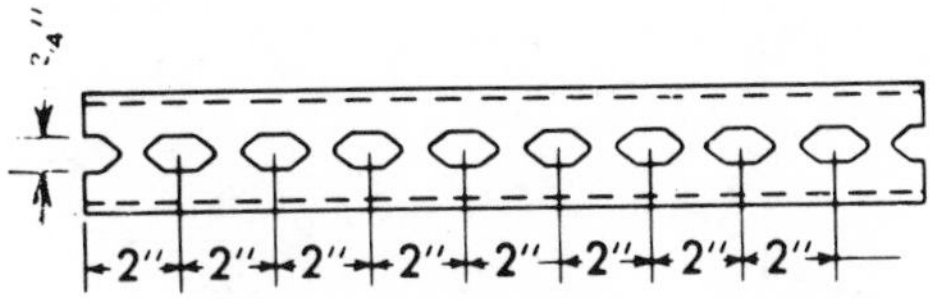

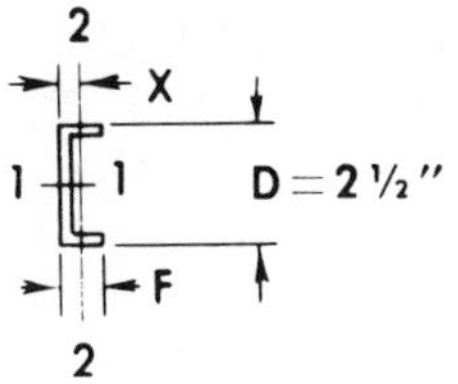

2½" PUNCHED SCREW STUDS AND CHANNEL STUDS

Lengths

Screw CEE Studs, Screw Studs, Channel Studs, and unpunched Channels: 20', 24' and 32'.

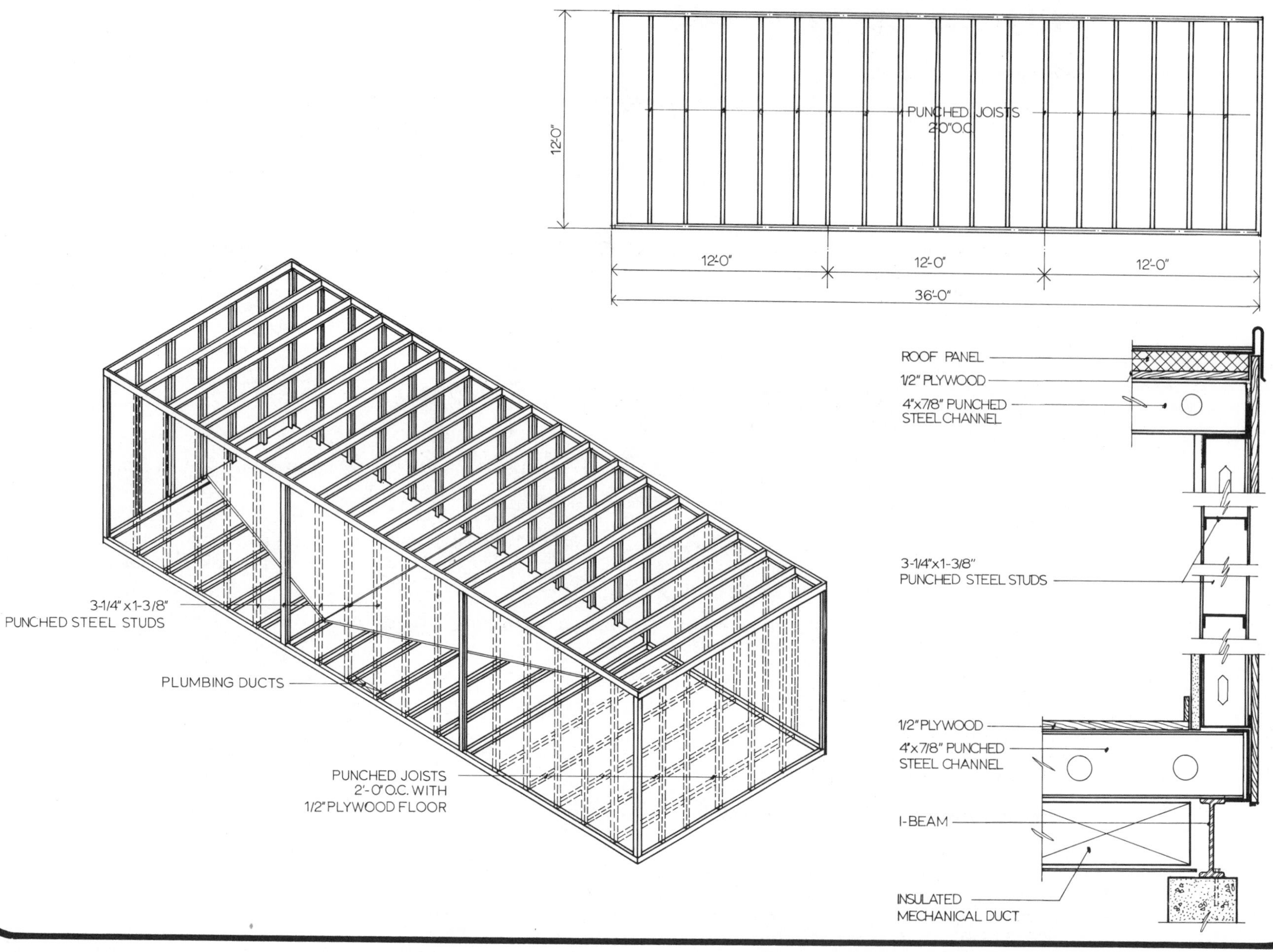
PUNCHED JOISTS
2'0" O.C.
12'0"
12'-0"
12'-0"
12'-0"
36'-0"
ROOF PANEL
1/2" PLYWOOD
4"x7/8" PUNCHED
STEEL CHANNEL
3-1/4"x1-3/8"
PUNCHED STEEL STUDS
1/2" PLYWOOD
4"x7/8" PUNCHED
STEEL CHANNEL
I-BEAM
INSULATED
MECHANICAL DUCT
3-1/4"x1-3/8"
PUNCHED STEEL STUDS
PLUMBING DUCTS
PUNCHED JOISTS
2'-0" O.C. WITH
1/2" PLYWOOD FLOOR

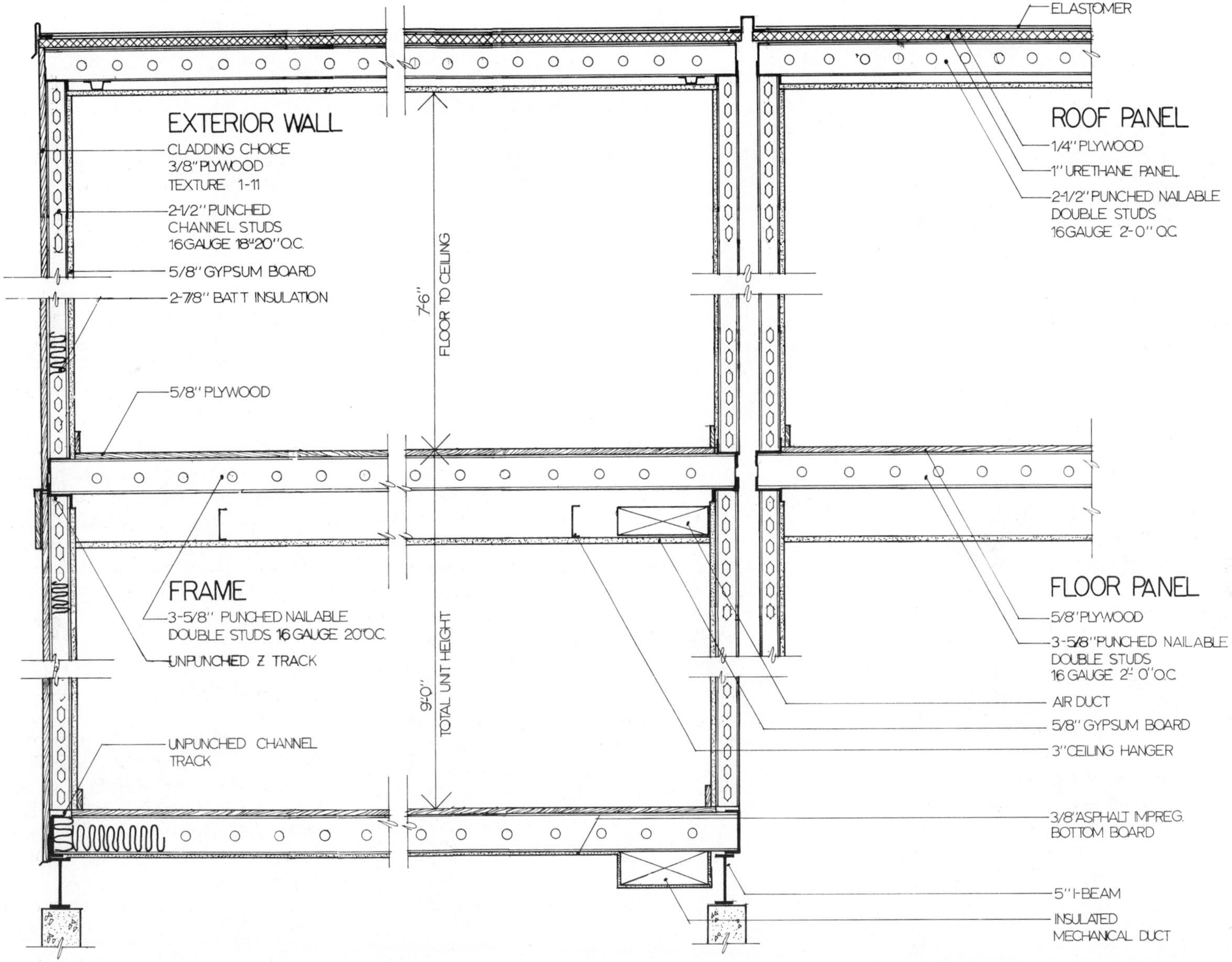
ELASTOMER
EXTERIOR WALL
CLADDING CHOICE
3/8" PLYWOOD
TEXTURE 1-11
2-1/2" PUNCHED
CHANNEL STUDS
16GAUGE 18"-20" O.C.
5/8" GYPSUM BOARD
2-7/8" BATT INSULATION
5/8" PLYWOOD
7'-6"
FLOOR TO CEILING
9'-0"
TOTAL UNIT HEIGHT
ROOF PANEL
1/4" PLYWOOD
1" URETHANE PANEL
2-1/2" PUNCHED NAILABLE
DOUBLE STUDS
16GAUGE 2'-0" O.C.
FRAME
3-5/8" PUNCHED NAILABLE
DOUBLE STUDS 16 GAUGE 20"O.C.
UNPUNCHED Z TRACK
UNPUNCHED CHANNEL
TRACK
FLOOR PANEL
5/8" PLYWOOD
3-5/8" PUNCHED NAILABLE
DOUBLE STUDS
16 GAUGE 2'-0" O.C.
AIR DUCT
5/8" GYPSUM BOARD
3" CEILING HANGER
3/8" ASPHALT IMPREG.
BOTTOM BOARD
5" I-BEAM
INSULATED
MECHANICAL DUCT

and plumbing. To obtain equipment ratings with masonry construction would require two to three times the thickness—this means the loss of vital floor space.

With nailable lightweight steel framing as one of our "stable" of structural systems, we have compatibility with standard steel for heavy construction uses such as lower levels of commercial or parking below housing, as well as compatibility with wood finish components and other closure materials and lightweight concrete decking. These framing sections are applicable to both low- and medium-rise structures. The collateral material compatibility and the inherent versatility of the system provide the maximum in design flexibility.

ECOLOGIC III—CONCRETE

Concrete Structure

ECOLOGIC's concrete system is an on-site casting technique that uses T-shaped forms to adapt hollow-core, "flexicore," "spandeck," "spancrete," and other common extruded concrete floor planks into a monolithic poured concrete bearing wall structure which is capable of integrating doors, duct and stair access, electrical, mechanical, and plumbing on-site.

This factory on-site technique operates by virtue of story-height T-shaped insulated steel-forms. These forms or shutters are precision-built in 12′ segments or full 36′ sections with 4′ wide "wings" which form the wall and the "integrating section" of the floor. The "wings" also support the floor slabs during casting.

Concrete "distance blocks" or formed kicker strips are used to set the wall thickness dimensions (generally 6″–8″) and to maintain alignment from floor to floor. Steel strip shutter clamps held by wedges give additional dimensional control and protect the shutters against uplift.

The kicker former illustrated also acts as a casting funnel at the top of the forms to facilitate the introduction of concrete.

Electrical wires or steam pipes set in the insulated forms accelerate the curing and hardening of the concrete in all temperature conditions and reducing curing from several days to only thirteen hours, or six hours for stripping.

Construction Sequence

1. A crane places the forms which are set dimensionally as described above and jacked into place vertically. Doors, ductways, and other opening forms are bolted to the forms in any desired location (unlike prefab panels, these locations can change economically). Prefabricated reinforcing mats containing electrical conduits, heating pipes, etc. are set between the forms.

2. Hollow-core panels, the lengths of which are determined by the desired span and its point of zero moment, are laid out accordingly on the form "wings" with reinforcing extending through the cores. The casting tunnel/kicker shutter is set across openings resting on the floor panels.

3. Concrete is then discharged by bucket or pump along the linear tunnel. The cores are filled back to the paper plug. High-frequency vibrators pack the concrete evenly and firmly. The kicker former resting on the floor panels also acts as a scaffold or platform from which the vibrator can be operated, opening forms can be set, and other operations performed.

4. As weather conditions require, the concrete is heated by steam, hot water, or electricity, or cooled by cold water in the heating pipes, to accelerate or retard curing as required. In 13 hours or less, the shutters can be stripped and lifted into position above to form the next floor. The steel forms provide a high-quality concrete finish which is ready to paint or paper directly.

5. Prefabrication of closure panels and mechanical, electrical, systems in reinforcing mats, rapid handling and placing of forms and components by heavy equipment, prepackaged components, systematized allocation of labor, good management control, etc., all operate to bring factory speed and precision to the building site.

Concrete Half-Tunnel

This special on-site construction technique for medium- and high-rise buildings is a further sophistication and synthesis of various types of forming methods that have been used successfully and economically for over 15 years in England, France, Sweden, Belgium, and the Netherlands.

The concrete half-tunnel was devised to obtain the economies of systematization while retaining the advantages of on-site construction. Some of the retained advantages inherent in on-site systems are:

* Monolithic pour structure cannot be affected in a precast panel system even with wet joint. With monolithic construction, intricate connections and joining techniques are eliminated. Subsystems can be totally integrated. Acoustics are improved. Seismic and additional height conditions can easily be adjusted for.

* Flexibility is provided in location of door, duct, and stair access. Unlike many large panel systems, on-site adjustments are facilitated by the on-site technique (i.e., no damaged special panels holding up the job).

* All the mechanical and electrical is integrated into the superstructure without violation of union and code regulations.

* Many panel systems and even other forming systems impose considerable dimensional limitations in spans or in building height (because of inability to vary wall thickness). With the T-forming system, the flexibility is as total as the generic extruded hollow core slab and wall thickness can be varied even from floor to floor for economies in thinner wall sections in the upper floors of high-rise construction. This is done simply by setting forms apart with the appropriate-size "distance blocks."

* Complete flexibility in adaptation to unions and code requirements is possible only with systems such as the half-tunnel, which is a method of rationalizing and speeding on-site construction because the problems of transferring certain union activities to a factory do not exist. Even situations which might arise because of increased efficiency in placing subsystems can be solved by manipulating the size, hours, and makeup of the work force, etc.

* User prejudice against precast homes, especially after the well-publicized collapse at Ronan Point, in England, would not be a problem with an on-site technique.

Perhaps the most significant advantages are freedom from "economic factory radii" and "minimum market volume demand."

* No "economic factory radius" restrictions mean that a system is *not* restricted to heavily urbanized areas.

* With on-site techniques, there is no limitation geographically to the economic radius of a production facility. This radius, required by prefabricated systems, varies according to the highways serving the plant and traffic conditions, and depends on a density and need in the area to maintain continuous production rate—plus transportation costs and facilities. Precast systems operating in the states will depend on large urban or megalopolis locations for a market, and although our superior U. S. highways increase the European radii, the efficient distance from plants still ranges from 60 to 200 miles.

* These severe restrictions eliminate the areas of the U. S. which differ from the dense urbanization of Europe and of certain urban strips in the U. S., in that they are dispersed centers of population. In these areas, housing is needed but the area cannot supply a plant with a permanent market. Only on-site techniques can answer the needs of these areas.

* Although on-site forming techniques can operate as high-volume production systems and can adapt perfectly to answering the needs of heavily populated areas, it is to the more "unreachable" less densely populated areas described above that we have directed the three prototype systems of our stable of systems.

* There is a minimum market volume demand per year, and minimized need for aggregating the market. The on-site factory is not limited to any particular region or area. Because of this, the need for aggregating the market is a crucial factor inasmuch as a tight regionalized market is not required. The market volume demanded per year for successful economic operation is generally less than one-half of the required volume for a typical precast panel system (1000 units per year versus 2000 units per year). The market should not need to be limited to a particular region based on a factory location.

* Local contractors can be used, and the superstructure, with totally integrated mechanical and electrical, is so complete that it can, at the point of the closure and finishing phase, be sold or turned over to local union or non-union contractors for completion while the contractor or experienced licensee moves to other need areas.

ECOLOGIC III
CONCRETE

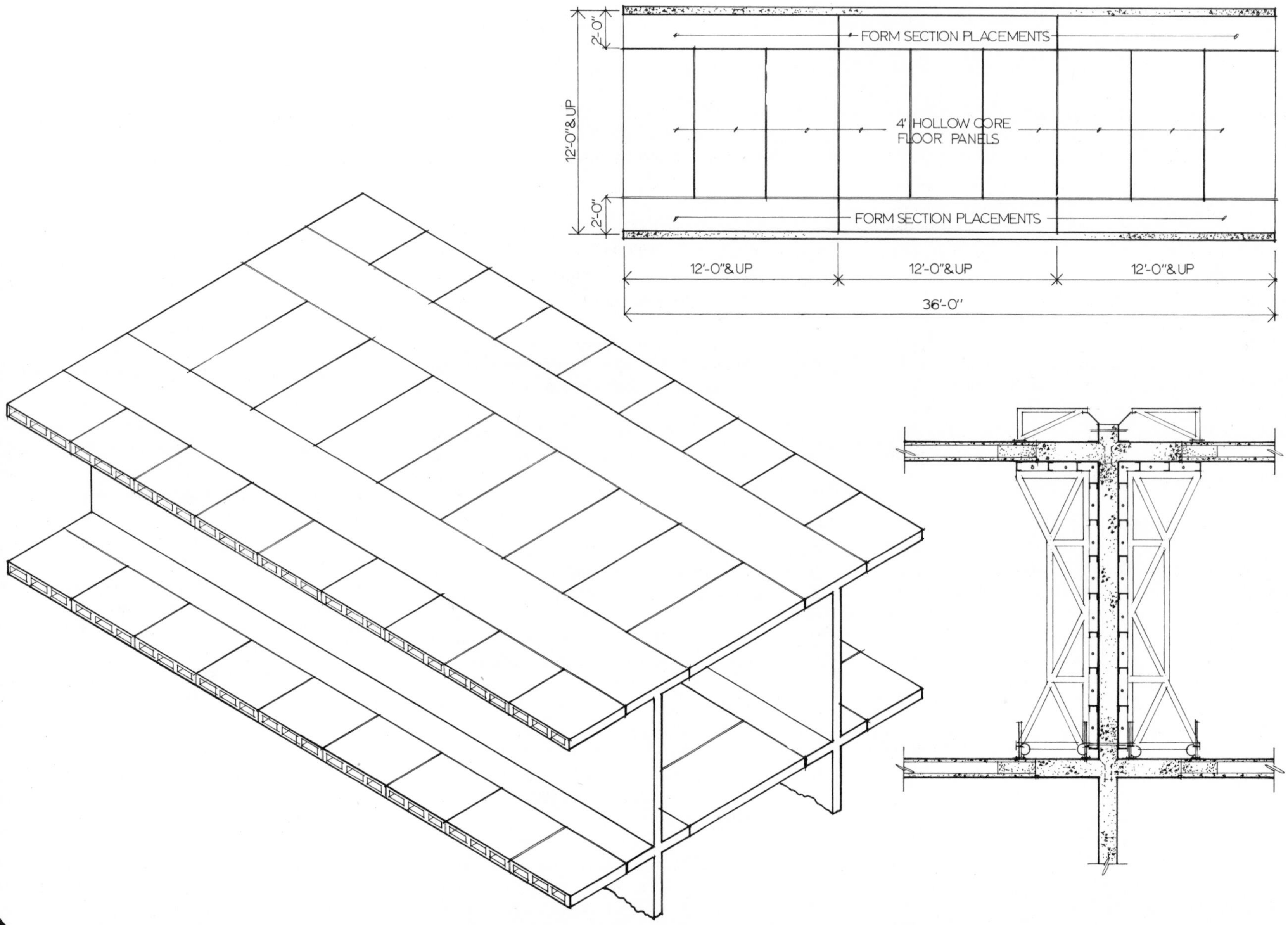

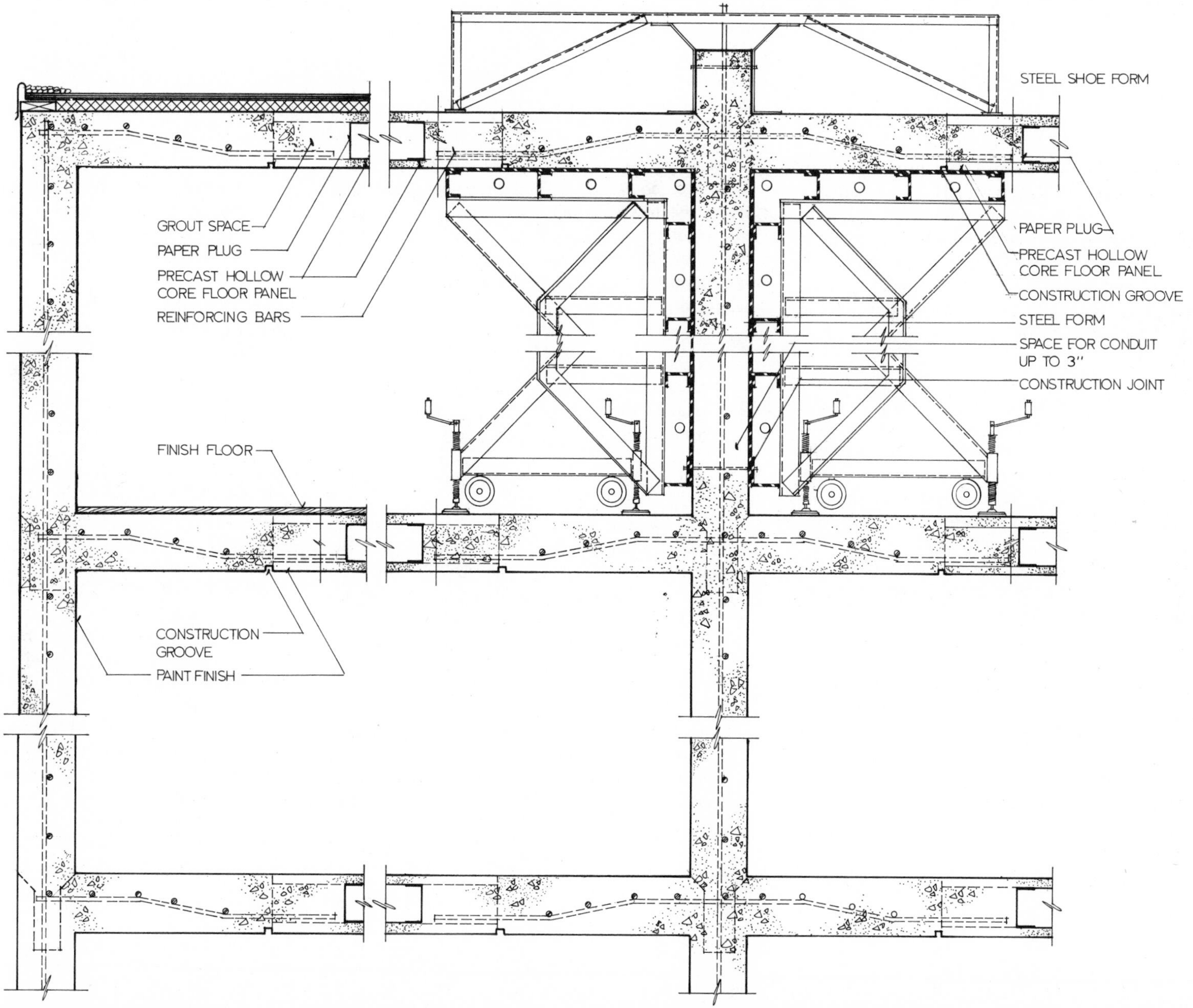
STEEL SHOE FORM
GROUT SPACE
PAPER PLUG
PRECAST HOLLOW CORE FLOOR PANEL
REINFORCING BARS
PAPER PLUG
PRECAST HOLLOW CORE FLOOR PANEL
CONSTRUCTION GROOVE
STEEL FORM
SPACE FOR CONDUIT UP TO 3"
CONSTRUCTION JOINT
FINISH FLOOR
CONSTRUCTION GROOVE
PAINT FINISH

ECOLOGIC's half-tunnel, by using standard precast hollow-core panels which are available in every part of the country, eliminates elaborate table or tunnel-forming required to cast floor but retains the flexibility of on-site casting in the walls and the 4' section of floor and ceiling immediately adjacent to the wall. This is the area where 90% of all door, duct, and stair areas occur.

The structure, code conformances, and finishing operations are in no way dependent on the prefabricated cladding and partitioning elements as is often the case with large panel and frame systems (i.e., inter-unit fire and acoustic insulation in frame systems, or mechanical equipment and sometimes facade installation in panel systems).

"Wet" trades are practically eliminated because of the high quality of concrete finish obtained by using the steel shutters; there is no necessity to plaster walls or ceilings—these surfaces can be papered or painted directly.

Because of the dimensional precision of this system of construction, it is possible to make wider use of prefabricated cladding partitions and mechanical units, and the installation is greatly simplified.

When the construction proceeds quickly and on schedule, it is also possible to schedule the finishings completely. This enables maximum use to be made of prefabrication and mechanical handling and results in earlier occupancy. This type of scheduling guarantee follows from two important aspects of the concrete half-tunnel forming system:

* *Year-Round Operations*—Constant thermal treatment of concrete in the steel T-forms ensures that the hardening process is independent of external temperature and eliminates delays often caused by extreme cold and results in year-round employment of labor.

* *Few Simplified Operations and Good Management Control*—The control imposed by the steel T-forms plus the clarity of operations and management through use of templates and simple coding allows use of less highly trained labor, while at the same time improving quality of construction.

* Construction takes place in a daily cycle and is broken down into basic operations each of which has been specially planned, and the men necessary to run the T-shutter system have every day, at the same hours, the same operations to perform. Productivity is thus appreciably improved and the labor content reduced to a minimum.

ECOLOGIC offers a medium- to high-rise building system, with all the cost and labor saving which the word implies, that is within the financial and market situation and capabilities of the typical small developer. It is possible for him because of the *low initial capital expenditure.* There is no million-dollar factory to finance; once licensed, he has only the forms (which could be as little as $25,000) to purchase or rent. The savings represent the immediate return on an extremely low investment.

Finally, although the T-forming system is capable of a very high rate of production (dependent on the number of sets of shutters in operation) and although its efficiency would increase with a greater volume of units, it can be operated economically with the more typical U. S. project under 100 units. This low minimum total order plus high-volume production is unique in that most precast systems are not competitive until they reach 300—500 units as their minimum total order.

6 SYSTEM ECOLOGIC

A KIT OF PARTS FOR THE HABITAT

There are two ideas hidden in the word "system": One is the idea of the *system as a whole,* the other is the idea of a *generating system.*[1]

The notion of a total system is that of a "complete object," and it is appreciated as a product of interaction among its parts. A generating system is a kit of parts with rules about the way the parts are able to be combined into a variety of related "wholes." The properties and parts of the system, in fact, define the potential of the whole.

1. Christopher Alexander, Systems Generating Systems, Architectural Design, December 1968.

SYSTEM ECOLOGIC is such a kit of parts. It is developed as both a design kit and as a construction kit. It is a design technique and a three-part building system capable of performing immediately within the present range of building codes, union constraints, and economic considerations. It is adaptable as wood/foam panels, steel frame, or concrete monolithic construction, and applicable to low-, medium-, and high-rise construction.

The basic parts of the ECOLOGIC kit are large spatial elements which form the habitat of man in our civilization, *sleeping units* and *living units.* These, when combined, form *total units.* Each of the spatial elements (i.e., *sleeping* and *living*) is made up from a select list of subcomponents such as: closure components, flooring components, interior components, and roof components. These are selected and placed within the spatial elements according to the particular client's program requirements. A seemingly limitless number of *total units* can be combined from the selection and joining of the various *living* and *sleeping units.* The total units demonstrated in this section are primarily those based on some of the generic housing layouts which have proven to be the most functional and desirable. These total units are further defined by a unit enumeration which includes the letter prefix A, B, or C, indicating the *building types* with which each individual layout is most compatible:

- A = Single-family
- B = Row house, Low-rise, garden apartment walkup
- C = High-rise

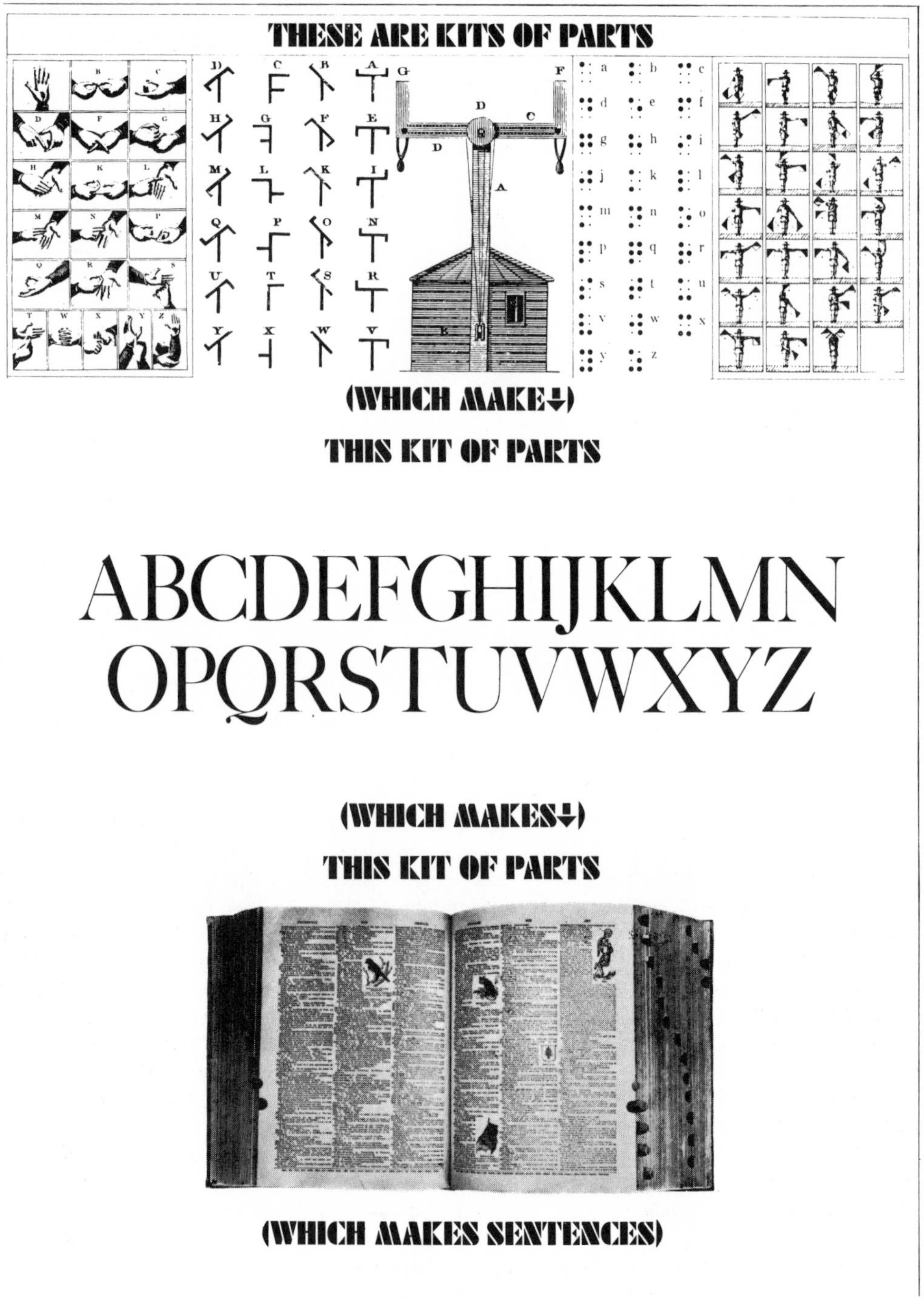
THESE ARE KITS OF PARTS
(WHICH MAKE↓)
THIS KIT OF PARTS
ABCDEFGHIJKLMN
OPQRSTUVWXYZ
(WHICH MAKES↓)
THIS KIT OF PARTS
(WHICH MAKES SENTENCES)

6-1
A KIT OF PARTS *From AD December 1968, "Systems Generating Systems" by Chris Alexander, first printed in Inland Steel Products Company Journal; Illustrations from Systemat Exhibit by James Robertson, San Francisco*

The numbers which follow the letter prefix indicate the particular spatial unit combinations involved. By these code numbers, the total units can be quickly identified and programmed, and even ordered. (See Chapter 8).

Design Standards for Combining Parts

The ASTM defines standardization as "the process of formulating and applying rules for an orderly approach to a specific activity for the benefit and with the cooperation of all concerned."

A mandatory standard is a standard whose use is compelled by law. In the U. S., most mandatory standards, except for the fundamental units of weight and measure, are written to protect health and safety.

The standards set for SYSTEM ECOLOGIC are a procedure established to evolve a rule of behavior. This rule is described in the following paragraphs, and it is demonstrated in the graphic *kit of parts*—which limits the number of elements, yet widens the choices and alternative solutions.

The main objective in standardizing choice is to develop a building system which is able to be assembled by either skilled or unskilled labor and which can be universally applied to both minimal and expanded housing projects, in a variety of geographic and social situations.

The general and specific standards which became the criteria for controlling the parts of our kit have already been discussed in detail in Chapters 1–5. *General standards* cover human criteria—sociological change and needs, function and flexibility, multipurpose use, individualism yet economy, light, space, and so on; and tech-

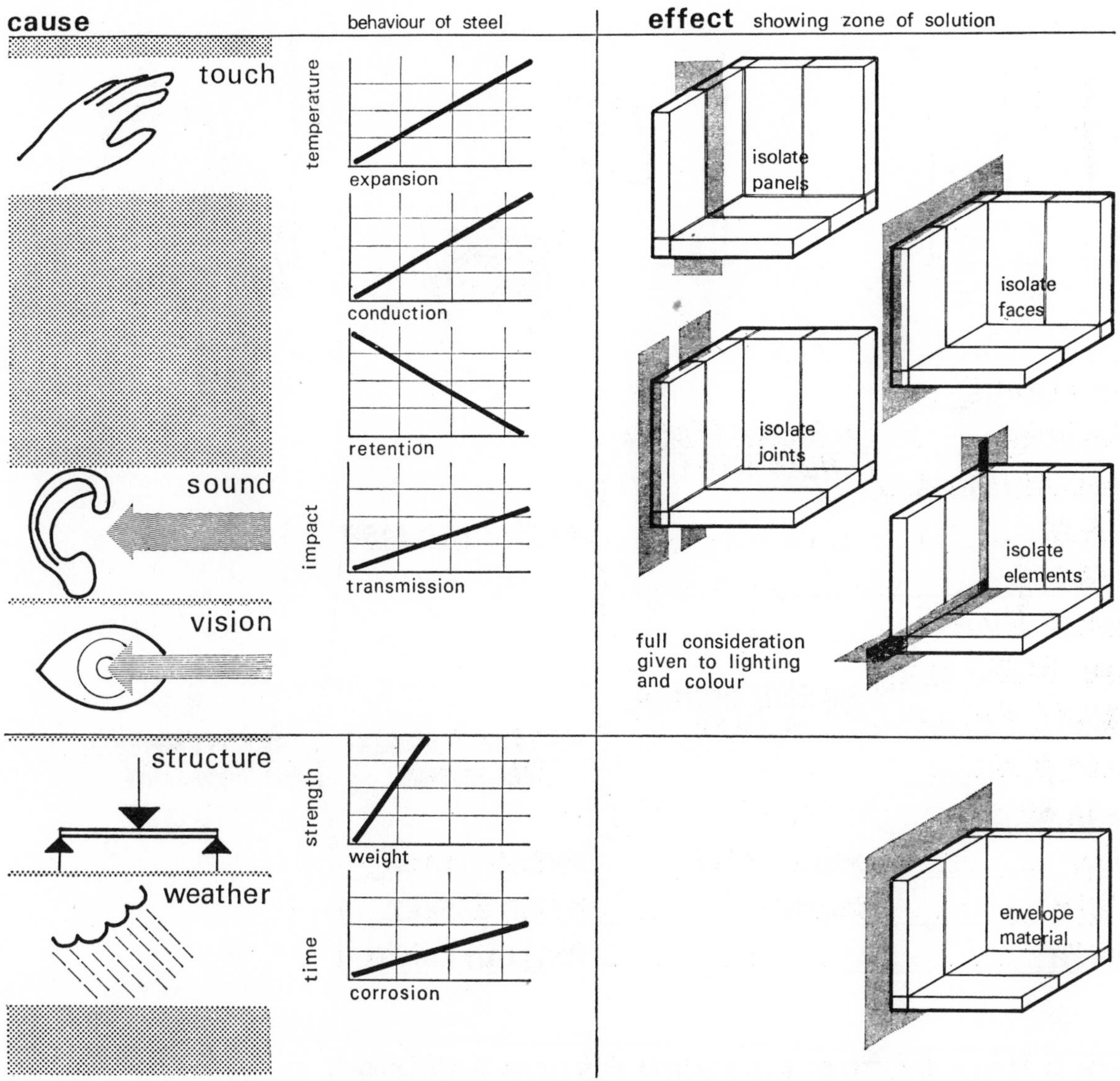

6-2
COMBINING PARTS This drawing is intended to indicate the characteristics of sheet steel when used as a building material. The left-hand column indicates certain requirements demanded of the component. These requirements have to be defined in terms of the material, as indicated in the center column. The resulting principles effecting the component design (or relationships) are indicated in the right-hand column. *Ibis Systems, London, England*

6-3
MODULARIZATION In order to standardize and thus simplify space for use within an industrialized building system, it is imperative to emphasize the need to formulate a viable basis for dimensional control. *"Selected Technological Aspects of the American Building Industry," by Richard Bender for the National Commission on Urban Problems, 1966*

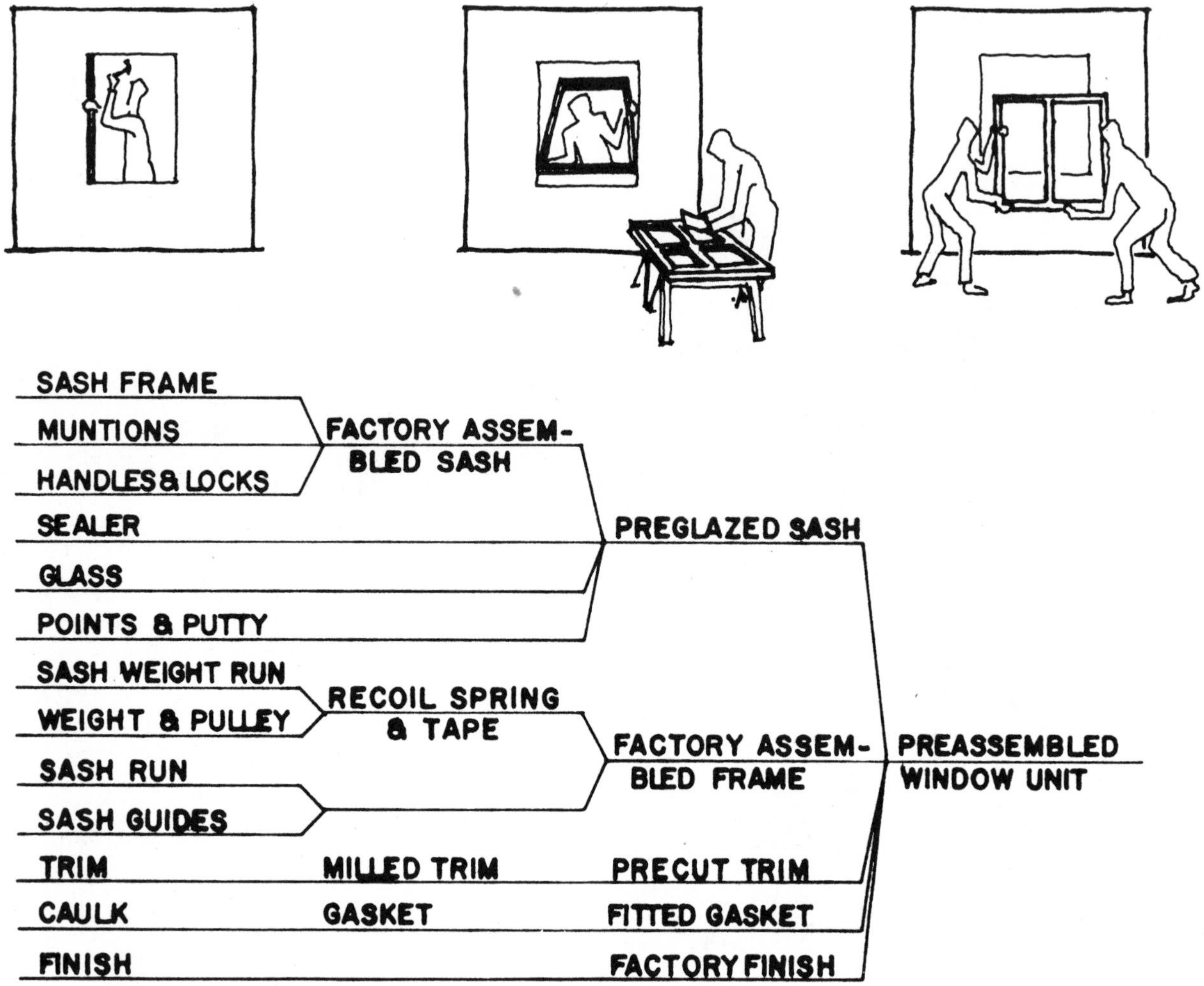

EVOLUTION OF THE FINISHED WINDOW ASSEMBLY

nological criteria—the urban plan, mass production, prefabrication, combinations and multipurpose flexibility. The *specific standards* of technology (as covered in Chapter 2) require that the spatial units form a total assembly which integrates all building functions (structure, enclosure, mechanical, electrical) into one system; that the elements must be related in detail, line, surface, and form, to all other elements of the assemblage; that the elements be capable of being factory-fabricated, machined, prefinished, and assembled, either on-site or preassembled; that a minimum of pieces require assembly by skilled workmen; that the total units and individual components must satisfy the basic code requirements, plan flexibility, and ease of addition and alteration for future expansion.

MODULARIZATION

In order to standardize and thus simplify space for use within an industrialized building system, it is imperative to emphasize the need to formulate a viable basis for dimensional control. In the case of SYSTEM ECOLOGIC, there are two systems of modular coordination which are based on (a) a planning module of 3'0" (PM), and (b) a structural and closure module of 4'0" (SM). Both modules, PM and SM, are based on four inches, which follows the international recommendations[2], and also conforms to recent findings concluded by ECODESIGN for HUD's In-Cities program.[3]

2. Second Report of European Productivity Agency, Project 174, Published by O.E.E.C., Paris 1961, obtainable from H.M.S.O.
3. In-Cities Experimental Housing Research and Development Project, Contract No. H-969, Abt/DMJM Consortium, July 1968.

6-4
TOLERANCE: STANDARD CONVENTIONS *Royal Institute of British Architects*

The use of a dimensional coordination for the design of spaces, while providing the basis for a consistent pattern of standardization, will also provide sufficient flexibility to cater to particular user needs, allow variety of forms, and enable adaptation to materials and components.

Why such an obvious tool as modularization, which is indeed a *prerequisite* for building and designing economically on a mass scale, has had such an acceptability problem is a question.

Traditionally, American building industry manufacturers have never coordinated their products dimensionally, this being a direct result of the free enterprise system. A business would start because grandfather had an extra load of lumber in his yard and, as an example, decided to fabricate windows. Depending upon the lengths of lumber at hand, the windows would conform to the most economic dimensions. The family would, out of ignorance or habit, continue to manufacture their products and variations based on the same lengths. Today, we have an untenable situation where few manufacturers of any building products have compatible dimensions, thereby causing everything built between the building elements to be custom-made. In some European nations, governmental controls are now exercised to establish dimensional coordination. But the American situation is accentuated by the multiplicity of makers of any given product.

The following quote from "The Adoption of Modular Coordination in Denmark," by Marius Kjeldsen for the Ministry of Housing in Copenhagen, thoroughly illustrates the present dilemma in this as well as other countries.

Tolerances: Standard Conventions

Procedure, in principle, for calculating the Maximum and Minimum sizes for a modular component.

1 Define the Modular Size (nM) of the component.

2 Define the Minimum Gap* (g).

3 Define the Positional Tolerance (p).

4 Define the Manufacturing Tolerance (t).

When the above have been settled, make the following calculations:

5 Determine the Minimum Deduction: $g + p + g$

6 Determine the Maximum Size: $S = nM - (2g + p)$

7 Determine the Minimum Size: $s = S - t$

8 Check the Maximum Gap (G): $G = g + p + \frac{t}{2}$

9 Check the Minimum Gap (g): $g = \frac{1}{2}(nM - S - p)$ and check against 2, above.

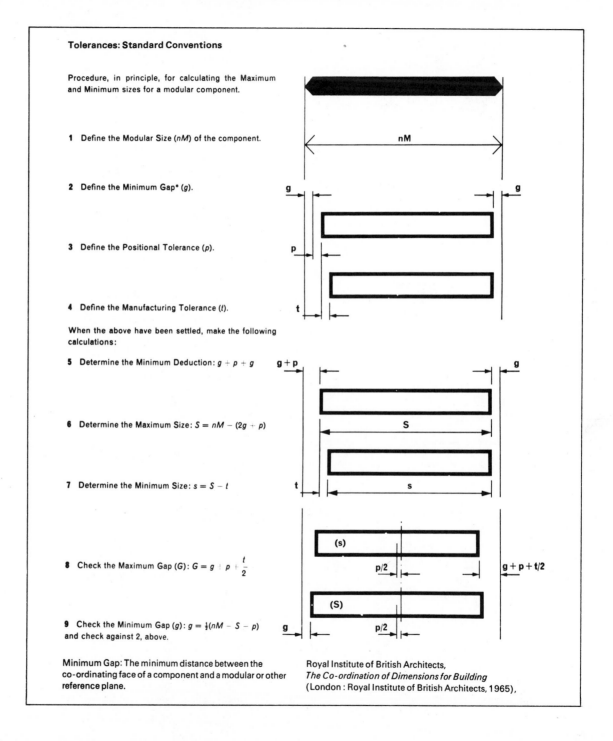

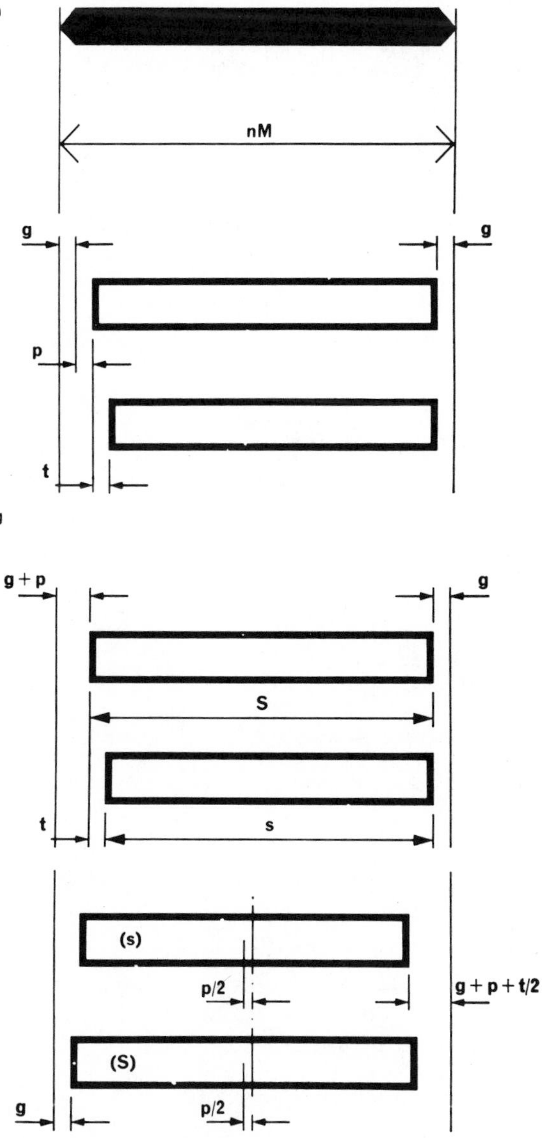

Minimum Gap: The minimum distance between the co-ordinating face of a component and a modular or other reference plane.

Royal Institute of British Architects, *The Co-ordination of Dimensions for Building* (London: Royal Institute of British Architects, 1965),

6-5
TOLERANCES: FLOOR-TO-FLOOR COMPONENTS *Royal Institute of British Architects*

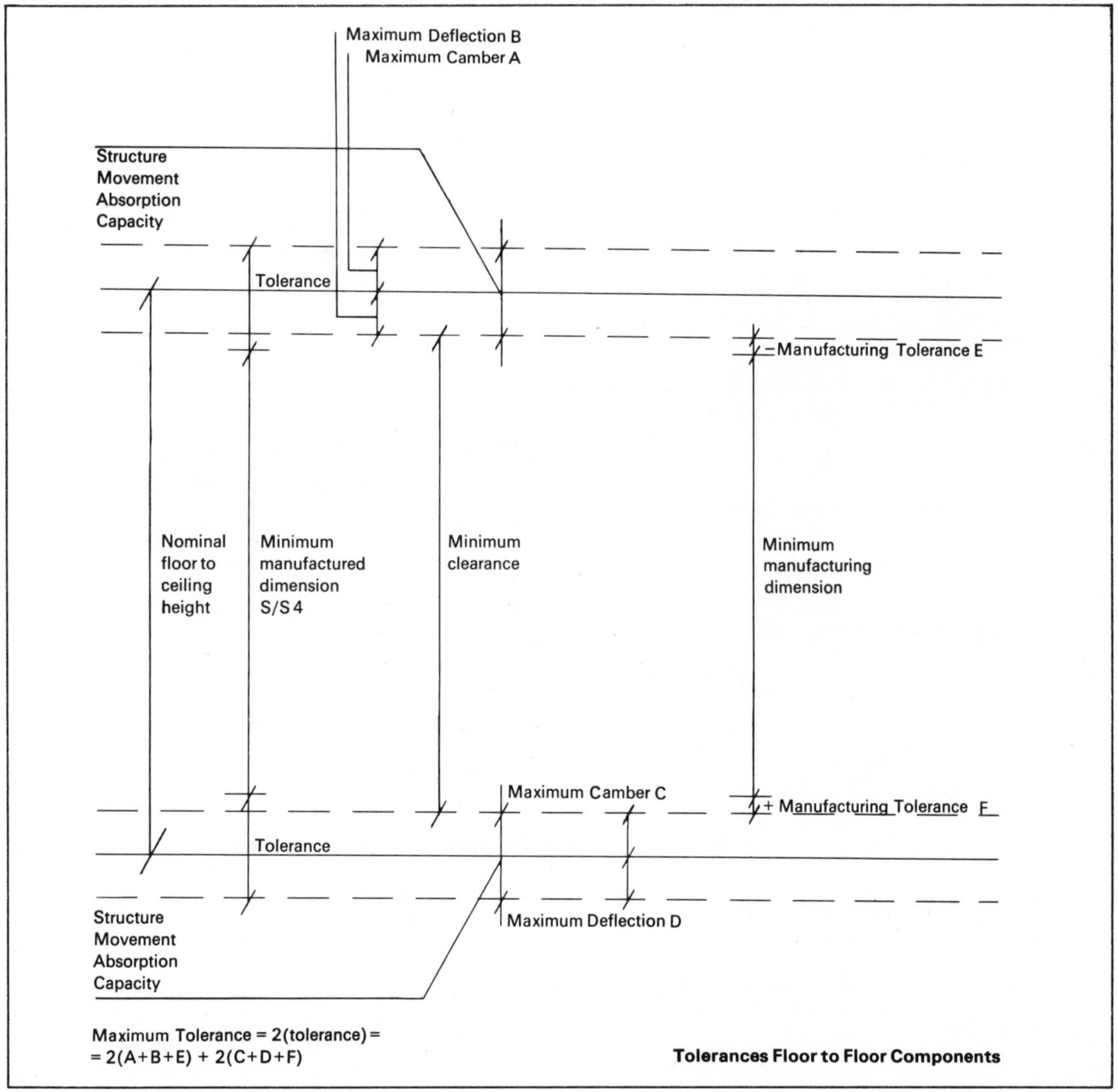

Tolerances Floor to Floor Components

"It is not difficult to arouse what seems to be a genuine interest in the matter, because it is quite easy for everybody to realize the many advantages which in the long run will result from modular coordination; but due to the slowness with which the effect manifests itself, it is difficult to get anybody to begin. This may rapidly lead into that most unfortunate situation in which everybody waits for somebody else to do something. The building material industry would very much like to produce building components of modular dimensions, but not until the demand is sufficiently large. Architects, engineers, and building owners would very much like to use modular building components, but since they are not as yet in production, their use is not prescribed. In this way, the vicious circle is set up, and it will have to be broken in some way or other."

The new awareness of this need for modularization is illustrated in a quote from "Modular Design of Low Cost Housing" by Mark Hartland Thomas:

"Hitherto, the prefabricated standardised components have been comparatively few in relation to the other materials that go into a building and are fabricated manually on the site. For example, a prefabricated window has a handmade brick wall built around it."

"Now, the prefabricated components are beginning to occupy so much of the construction that they jostle each other: the standard window has to be set between prefabricated panels, not brickwork or timber weather-boarding cut on site. At once a dimensional problem arises. How does one choose sizes for prefabricated wall panels when one realizes, as is too often the case, that the standard wooden windows are not made to the same sizes as the standard metal windows?"

Modular coordination offers a dimensional common denominator; the standardized module states to all who agree to utilize it that they will have a better prospect of market acceptance and a broader market by suiting the sizes of other manufacturers' components.

In Denmark, the size of the module agreed to is 100 mm. The inch-foot countries usually find the analogous size of 4" as the basic unit *(m)*. In one country, West Germany, a second module of 125 mm for brick construction is in parallel operation with the International Module of 100 mm for prefabricated components. Some 25 countries have issued national standards for modular coordination on the basic module of 100 mm or 4", according to their system of measurement.

When two close module dimensions are being used in order to accommodate both planning and existing structural efficiencies, they can be coordinated by use of a maxi-module or "critical number." It is easily seen that using $3m$ and $5m$ as modules, $8m$ is the critical number *(CN)*. The *CN* can be found by the formula:

$$CN = (a\text{-}1)\ (b\text{-}1)$$
$$= (3\text{-}1)\ (5\text{-}1) = 2 \times 4 = 8$$

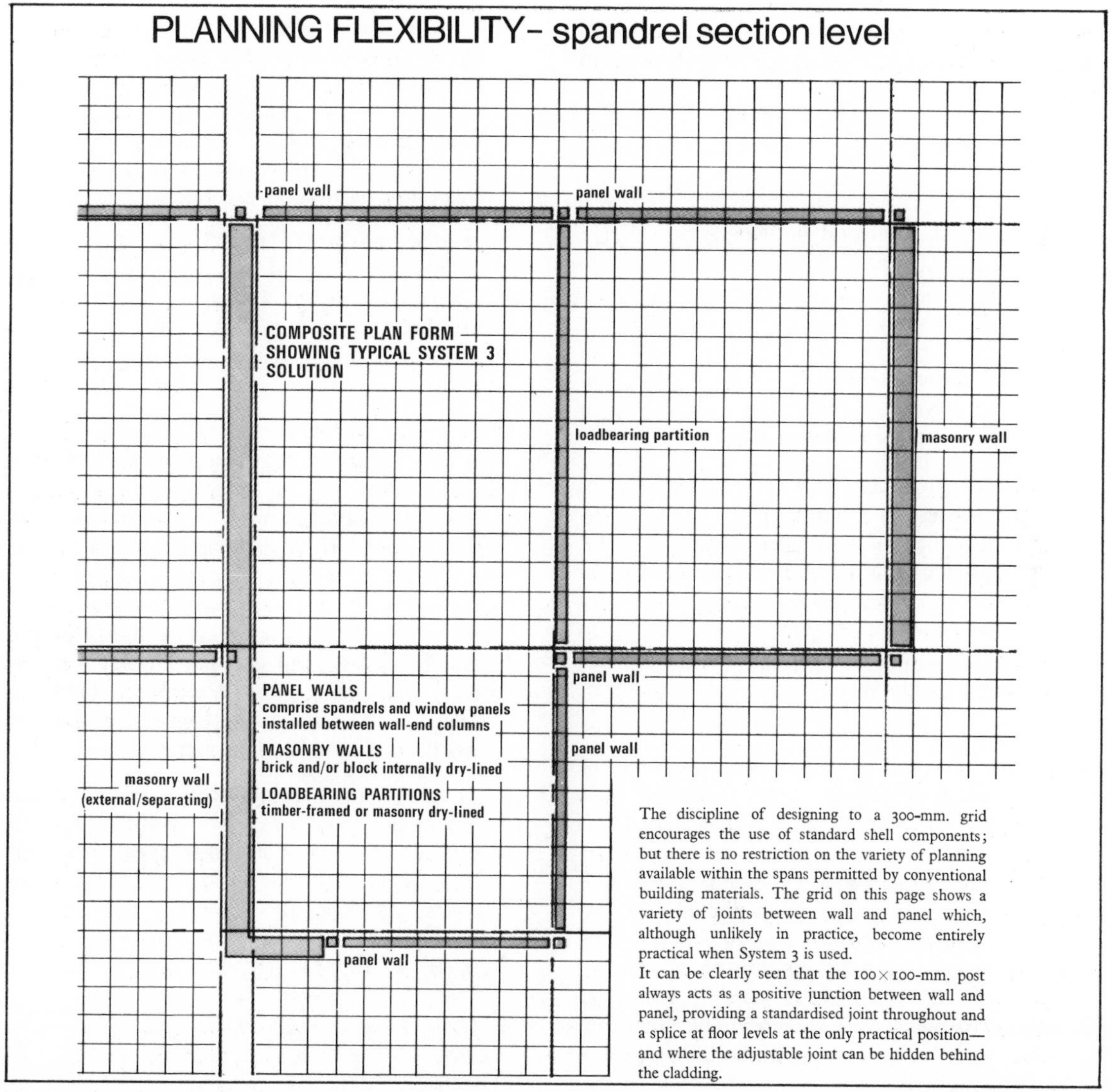

PLANNING FLEXIBILITY – spandrel section level

panel wall

panel wall

COMPOSITE PLAN FORM
SHOWING TYPICAL SYSTEM 3
SOLUTION

loadbearing partition

masonry wall

panel wall

PANEL WALLS
comprise spandrels and window panels
installed between wall-end columns

MASONRY WALLS
brick and/or block internally dry-lined

panel wall

LOADBEARING PARTITIONS
timber-framed or masonry dry-lined

masonry wall
(external/separating)

panel wall

The discipline of designing to a 300-mm. grid encourages the use of standard shell components; but there is no restriction on the variety of planning available within the spans permitted by conventional building materials. The grid on this page shows a variety of joints between wall and panel which, although unlikely in practice, become entirely practical when System 3 is used.

It can be clearly seen that the 100×100-mm. post always acts as a positive junction between wall and panel, providing a standardised joint throughout and a splice at floor levels at the only practical position—and where the adjustable joint can be hidden behind the cladding.

6-6
MODULAR COORDINATION Offers a dimensional common denominator. This diagram illustrates the various wall types working with the 100 mm module. *Drury Building Service Ltd.*

There is also a formula for three sizes, provided that they are consecutive. If the first is an even number,

$$CN = \frac{a^2}{2}$$

and, if the first is an odd number,

$$CN = \frac{a(a\text{-}1)}{2}$$

When the three numbers are not consecutive, there is no formula to help and one must have recourse to a table of critical numbers.

SYSTEM ECOLOGIC has a planning module of 3′ and structural module of 4′ with the *CN* being 6. Evidence the overall dimensions of 12′ by 36′ and the compatibility with the 4′ closure widths imposed by existing structural efficiencies of panels and with circulation 3′ in width, which is the efficient space planning size for domestic passageways.

1. *Modular Combinations*—In low-cost housing, architectural planning often demands the flexibility to vary spaces by one module. If it is not possible to achieve 1*m* flexibility by combining components of dissimilar sizes, it would be necessary to manufacture components of every modular size. It is clear that components of almost an infinite number of sizes exist on the present market. The problems exist when the dimensioning does not relate to 4″—the module which we suggest is the most suitable.

Because the smallest number of components makes generally for the least expensive and fastest method of construction, certain critical dimensions become most obvious and prevalent in design, especially if one is attempting to create plan and building type flexibility with good architectural design.

It does not make much sense to have the simplistic notion of one module with incremental growths of 1*m* and 1*m* and 1*m*. It is infinitely more logical to work an increase based on something like the famed Fibonnaci series of 1*m* + 2*m* + 3*m* + 5*m* + 8*m* + 13*m*.

Choosing particular numerically related sizes, as for example 3*m* and 5*m,* works! Using two widths of 3*m* only gives the designer 3*m* flexibility. Usually, a manufacturer will double and triple his basic sizes as in panels—a width of 3*m* will have a second size of 6*m* offered. This is common practice for manufacturers, but we lose flexibility because two 3*m* panels will give 6*m* already and 12*m* with four threes, etc. A common factor between two sizes is self-defeating and not worthwhile as a module system.

To avoid this "factor effect" one can choose 5*m* as a second size. Two widths of 3*m* and 5*m* will fill every modular space from 8*m*. That is to say that, with only two components, we have 1*m* flexibility from 8*m* onwards. 3*m* and 4*m* work similarly, as 3, 4, 6, 8, 9, 10, 11, 12, 13, 14, 15, 16, ...

2. *Manufacturers, Architects, and Modules*—In order for a manufacturer to decide what standard sizes his products should be, the critical number is a useful guide for him to work with. If he should want to convert fiberboard panels, commonly 12*m* wide, into panels for partitions, he may cut the 12*m* into two sixes, or three fours, or four threes, all lacking the critical number. Those dimensions have only 6*m* or 4*m* or 3*m* flexibility. However, if he cuts unequally into 5*m* and 7*m,* the critical number is 24, by the formula mentioned above. If, besides cutting the material into two small panels, he offers a large 12*m* panel as well, then the critical number is not reduced below 24 and the larger size is easier and less expensive to install. The quantity of waste in any particular subsystem on a module design is reduced to practically nothing by utilizing the critical number concept.

The critical number is not the panacea to the architect. He must not only identify number combinations which fill a required space but also take into account what the various alternative permutations of it bring about aesthetically: whether to arrange the three A's and two B's thus: ABABA, or thus: BAAAB, or thus: BBAAA, and so on.

It should be pointed out that in SYSTEM ECOLOGIC when we speak of additive components capable of different combinations, they include doors, windows, and panels, as well as skylights, etc. Some components are additive in one direction and single in the other.

THE PARTS THEMSELVES

The selection criteria for the parts of the habitat kit are straightforward:

* The smallest number which will create the greatest number and variety of spaces.
* Availability of parts and ease of maintenance and replacement.
* Ease of assembly and/or erection.
* Module compatibility.
* Efficient use of space inherent in components.
* Relevance to scale of project and human scale.
* Ease of transport.
* Economy.
* Adaptability and viability.

The following components or parts were selected for use within SYSTEM ECOLOGIC. Some are standard proprietary items, readily able to be chosen from existing catalogues and, thusly, providing the total system with an ever-increasing range and mixture of components.

By combining these subsystems components with the space units, one can truly exploit and utilize the entire "kit of parts." It becomes a tool for the user and everyone involved in the building process (architect, developer, user) to achieve successfully planned dwelling units for each individual site and situation without constant duplication of efforts or dimensional error.

In the next pages, the sub-pieces of this "kit of parts" are displayed.

1. Structural Frame

The most basic building element to SYSTEM ECOLOGIC is the structural frame in the following dimensions:

12′ x 36′: The largest size frame available, the proportions were determined, among other things, by the allowable transportation dimensions as set by the Department of Transportation for highway portage.

12′ x 24′: This elemental size is determined by users' requirements for comfortable living areas and service core combinations.

12′ x 12′: The logical basic room size; for instance, a 10′ x 12′ bedroom, allowing a 2′ wide x 12′ long storage closet area all within the

12′ x 6′: An *add* module which, when combined with the others above, gives the entire planning module (PM) greater flexibility, keeping in mind that the critical number (CN) in this system is *6′.*

2. Foundations/Flooring

Because all sites are not necessarily level, the footings and foundation piers are to be engineered for each particular site and are poured in the conventional manner, precast, or sonotube in-situ on a 12′ module in order to correspond with the bearing module of the component units. No additional excavation is necessary other than a trench for utilities and service lines. These are brought to the meters where hot water heater, electrical, gas, and waste lines are located.

There are four flooring options: wood (standard oak), full cover vinyl, concrete slab, or carpet. Floors are designed to support a uniform live load of 40 psf plus the dead load of the materials.

3. Service Components

Each dwelling could have its own thermostatically controlled forced-air heating and cooling system. The system can operate with any energy source or a combination of energy sources most economical in the building location. The capacity of the system will also vary with the climate conditions in the building location. The Chrysler gas-fired furnace with metal ductwork, registers, and flue is used as an example.

Plumbing components could include the use of ABS (acryonitrile butadeine styrene) pipe. This material cuts plant fabrication and shipping costs substantially, compared with the use of metal pipe. Each unit would have factory-installed molded bathrooms, as by Hytec, Inc., and a 60-gallon hot water heater.

4. Interior Components

Storage closets, partitions, and stairs are shop-fabricated and placed so as to act as space dividers wherever possible, offering themselves as useful functioning elements and also as sound barriers and for visual privacy. Interior partitions are 8′-0″ steel-studded structure (1-5/8 x 25 gage) with 1/2″ gypsum board coverage both sides. Doors are prehung in preframed wall panels and are hollow-core (interior) or solid-core (exterior).

5. Closure Components/Siding

The central theme is the interchangeability between the six basic degrees of closure that give the ten closure panel layout alternatives. Standard glass window units are used in temperate closure components. Standard louvre units in tropical closure components.

Siding alternatives for the closure components utilize the most practical "Sweet's Catalogue" products to insure precise cost estimates.

6. Roofing Components

There are three roof components: flat, raised, and sloped. Two slopes can form a pitch, and also the possibility of skylights is available. The pitched or shed roof type can also adapt itself to a clerestory window treatment. There is also a raised roof alternate which allows for clerestories as well. Roofing, depending on geographic location, would generally be built-up or silicone.

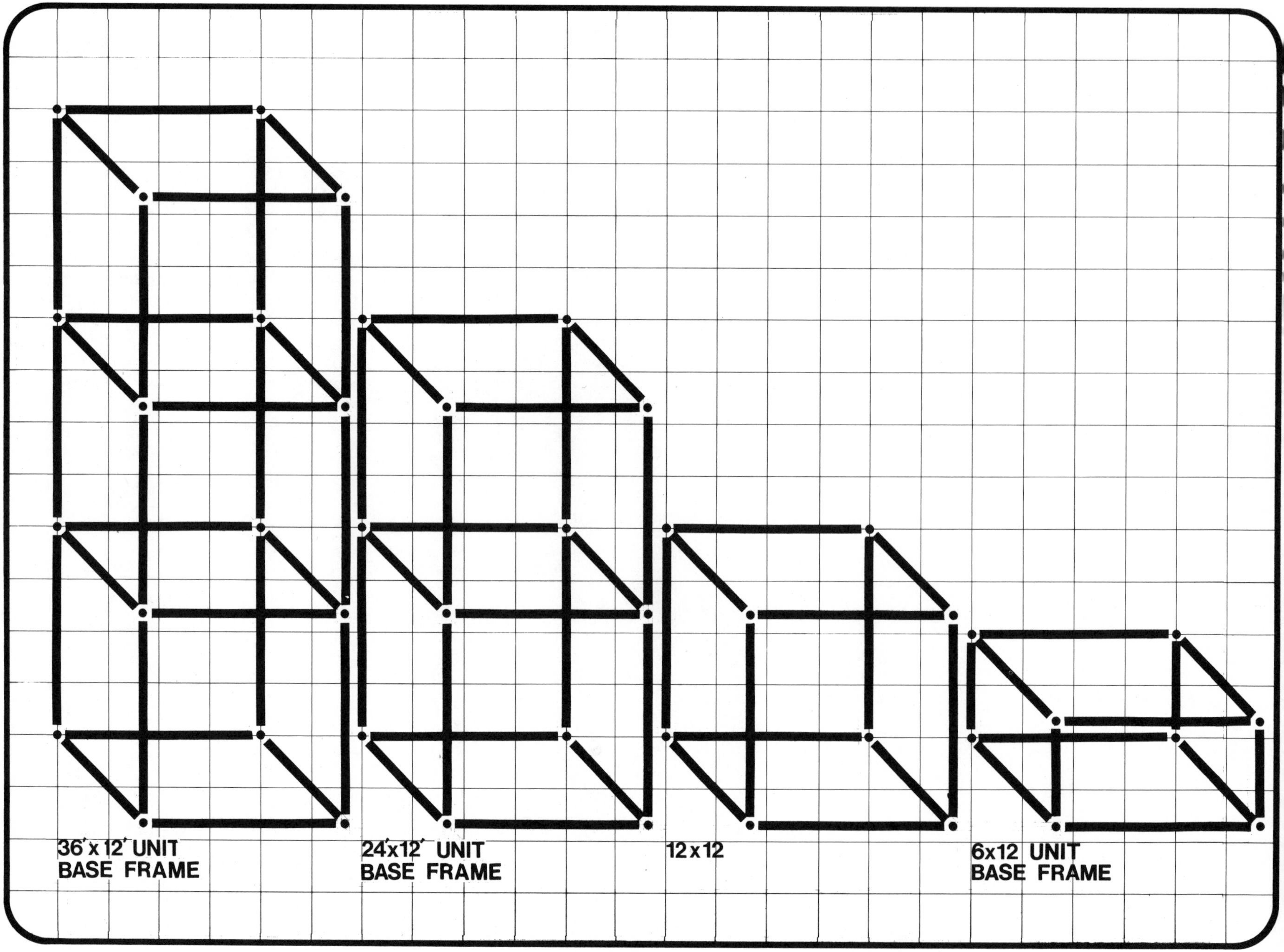
36' x 12' UNIT
BASE FRAME
24'x12' UNIT
BASE FRAME
12 x 12
6x12 UNIT
BASE FRAME

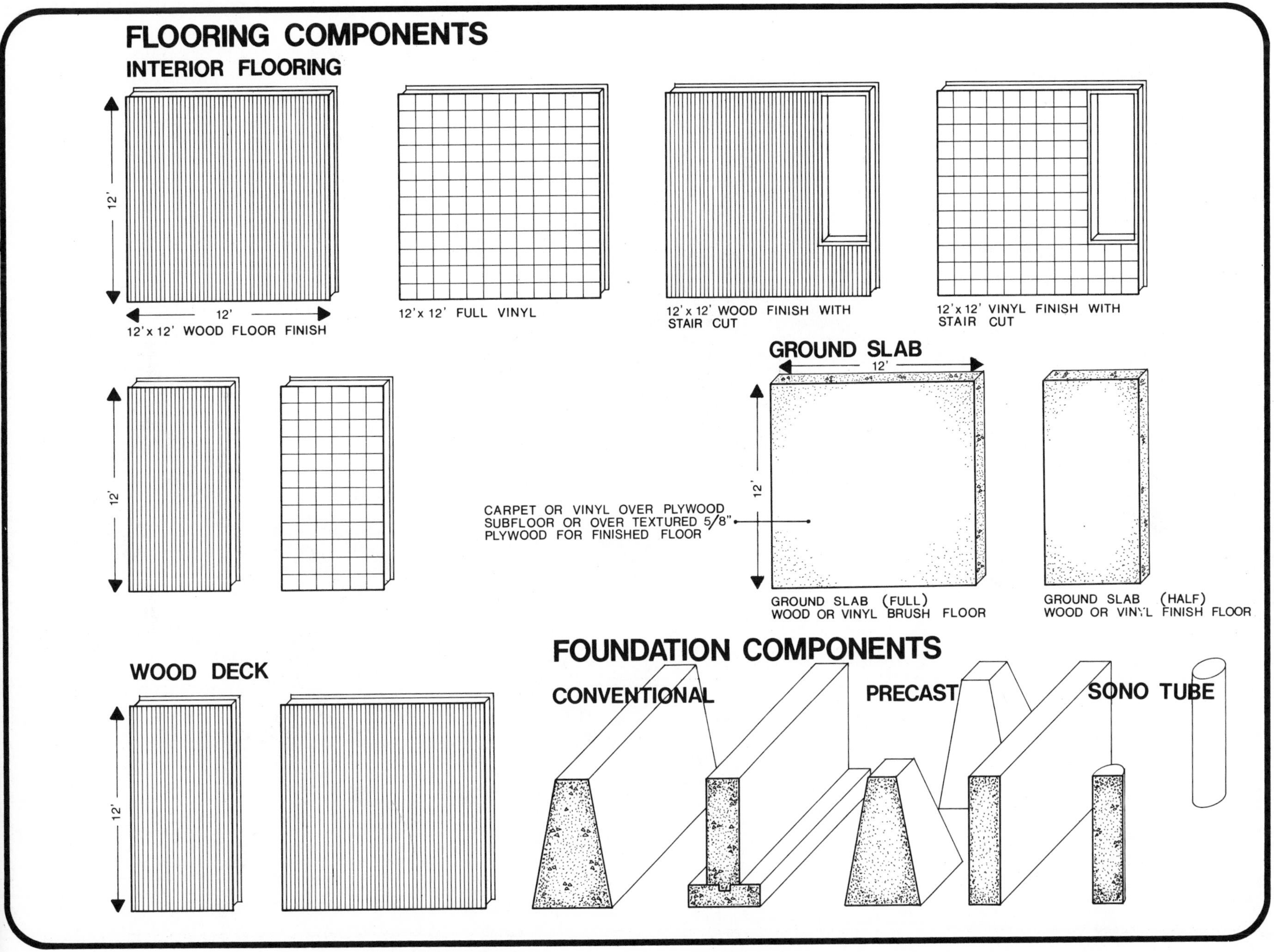
FLOORING COMPONENTS
INTERIOR FLOORING
12'
12'
12' x 12' WOOD FLOOR FINISH
12' x 12' FULL VINYL
12' x 12' WOOD FINISH WITH STAIR CUT
12' x 12' VINYL FINISH WITH STAIR CUT
12'
GROUND SLAB
12'
12'
CARPET OR VINYL OVER PLYWOOD SUBFLOOR OR OVER TEXTURED 5/8" PLYWOOD FOR FINISHED FLOOR
GROUND SLAB (FULL) WOOD OR VINYL BRUSH FLOOR
GROUND SLAB (HALF) WOOD OR VINYL FINISH FLOOR
WOOD DECK
12'
FOUNDATION COMPONENTS
CONVENTIONAL
PRECAST
SONO TUBE

SERVICE COMPONENTS

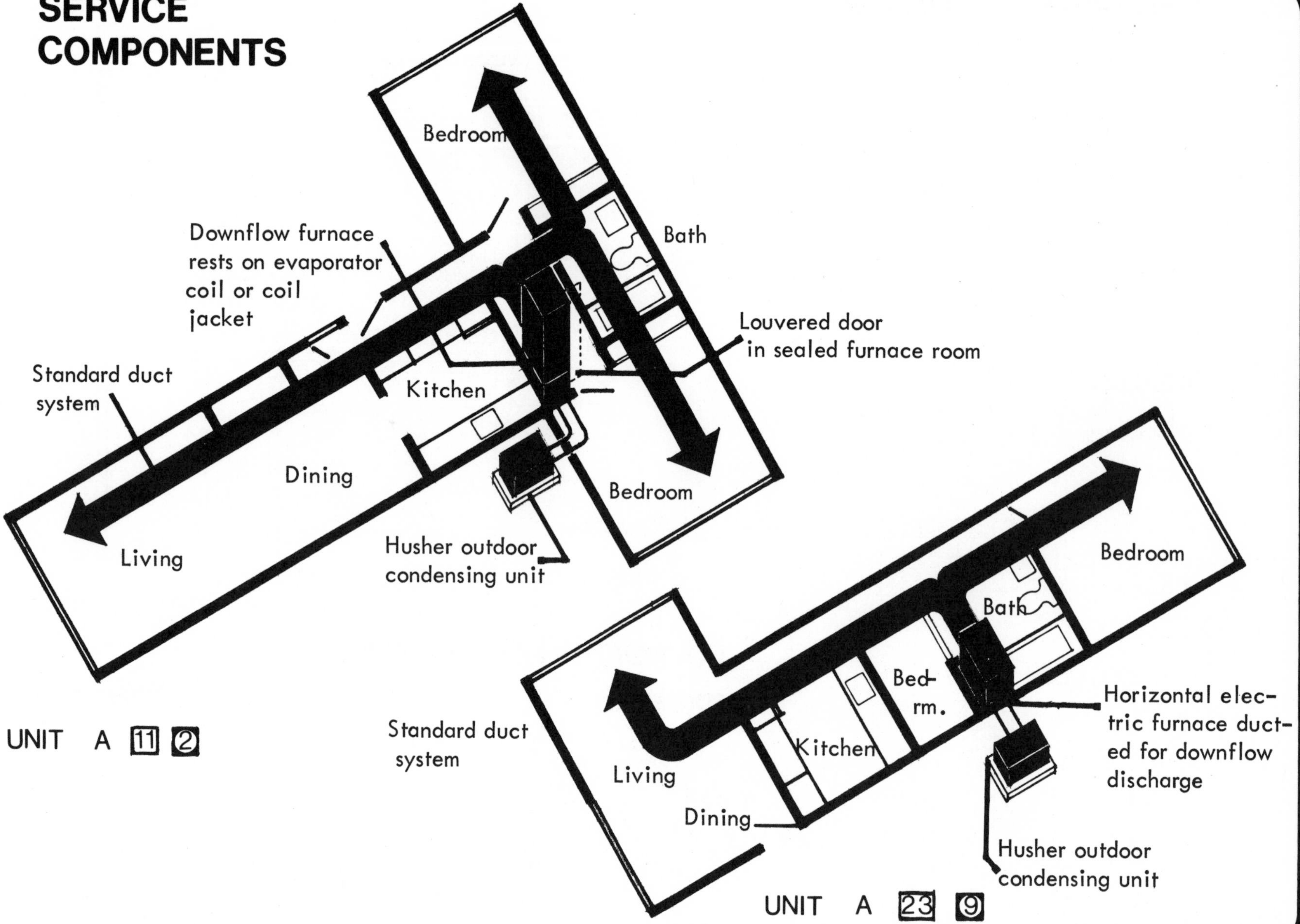
Bedroom
Bath
Downflow furnace rests on evaporator coil or coil jacket
Louvered door in sealed furnace room
Standard duct system
Kitchen
Dining
Bedroom
Living
Husher outdoor condensing unit
UNIT A 11 2
Bedroom
Bath
Bed-rm.
Horizontal electric furnace ducted for downflow discharge
Standard duct system
Kitchen
Living
Dining
Husher outdoor condensing unit
UNIT A 23 9

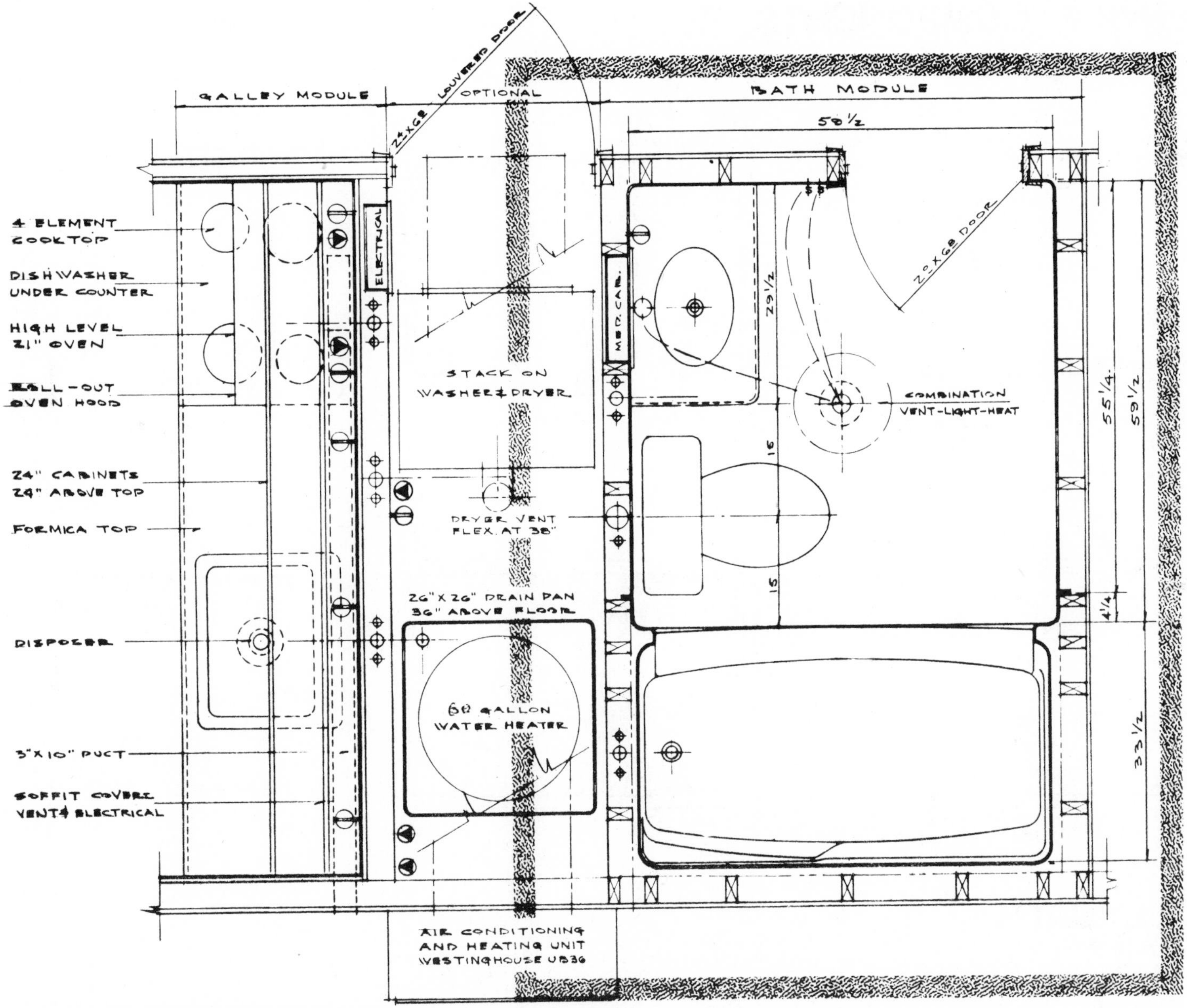
GALLEY MODULE
OPTIONAL
24 X 68 LOUVERED DOOR
BATH MODULE
58 1/2
4 ELEMENT COOKTOP
DISHWASHER UNDER COUNTER
HIGH LEVEL 21" OVEN
ROLL-OUT OVEN HOOD
24" CABINETS 24" ABOVE TOP
FORMICA TOP
DISPOSER
3"X10" DUCT
SOFFIT COVERS VENT & ELECTRICAL
ELECTRICAL
STACK ON WASHER & DRYER
DRYER VENT FLEX. AT 38"
26"X26" DRAIN PAN 36" ABOVE FLOOR
60 GALLON WATER HEATER
MED. CAB.
20 X 68 DOOR
29 1/2
COMBINATION VENT-LIGHT-HEAT
15
15
55 1/4
59 1/2
4 1/4
33 1/2
AIR CONDITIONING AND HEATING UNIT WESTINGHOUSE UB36

SERVICE COMPONENTS

SERVICE COMPONENTS

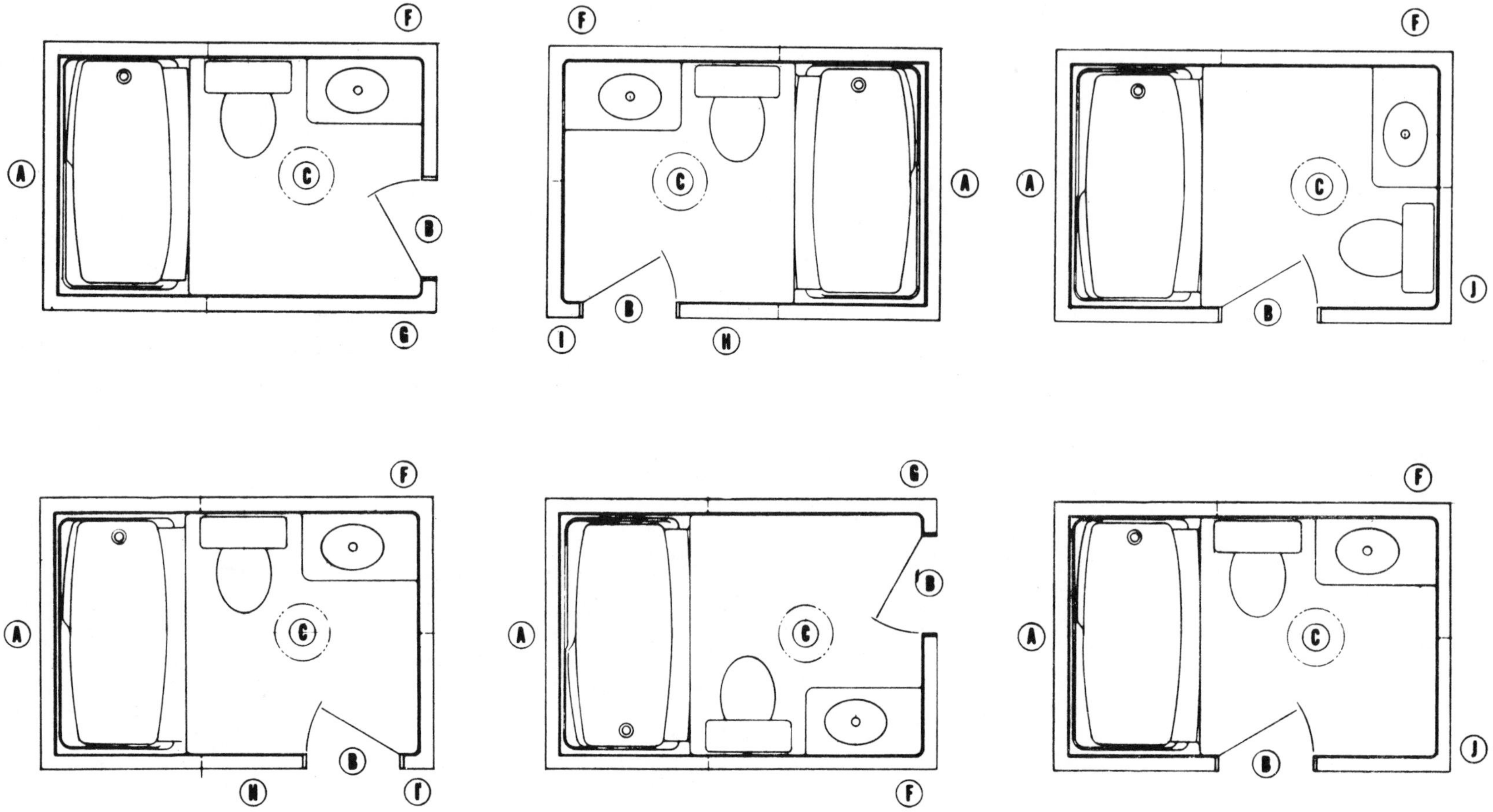

FULL BATH

INTERIOR COMPONENTS

KITCHEN TYPE 1
U-SHAPED (BASE 9' x 6')

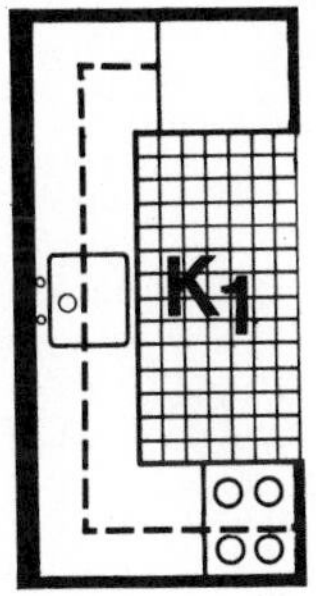

KITCHEN TYPE 2
PARALLEL (BASE 9' x 9')

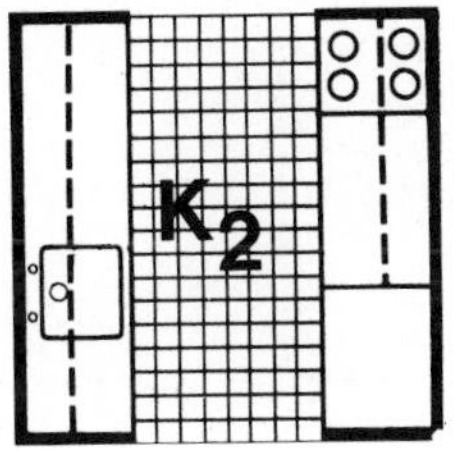

KITCHEN ADDS
EATING L'S

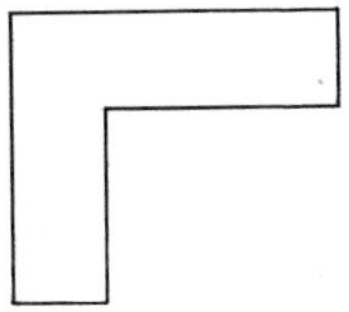

3' COUNTER SECTIONS

TOILET TYPE 1
HALF BATH (3' x 3')

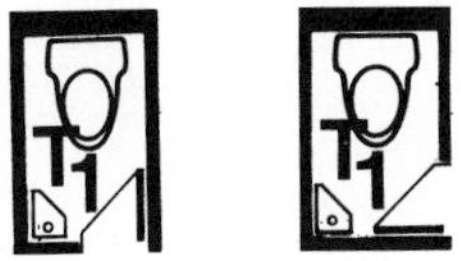

TOILET TYPE 2
FULL BATH (7'-6" x 6')

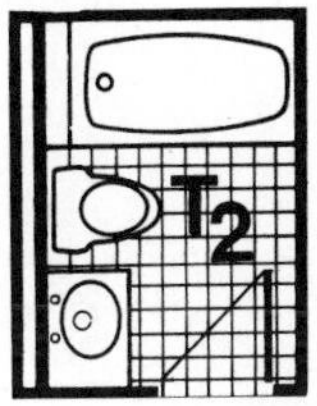

FULL BATH (7'-6" x 6') SHOWN w/ TOILET ADD

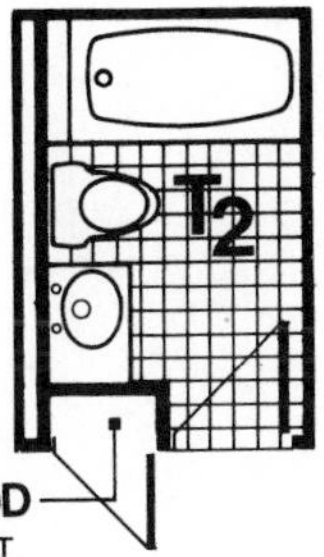

TOILET ADD
LINEN CLOSET (3' x 1'-1½")

STORAGE PARTITIONS, CLOSETS
(3' x 2'-6")　　(6' x 2'-6")

SPECIAL PURPOSE CLOSET
STACKED WASHER / DRYER (3' x 2'-6")

FIREPLACE

STAIR TYPE 1
STRAIGHT RUN (3' x 10'-3")

STAIR TYPE 2
SPLIT RUN (6' x 8')

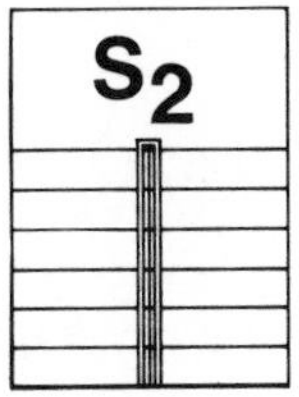

DESK - BENCH
DESK - BED
DESK / BENCH - MOUNTED AT 36"
DESK / BED - MOUNTED AT 20"

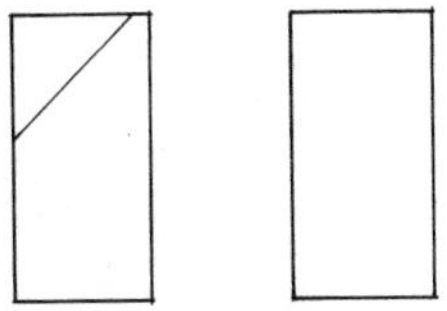

WET WALLS

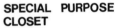

INTERIOR COMPONENTS

CLOSURE COMPONENTS – TEMPERATE

1. FULL GLASS

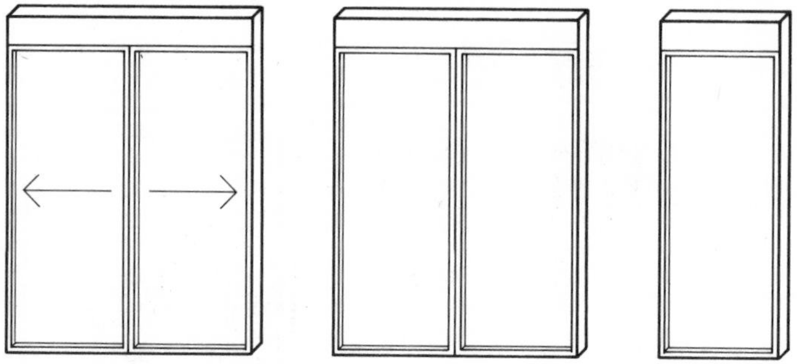

2. VERTICAL HALF GLASS

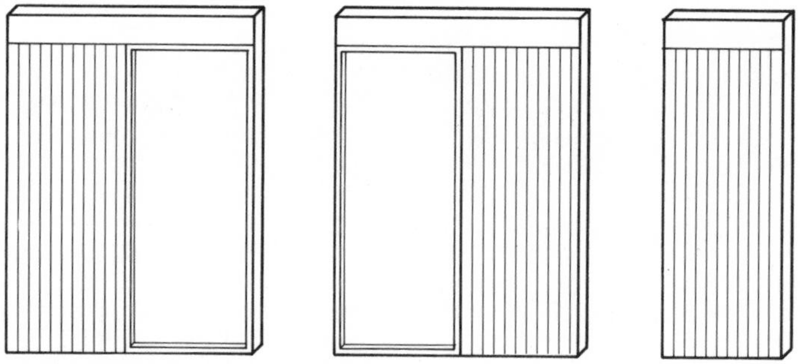

3. HORIZONTAL GLASS

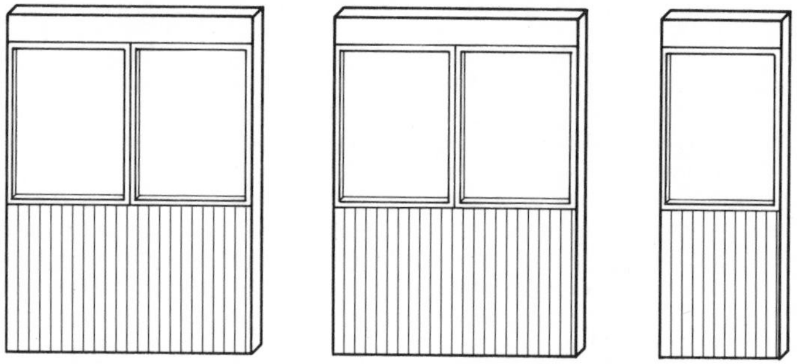

4. QUARTER GLASS

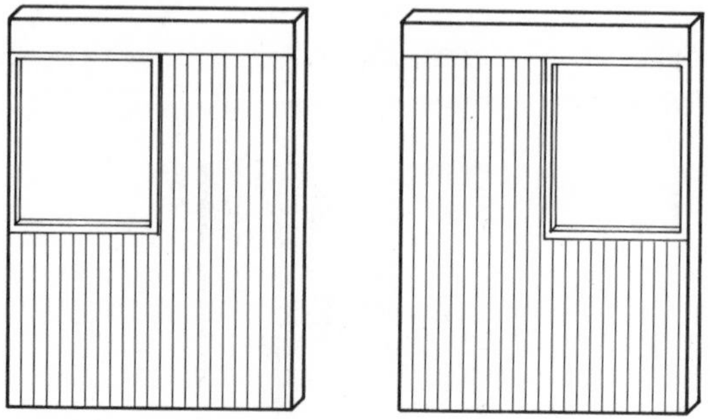

5. SOLID PANEL

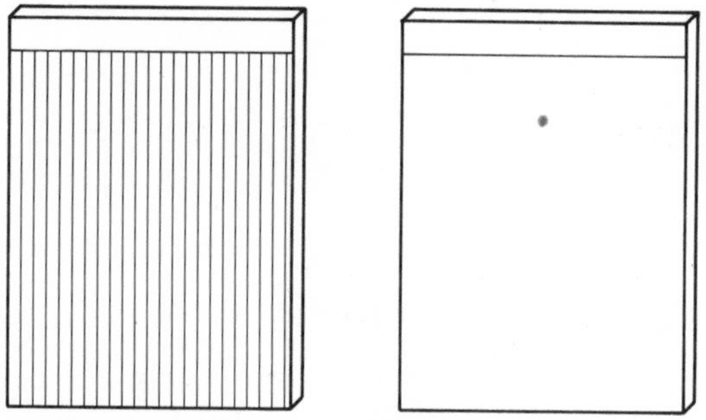

6. PROJECTED PANEL

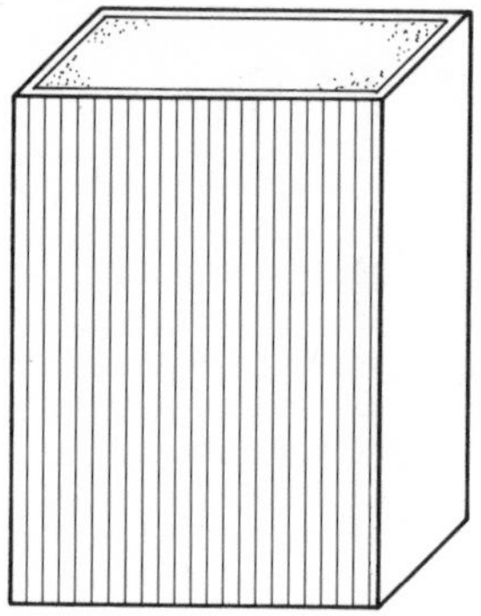

CLOSURE COMPONENTS – TROPICAL

7. FULL LOUVER

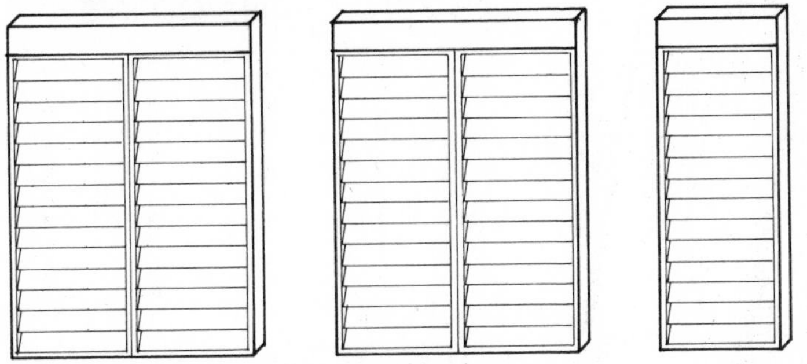

8. VERTICAL HALF LOUVER

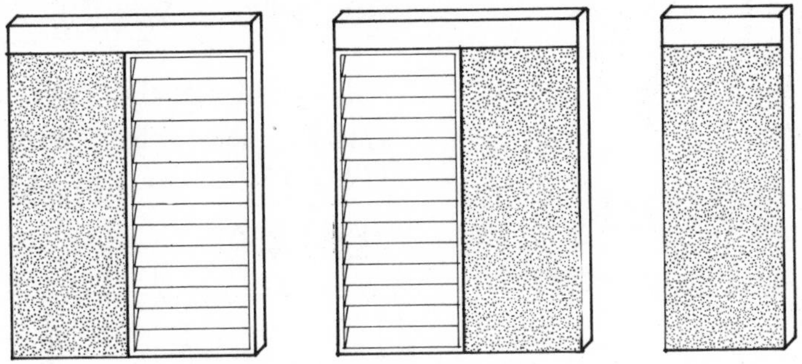

9. HORIZONTAL LOUVER

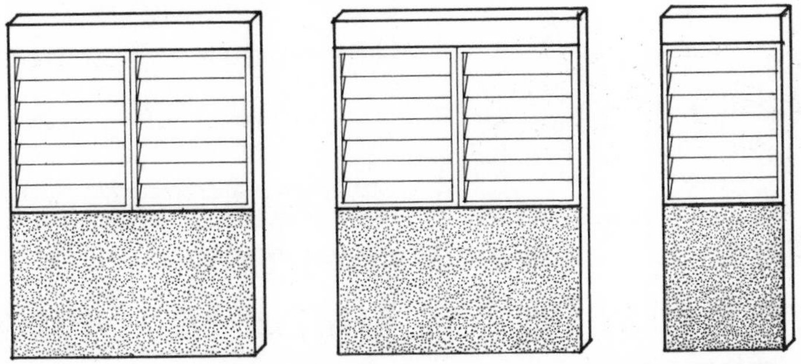

10. QUARTER LOUVER

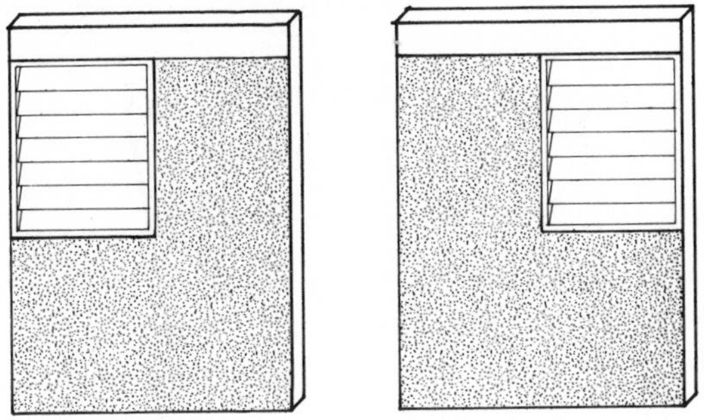

EXTERIOR SIDING
7/8"
8"
7/8"
4"
4"
HORIZONTAL
7/8"
5"
5"
2"
7/8"
10"
2"
VERTICAL
aluminum, vinyl siding
STUD OR RAFTER NAILING
HAND SPLIT WESTERN RED CEDAR BARN SHAKES
1/2" CEDAR FACED PLYWOOD SHEATHING
wood shakes
AGGREGATE FINISH
FIBER REINFORCING
EXPANDED FOAM
textured plastic foam panels
4"
4"
4"
PREMIUM GRADE ONLY
WIDTH 48-3/8" (48 face)
LENGTHS: 8', 9', OR 10'
VERTICAL
4"
4"
HORIZONTAL
textured plywood panels

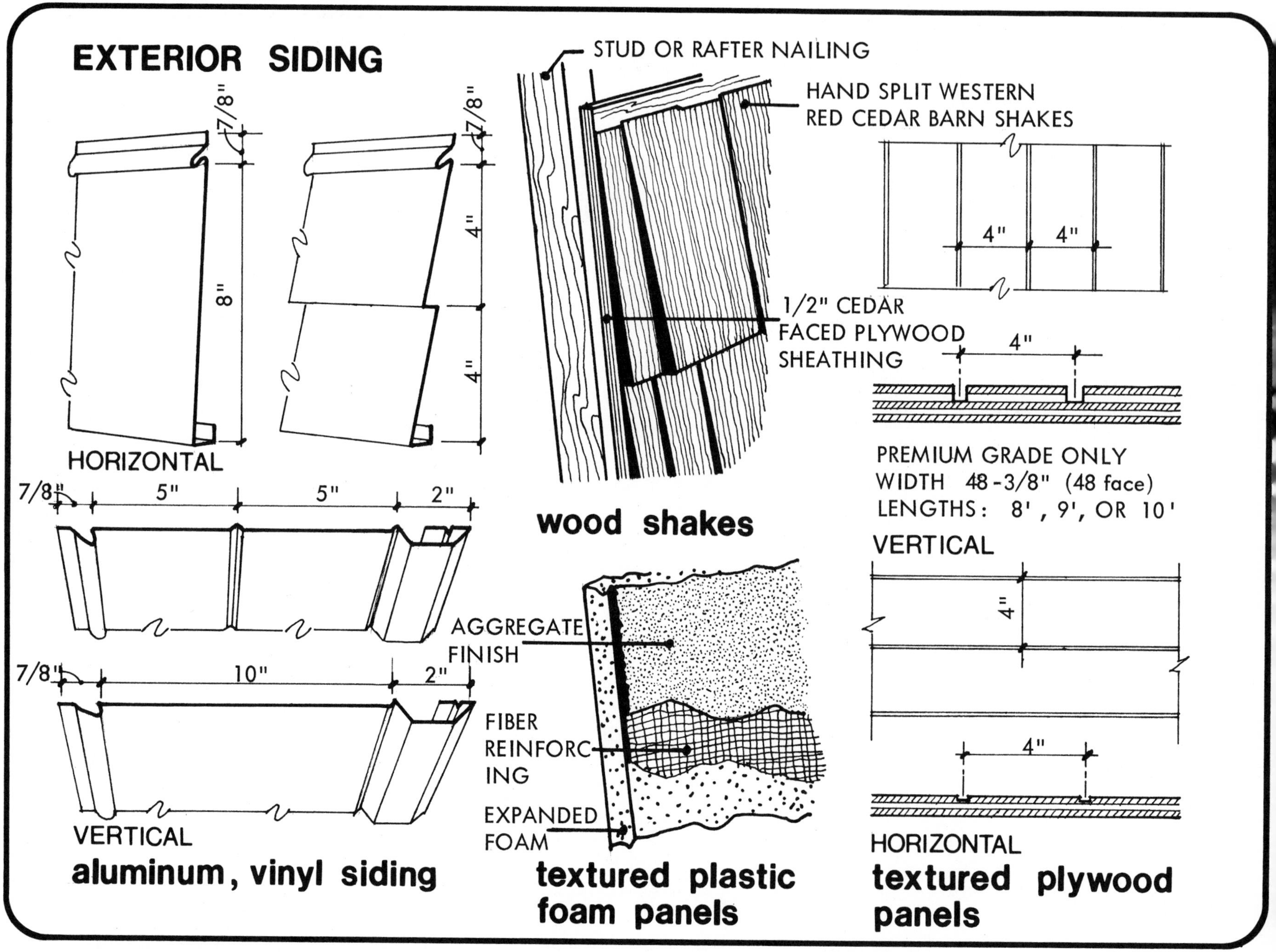

ROOFING COMPONENTS

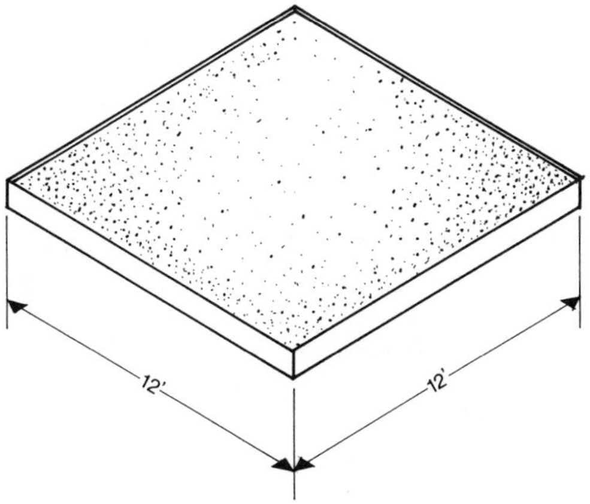

FLAT ROOF

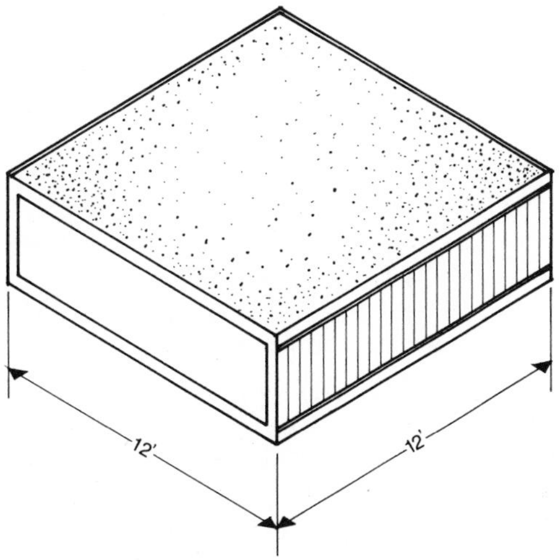

RAISED ROOF

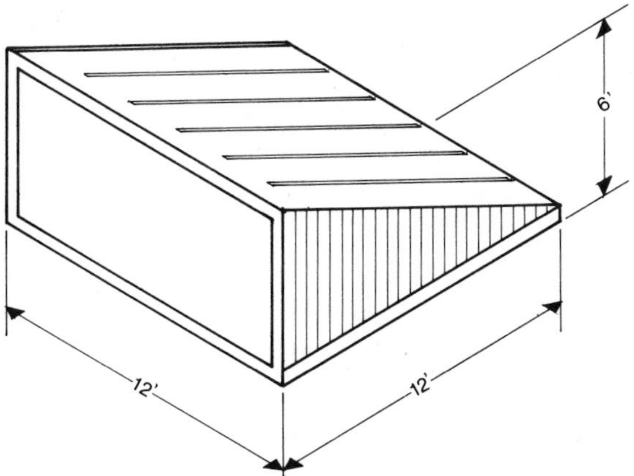

SKYLIGHT ROOF

7 HOW THE KIT WORKS!

BUILDING IN THE LIFE STYLE

Using a "kit of parts" as a design, planning, and construction tool results in advantages for all involved in the building process, but most importantly, it provides standards and "rules of the game" that allow for immediate dialogue.

Architect—It provides a technique of standardization which allows flexibility in design and gives more time to explore more innovative overall designs and site plan concept; a reduction in time spent on detail drawing; organized feedback from client and contractor.

Tenant/User—The kit of parts can be used as a game technique of discerning user needs. Component parts can be moved about within modules to show preferred layouts. Working with the architect, the user can readily see the parameters of size and function which control certain decisions.

Manufacturer and Supplier—Organized demand for components enables longer production runs, efficient programming for manufacture, simplified storage and transportation.

Contractor—More comprehensive, consistent, and better detailed information provides opportunities for earlier collaboration with architect, more efficient contract planning and more economical use of labor, materials, and mechanical equipment, better working conditions, earlier completion of a waterproof shell.

Engineers (Structural, Mechanical)—Time is reduced in detail preparations because the main concern is with the assembly of predetermined units; known performance data on structure are available at an early stage; extensive standard information and unit quantities, specifications, and cost data are provided.

Developer—A building which is designed and erected more quickly than is possible by traditional means, with high quality prefinished components produced under controlled factory conditions at less cost, means generally an earlier return on capital investment.

In utilizing the kit, it is proposed that factory-fabricated dwelling units of the three basic building types will be planned from the structural frame. The wood honeycomb, steel, or concrete shell would be selected according to the requirements of the site and situation as outlined in Chapter 5. Within these space modules, the "parts" or components would be organized in order to accommodate the "living and the "sleeping function" of the human habitat. These arrangements could vary almost infinitely in adapting to needs, desires, and budgets of the individual family. The following plates show some of the more common layouts in this first step of the kit's utilization—definition of *living units* and *sleeping units* layout alternatives.

Living Units

The essential space elements and functions which compose the living area of a dwelling unit are the living room itself, the dining area, and the kitchen. Of course, any number of additional rooms, such as dens, family room, or parlor can be included in this group, and some are shown in the layouts. The generic configurations, i.e., placement of the elements, one to the other, as well as the exterior form, are stated in these plans. It is implied in some cases that the sleeping units (when joined to certain living units) will be on another level, but this is obvious in the plans themselves.

Sleeping Units

As stated with respect to living units, these are examples of basic generic layouts of the essential space elements and functions which make up the sleeping area of a dwelling unit: the bedroom, the bathroom, and storage areas.

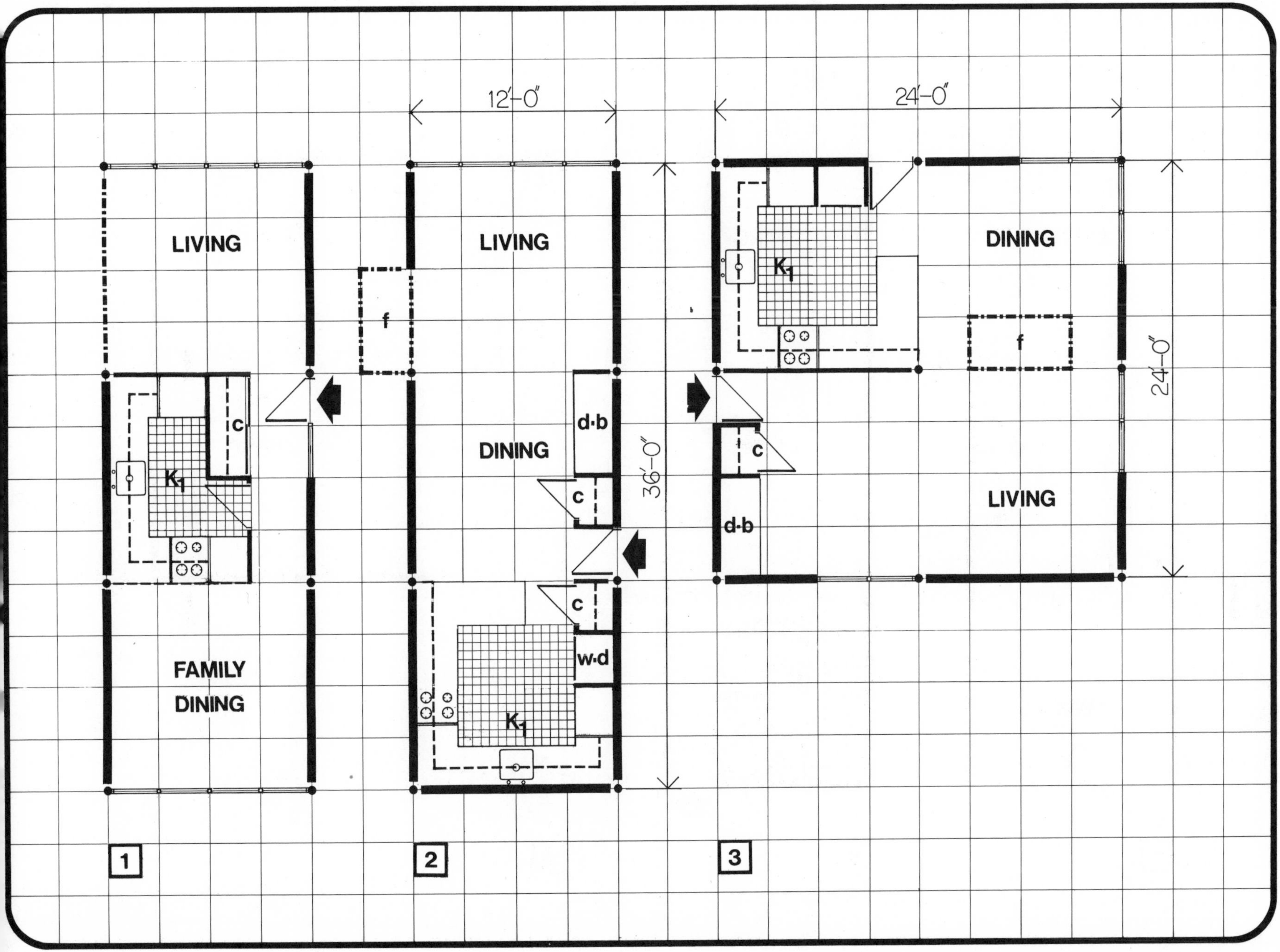
12'-0"
24'-0"
LIVING
DINING
FAMILY
DINING
c
K_1
f
d·b
w·d
36'-0"
24'-0"
1
2
3

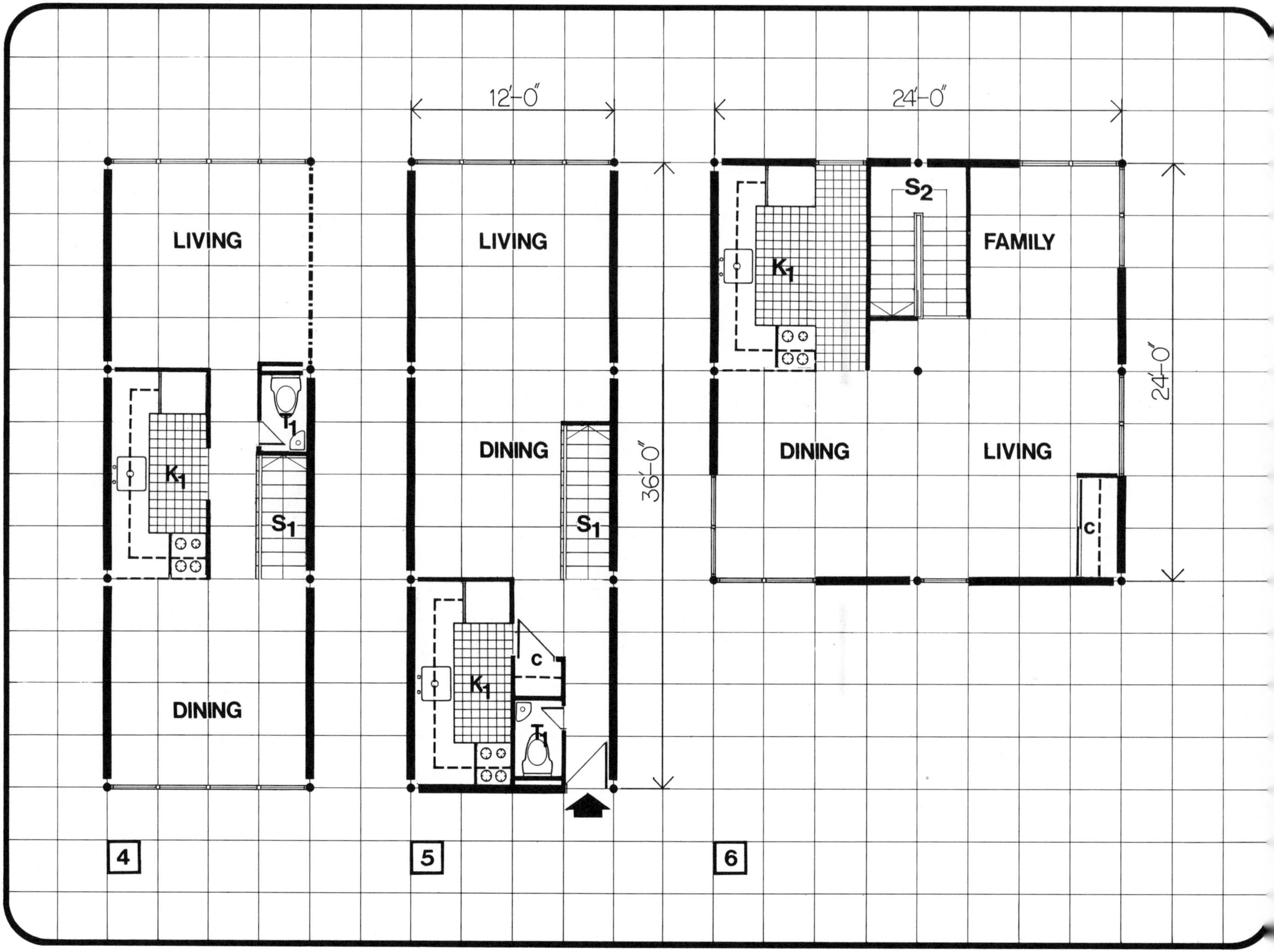
LIVING
K1
T1
S1
DINING
4
12'-0"
LIVING
DINING
S1
36'-0"
K1
c
T1
5
24'-0"
K1
S2
FAMILY
24'-0"
DINING
LIVING
c
6

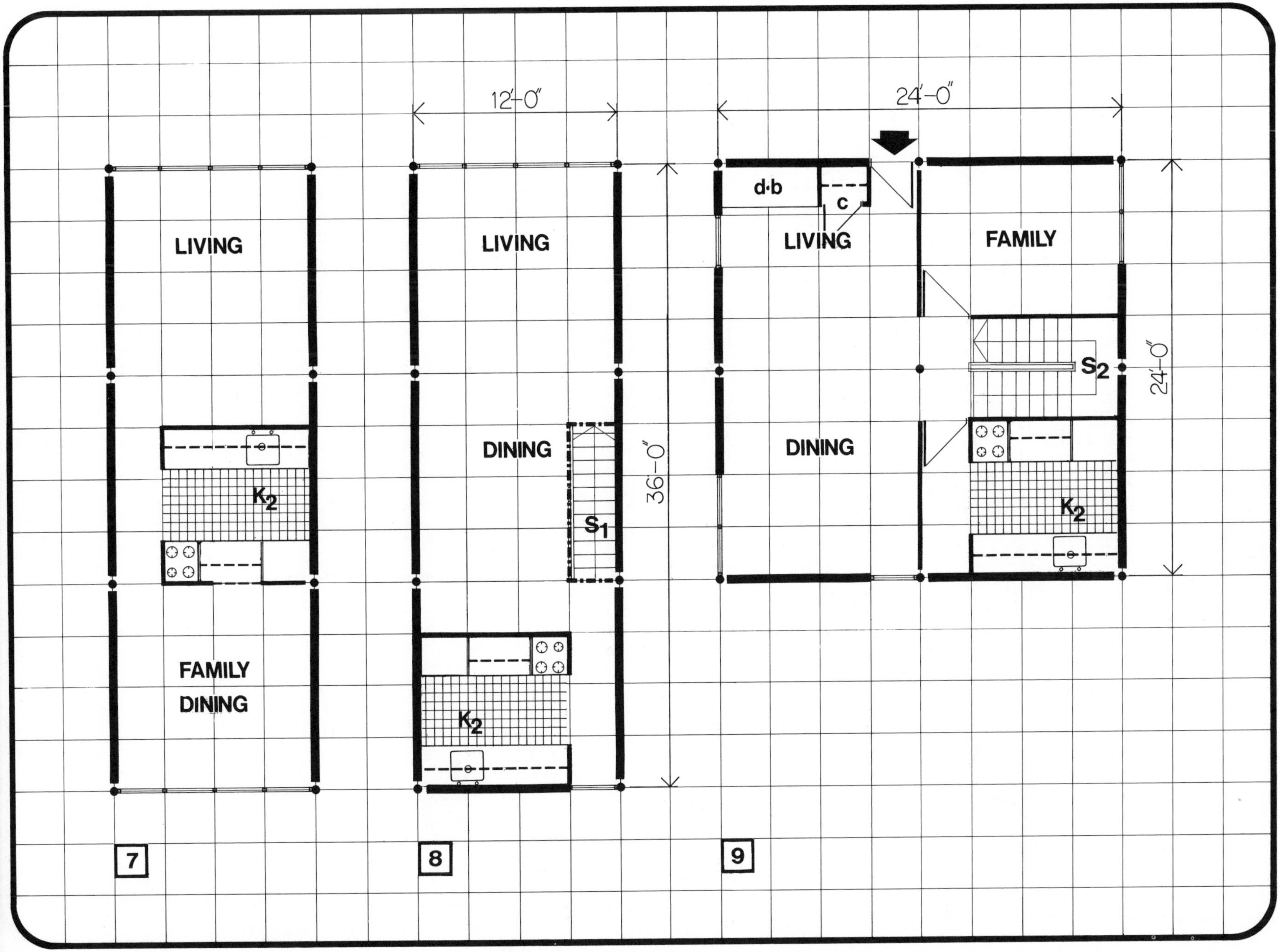

LIVING UNITS

12'-0"
24'-0"
d·b
c
FAMILY
DINING
DINING
S2
24'-0"
36'-0"
K2
S1
K2
K2
FAMILY
DINING
K2
7
8
9

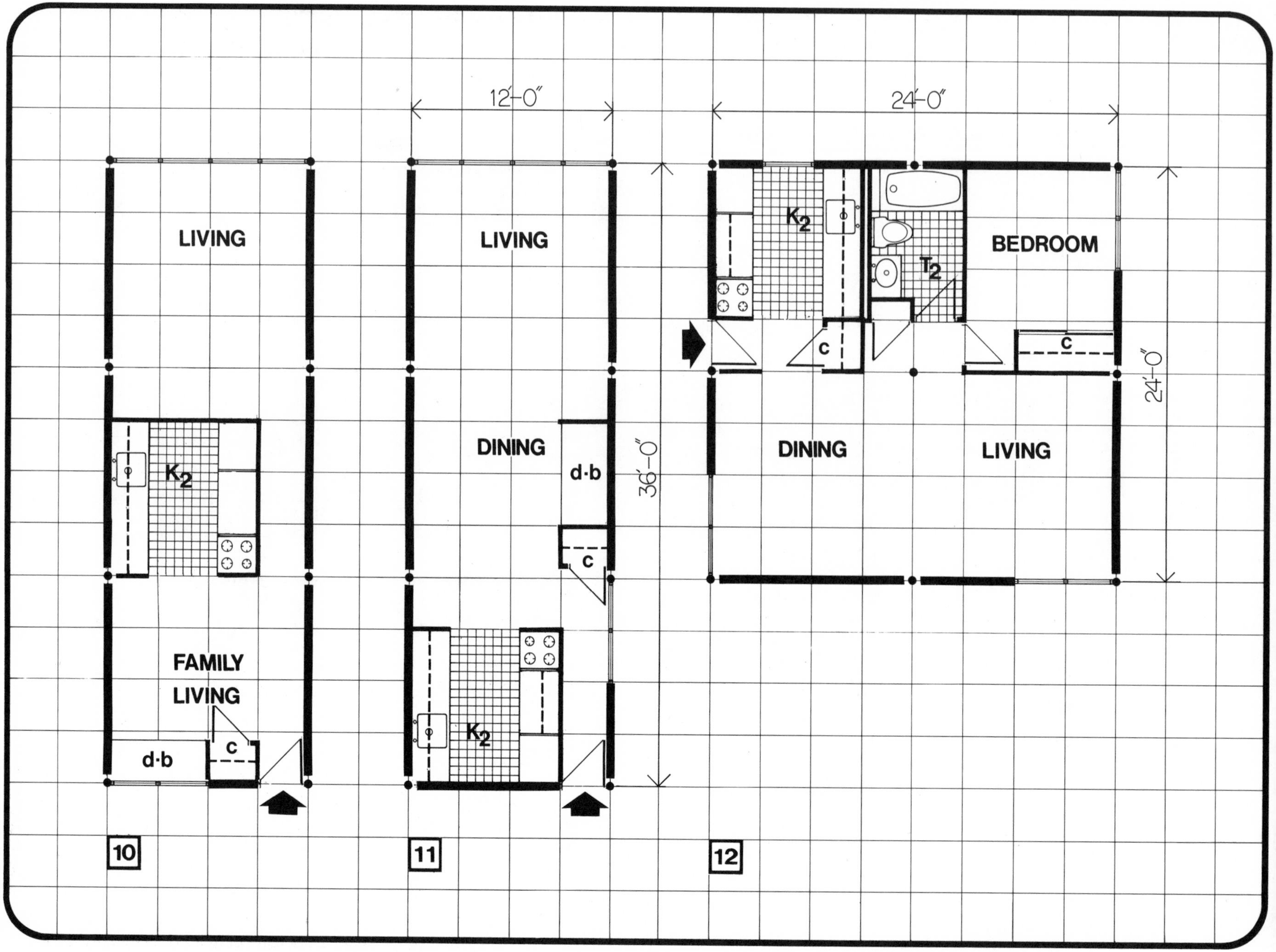
LIVING
K2
FAMILY
LIVING
c
d·b
10
12'-0"
LIVING
DINING
d·b
c
36'-0"
K2
11
24'-0"
K2
T2
BEDROOM
c
c
24'-0"
DINING
LIVING
12

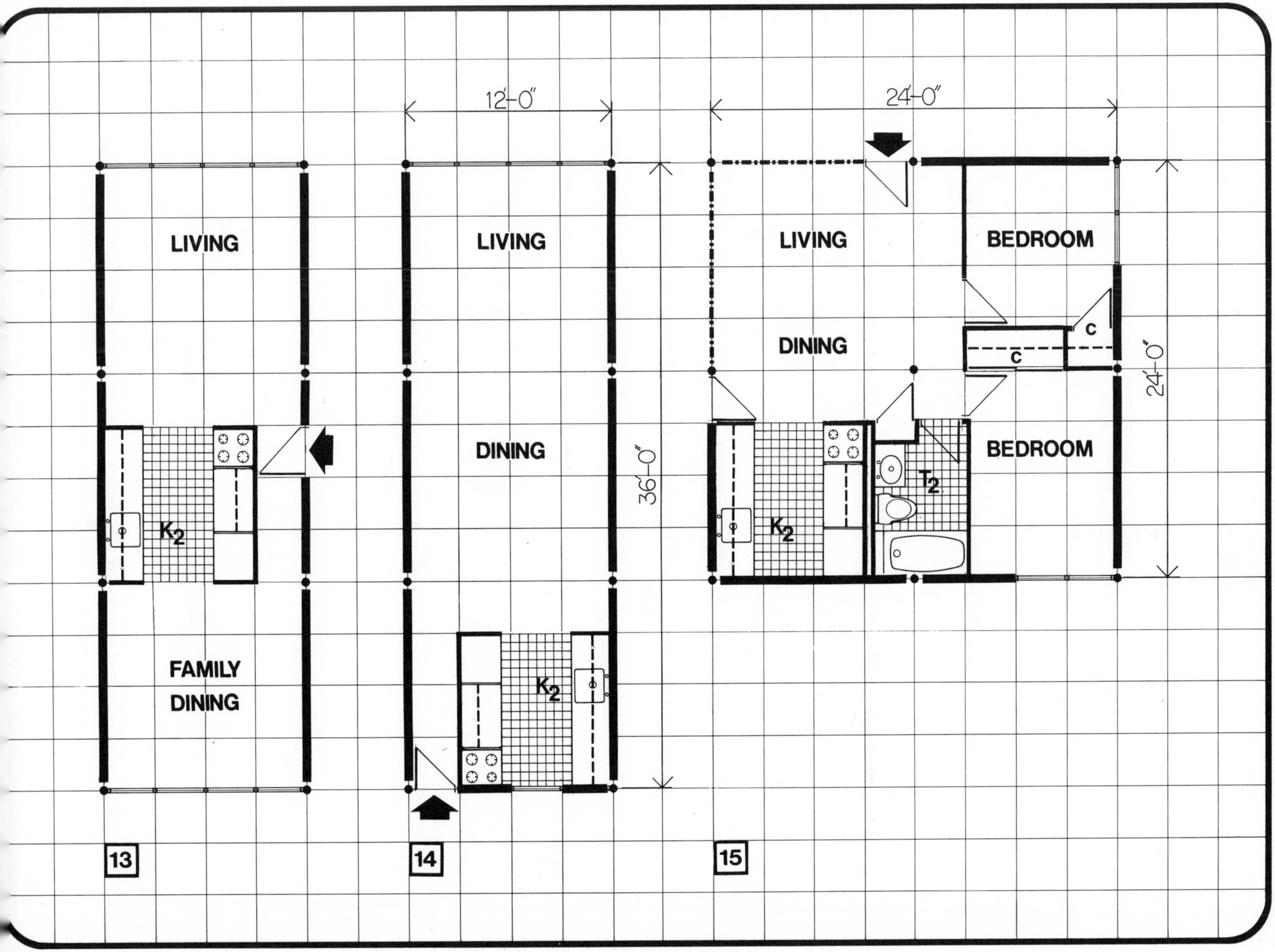

12'-0"
24'-0"
36'-0"
24'-0"
LIVING
K2
FAMILY
DINING
LIVING
DINING
K2
LIVING
DINING
BEDROOM
C
C
K2
T2
BEDROOM
13
14
15

LIVING UNITS

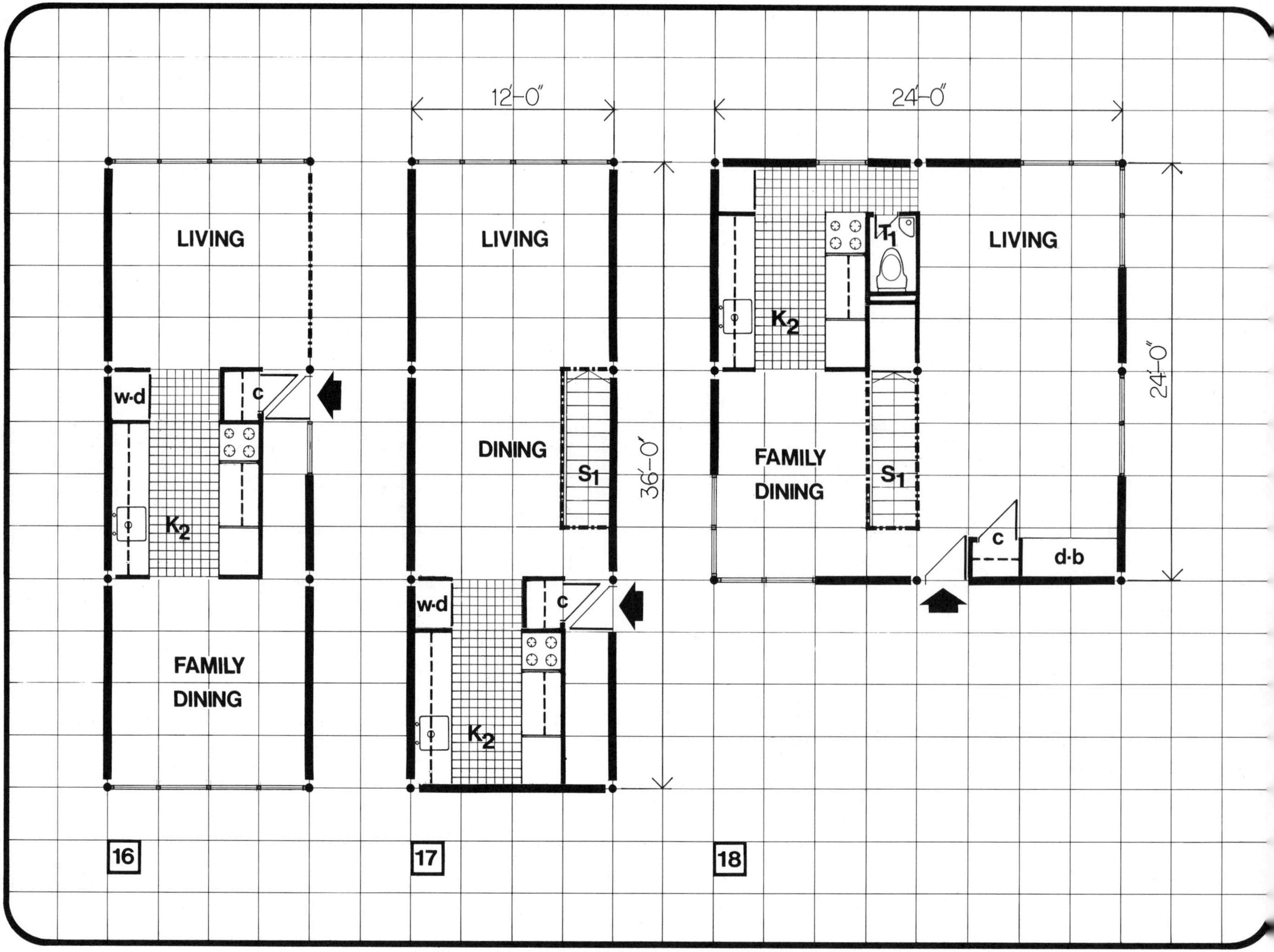

12'-0"
24'-0"
w·d
c
K2
FAMILY DINING
16
DINING
S1
36'-0"
w·d
c
K2
17
T1
K2
FAMILY DINING
S1
c
d·b
24'-0"
18

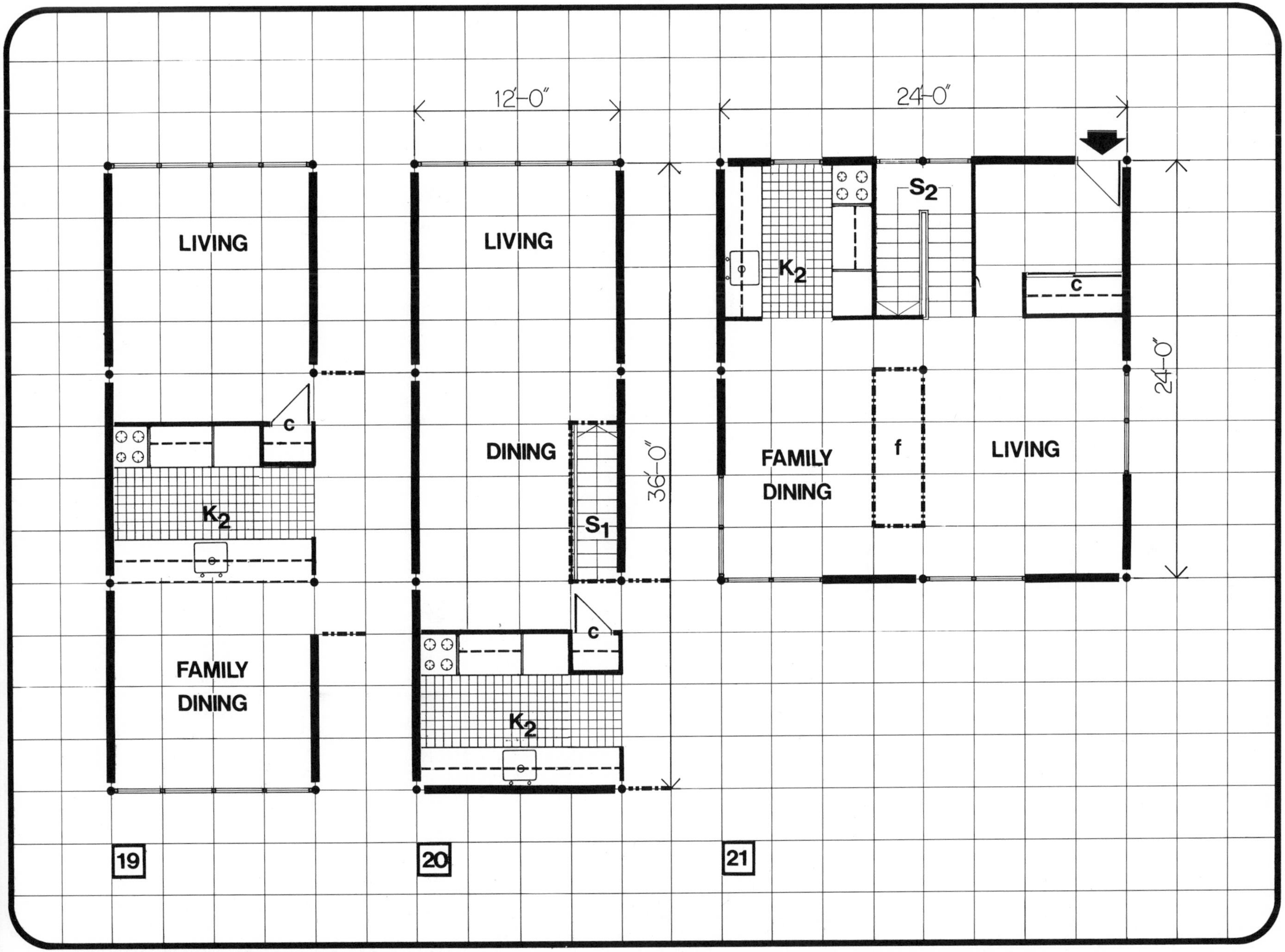
12'-0"
24'-0"
LIVING
LIVING
S_2
K_2
c
c
DINING
36'-0"
24'-0"
FAMILY
DINING
f
LIVING
K_2
S_1
FAMILY
DINING
c
K_2
19
20
21

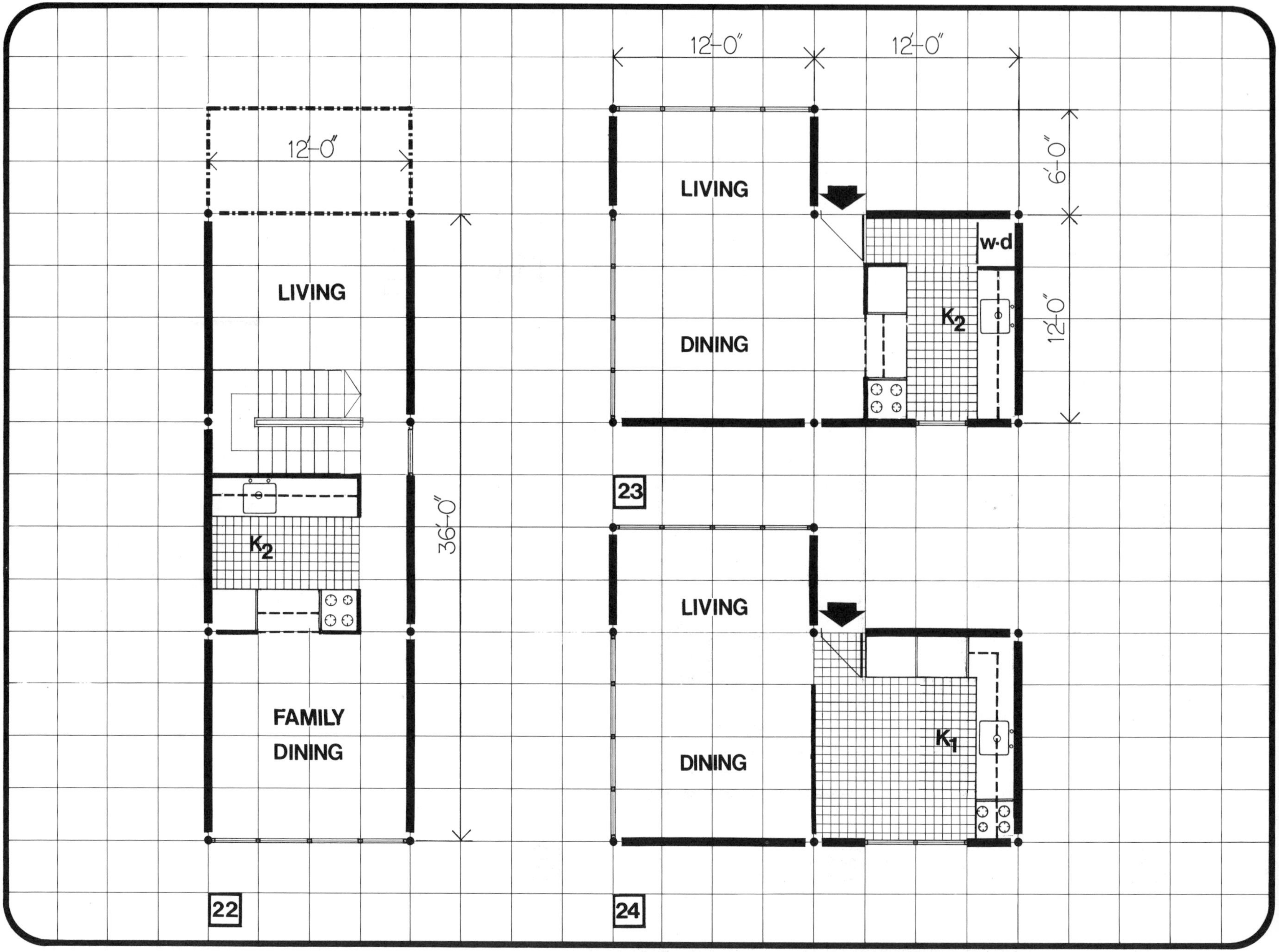
12'-0"
LIVING
K_2
FAMILY
DINING
36'-0"
22
12'-0"
12'-0"
6'-0"
12'-0"
LIVING
DINING
w·d
K_2
23
LIVING
DINING
K_1
24

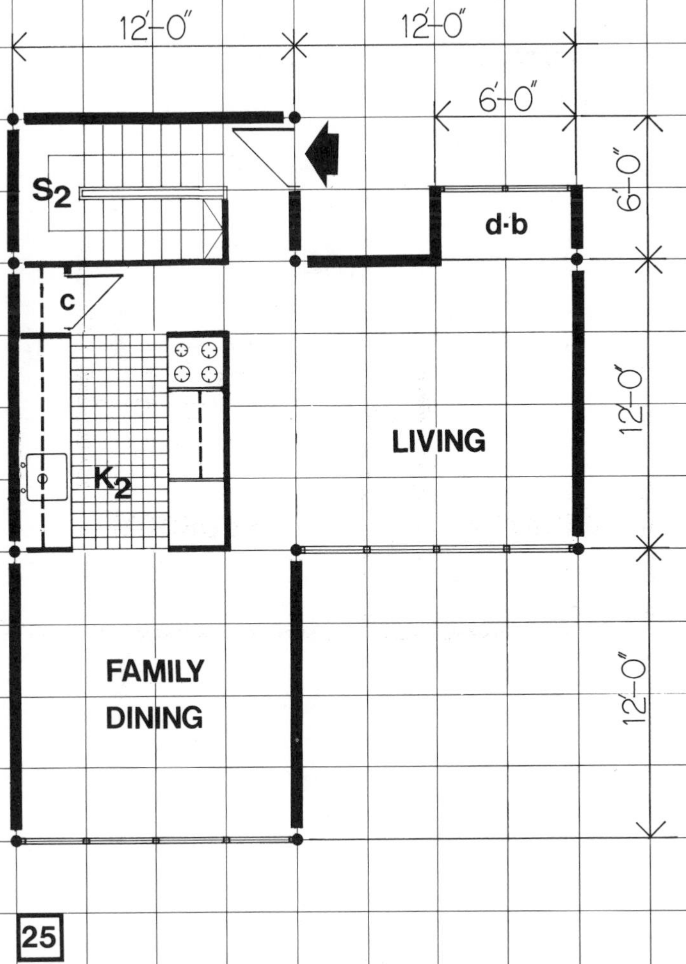
12'-0"
12'-0"
6'-0"
6'-0"
S2
d·b
c
LIVING
12'-0"
K2
FAMILY
DINING
12'-0"
25

SLEEPING UNITS

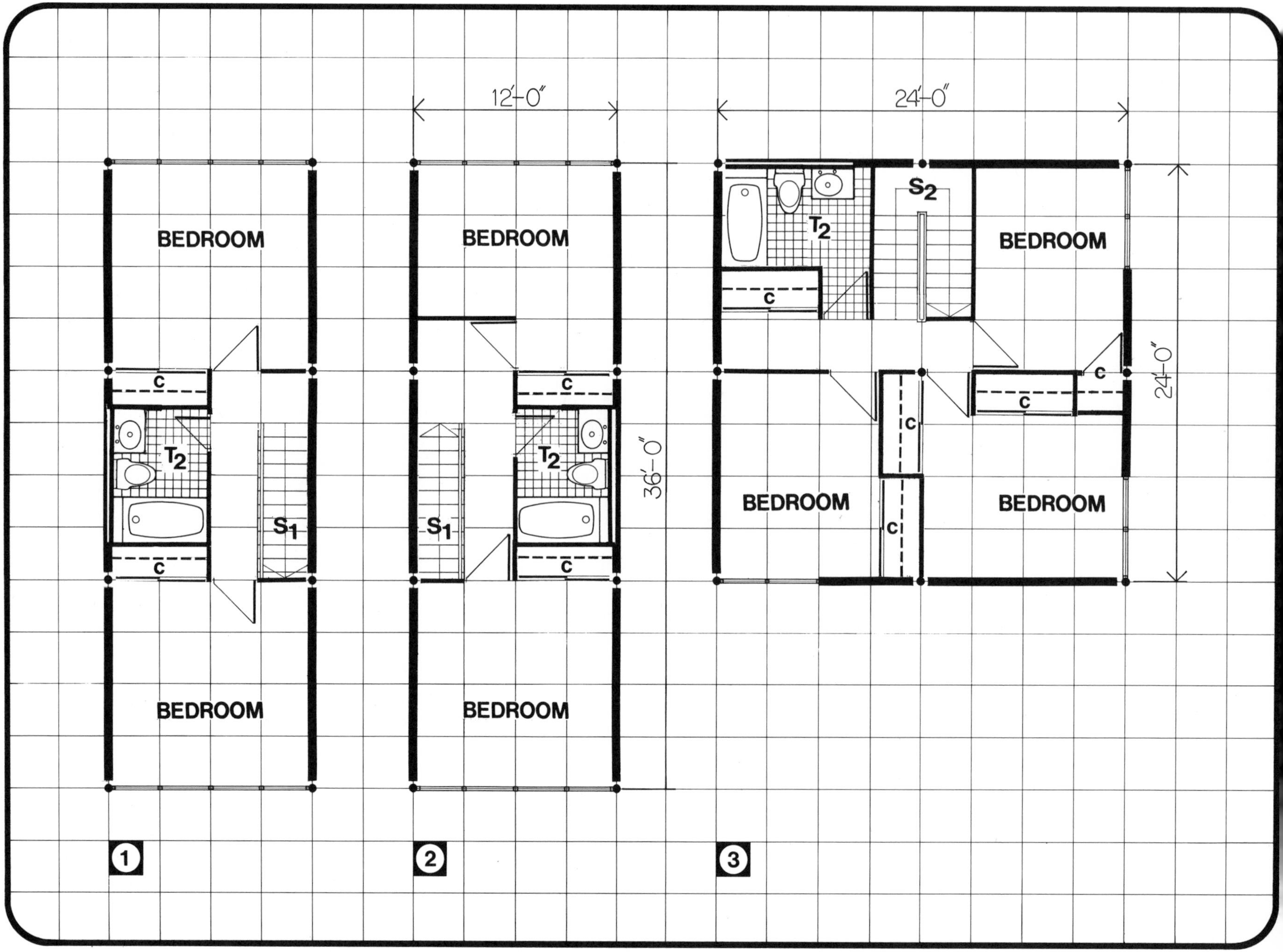

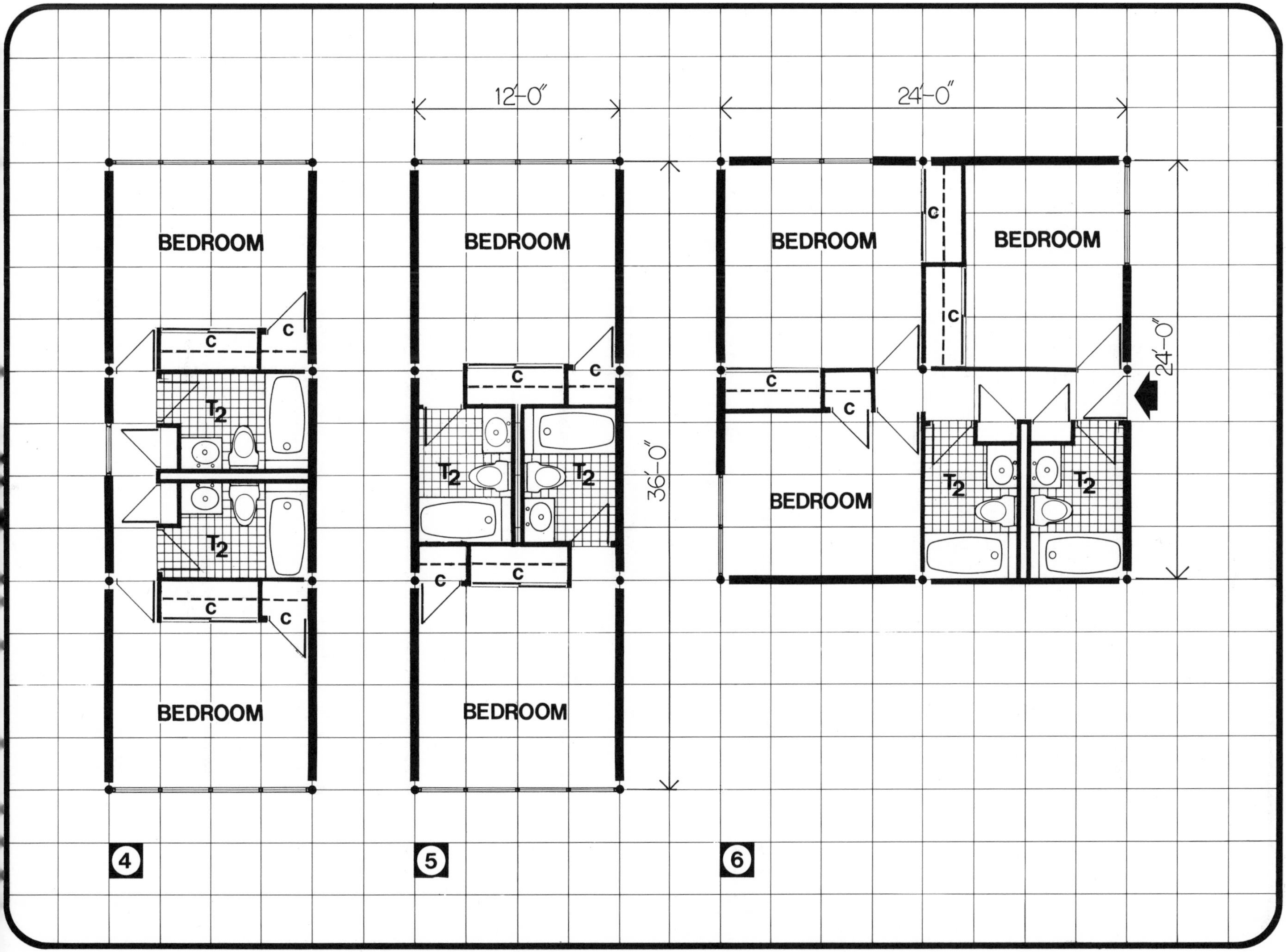

SLEEPING UNITS

SLEEPING UNITS

BEDROOM
c
c
S_2
T_2
c
c
BEDROOM
7
BEDROOM
c
c
S_2
T_2
c
c
BEDROOM
8
12'-0"
BEDROOM
c
c
T_2
b
c
BR.
c
d·b
c
d·b
BR.
b
c
36'-0"
9

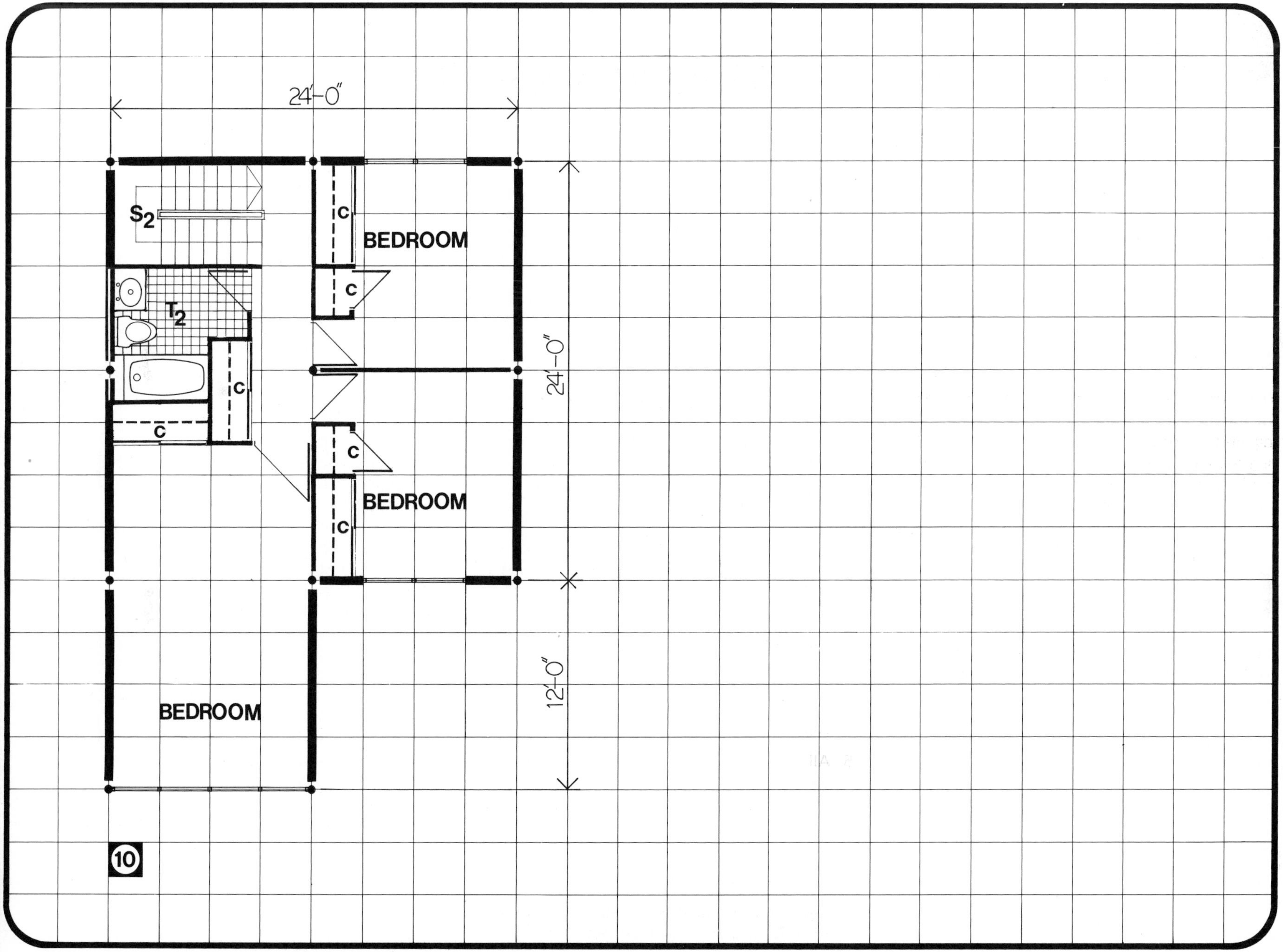

SLEEPING UNITS

8 THE TOTAL UNIT – GENERIC HOUSING TYPES

The final step in the utilization of the "kit of parts for the habitat" is planning and fabricating dwelling units of the three basic building types A = single-family; B = garden apartments/row house; C = high-rise), by combining the preceding *living units* and *sleeping units.*

The early implication of this "system" utilizing existing on-the-market building elements could make SYSTEM ECOLOGIC economically capable of resisting the so-called "invasion" of mobile homes into the housing market, while still being capable, with this small kit of parts, of adapting more gracefully and flexibly than the mobile home to local ecologies and individual needs of the human habitat.

ORIGINAL DESIGN CRITERIA

1. The structure itself can be produced on a modular basis *in factory* or *on-site.*
2. Pure geometry and dimensional standards are used in order to provide more flexibility and compatibility between manufacturers.
3. All assemblage can be done at the site—work on and off the job site is organized to promulgate reasonable work efficiencies.
4. Maximum size of the units is 12′ x 36′ and is prescribed by the need for easy transport under present-day standard road conditions and restrictions, while retaining the capability of adjusting to 14′ when state laws permit.
5. All environmental and climatic adaptations can be achieved through a minimum of alternate materials and space module combinations.
6. Use of existing components, subsystems, and materials presently available and selectively utilized will achieve lower total cost.
7. Low- and high-density situations can be provided for within the same system.

THE TOTAL UNIT

When the living units and sleeping units are eventually mated, the magical *total dwelling unit* appears. A large number of combinations is possible, but this study illustrates only a few—those that are almost classical in either their simplicity (the square, the rectangle, etc.), their historic antecedents (Philadelphia row house, colonial center entry, ranch style), their present-day competition (the mobile home, the prefab, the high-rise apartment, the fourplex, the garden apartment). All of these are constructed from the few components illustrated in the kit of parts in Chapter 6. These combinations which are possible demonstrate the great possibility of competing successfully with the blighted building industry, the speculative home and mobile home markets. Coupling the possible choices in materials, spatial add-on's, and functioning plans makes SYSTEM ECOLOGIC both a housing system for the developer and a design technique and tool for the architect and client.

ELEVATION AND PLANS

The elevations show typical adaptations of closure and roof components to plans. The plans are coded by letters indicating building type adaptability and numbers showing specific living unit and sleeping unit combinations utilized.

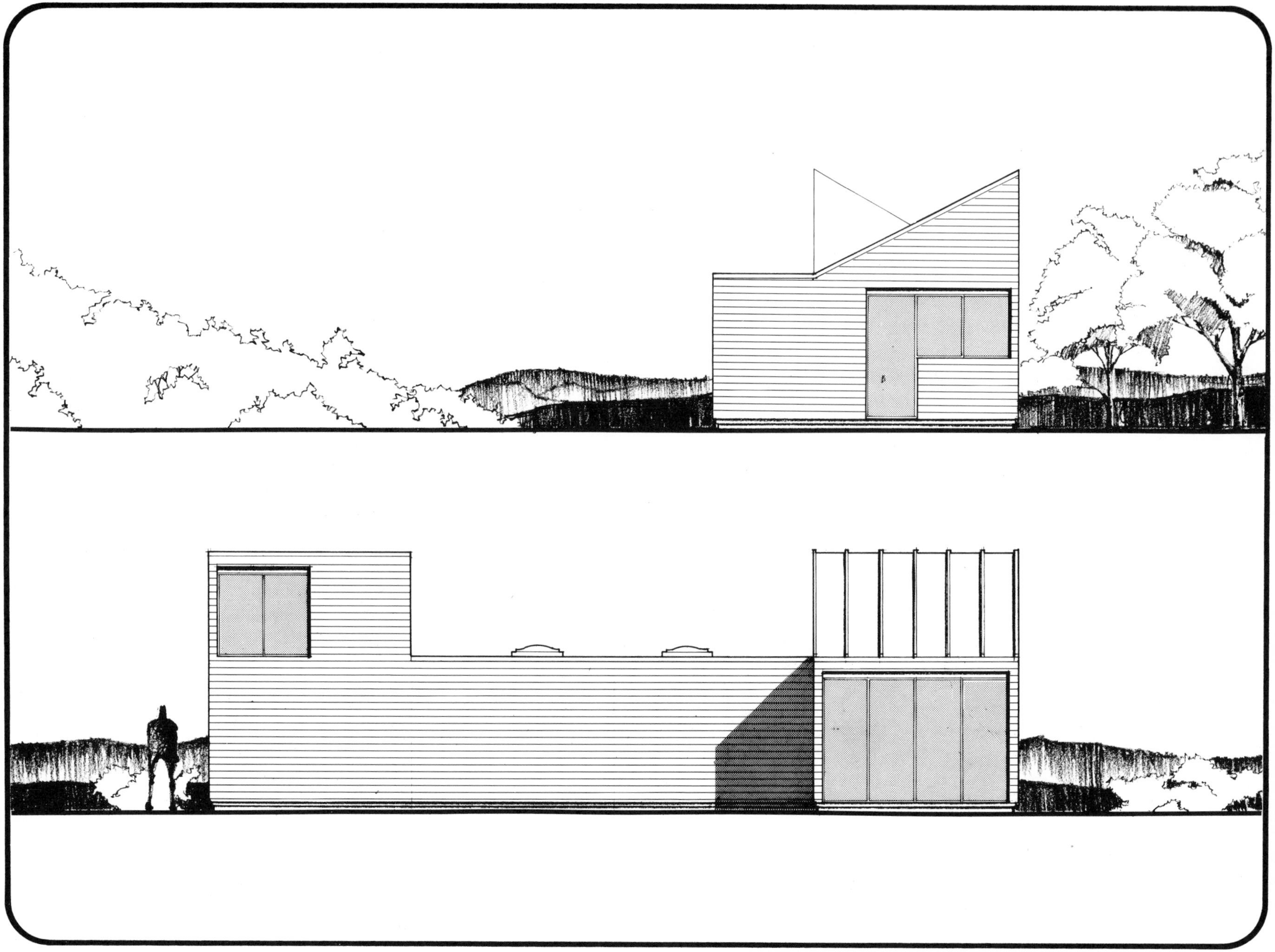

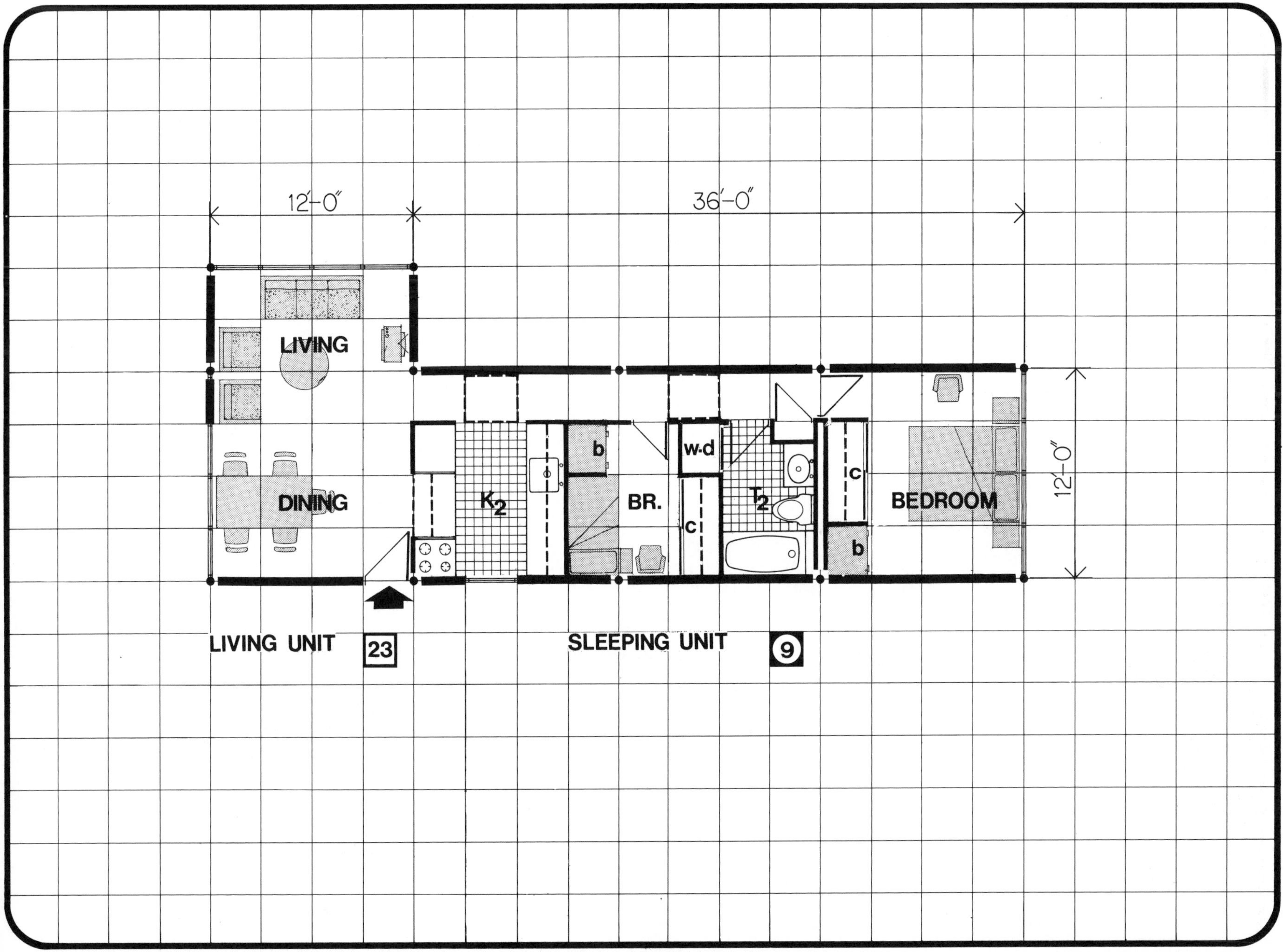

TOTAL UNITS

A 23 9

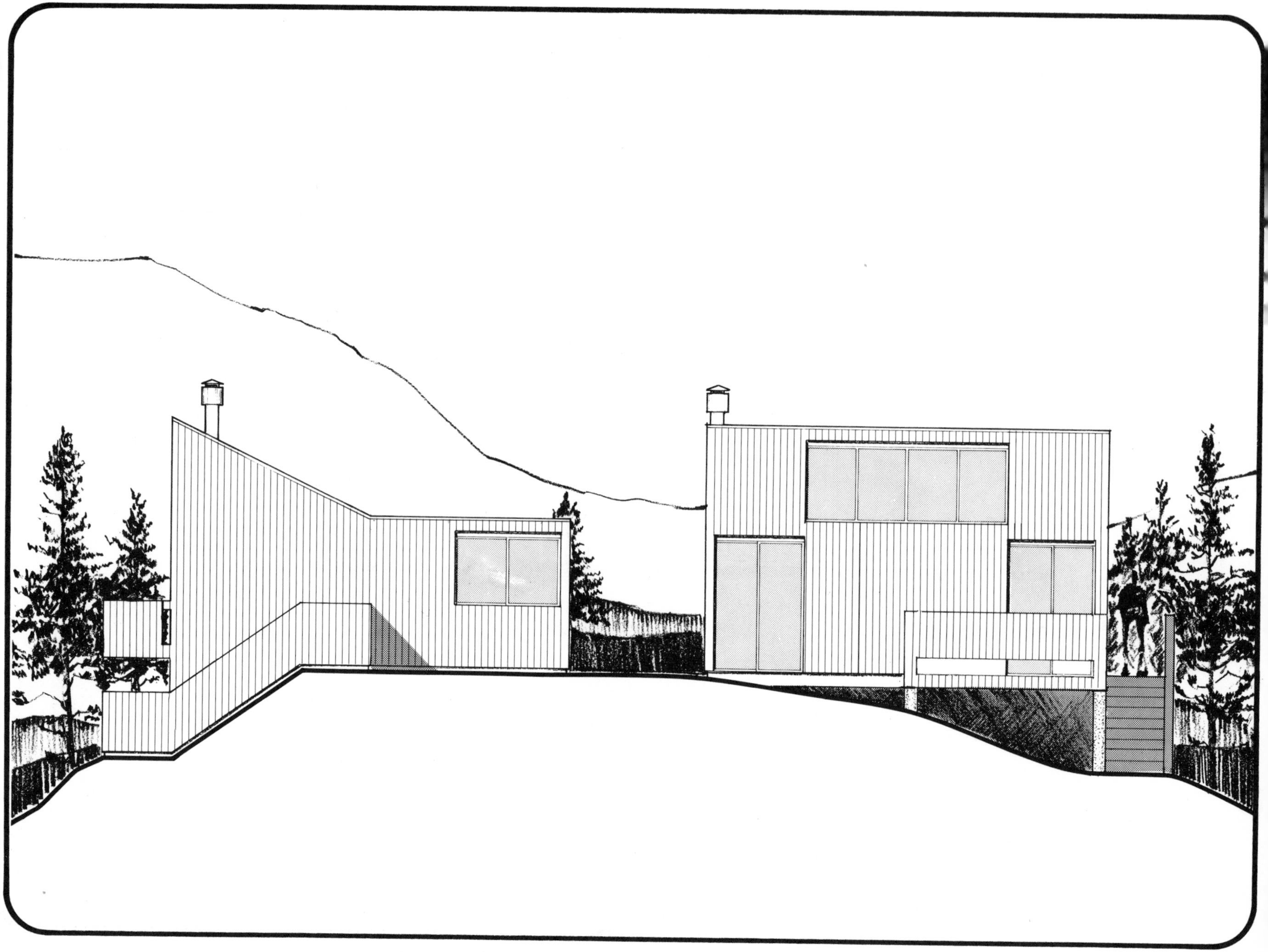

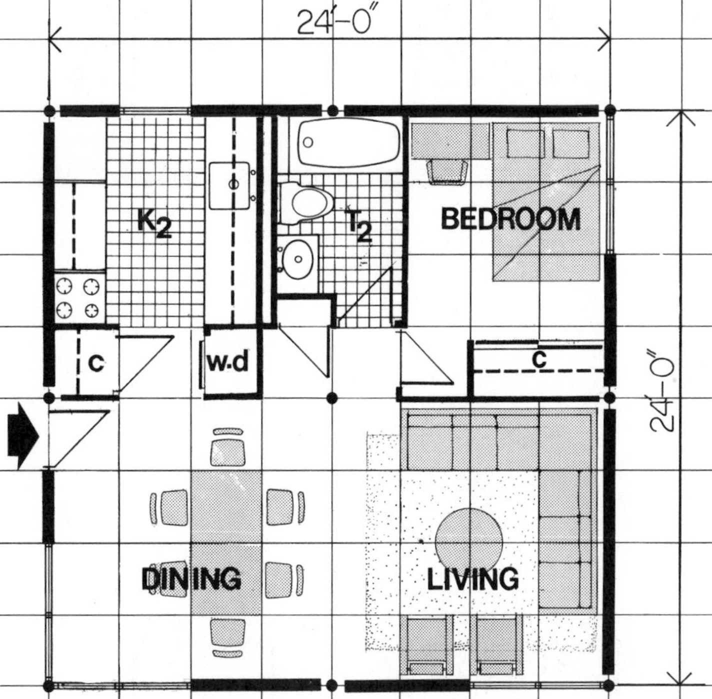

LIVING UNIT 12

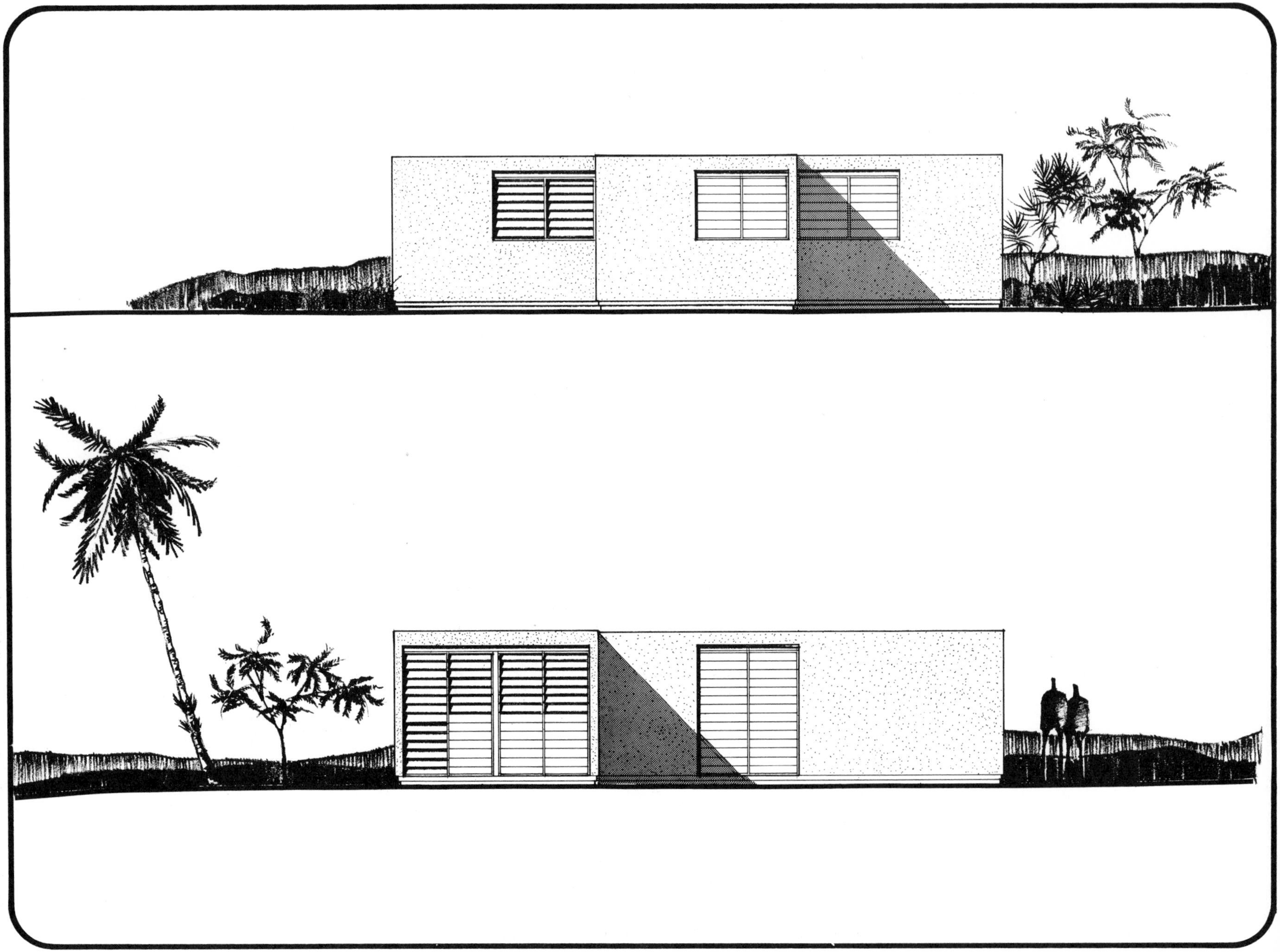

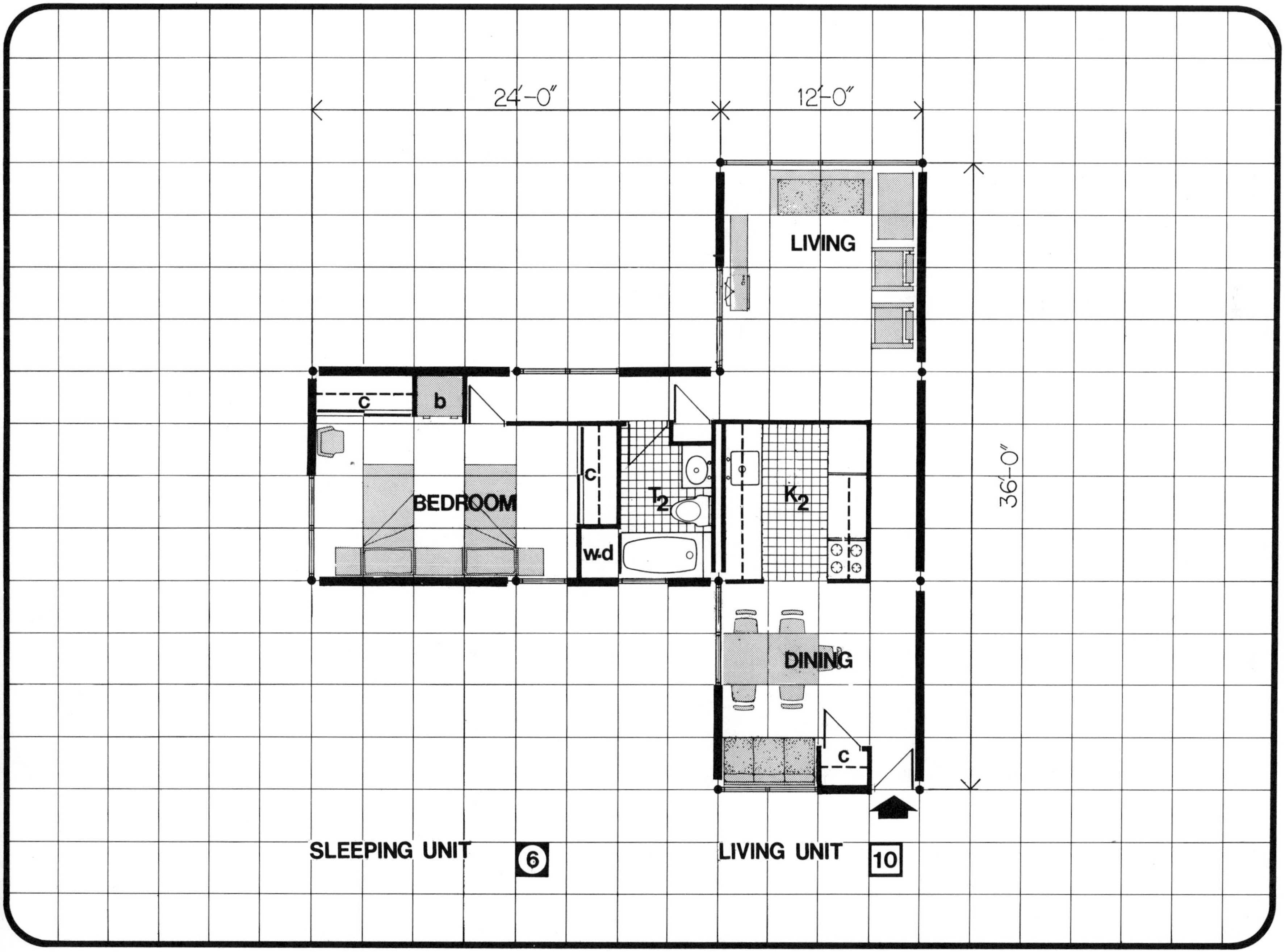

TOTAL UNITS

A 10 6

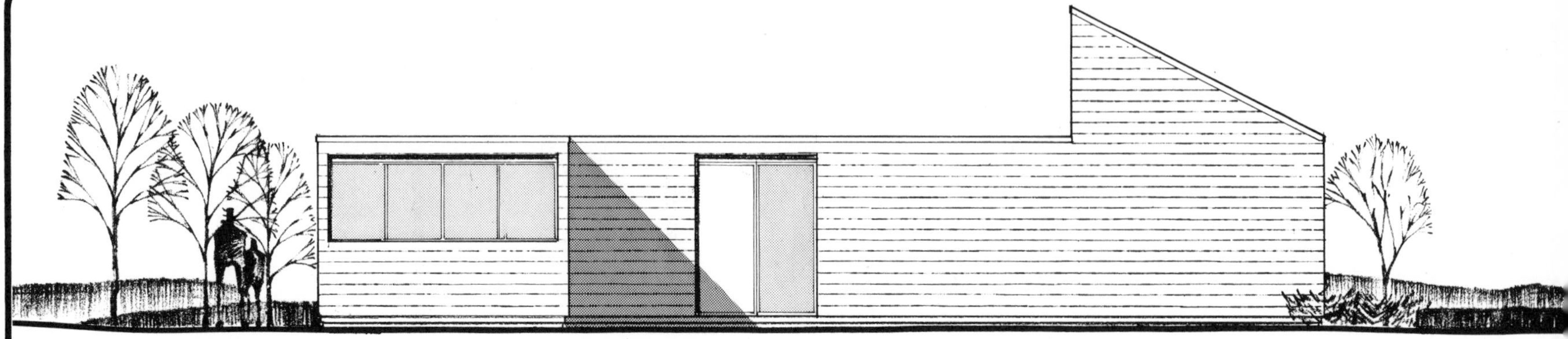

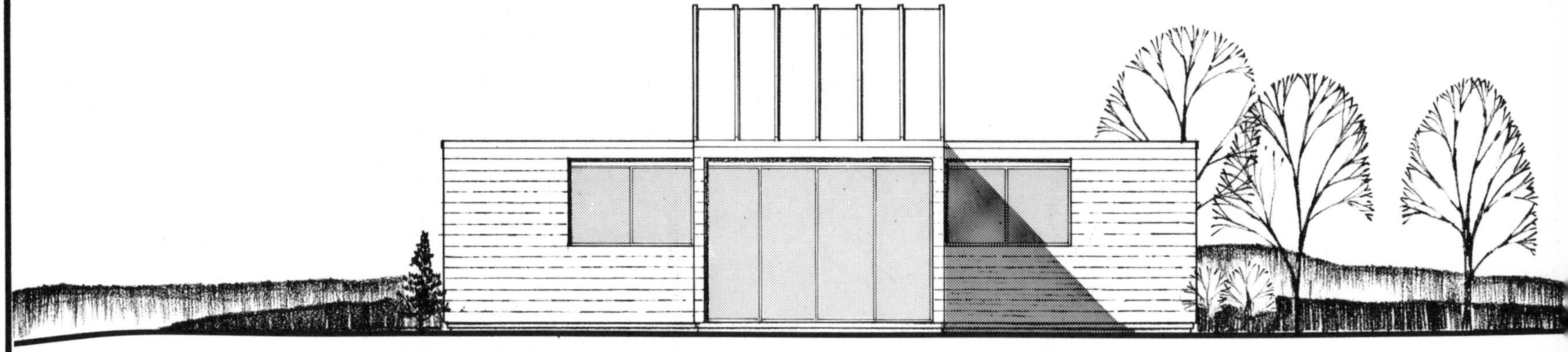

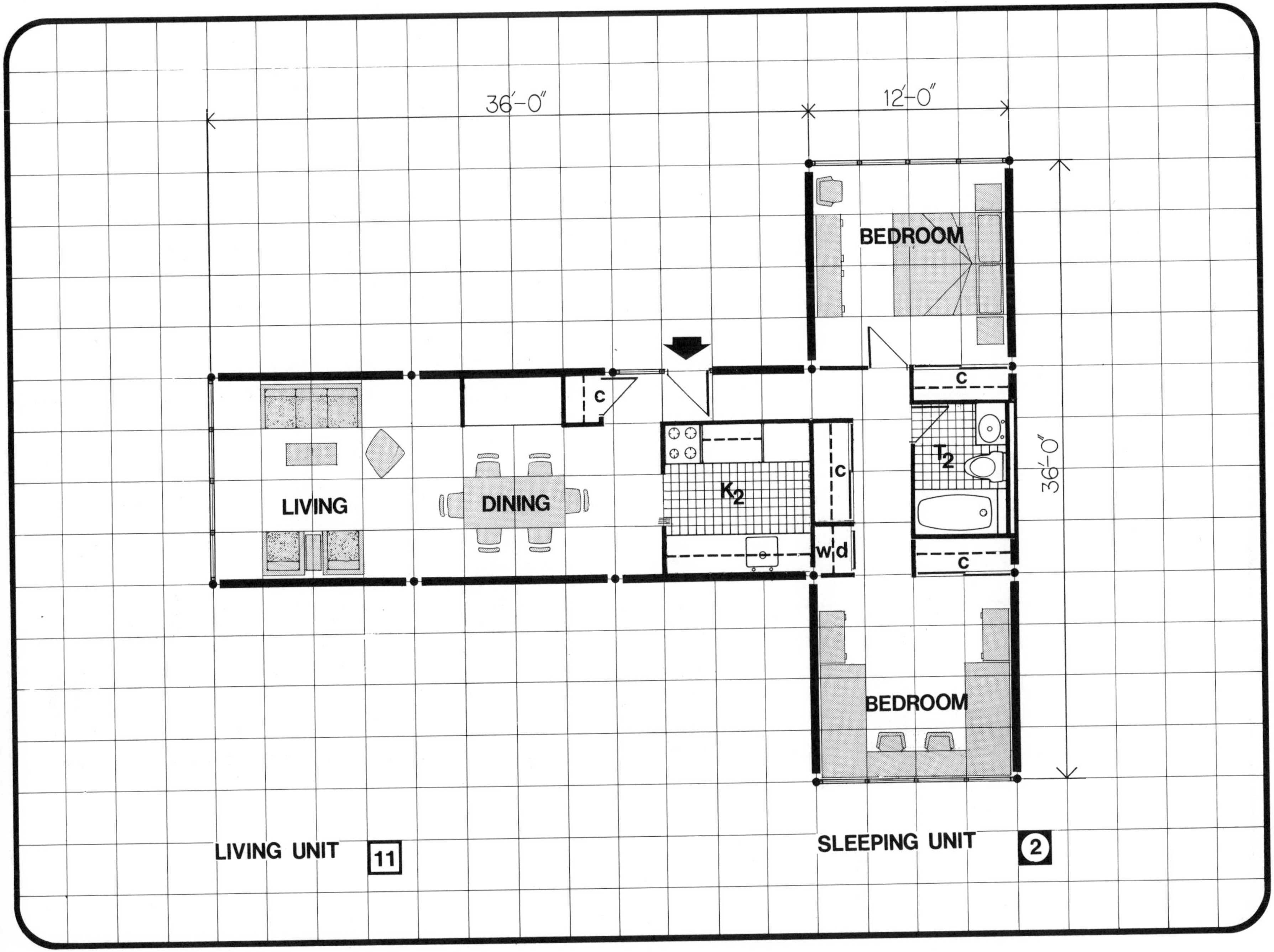

TOTAL UNITS

A 11 2

ELEVATIONS

A 23 9

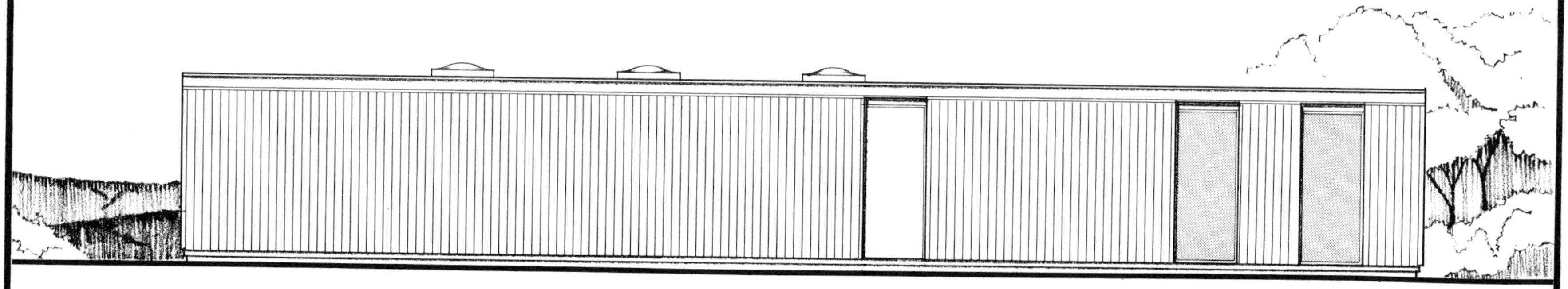

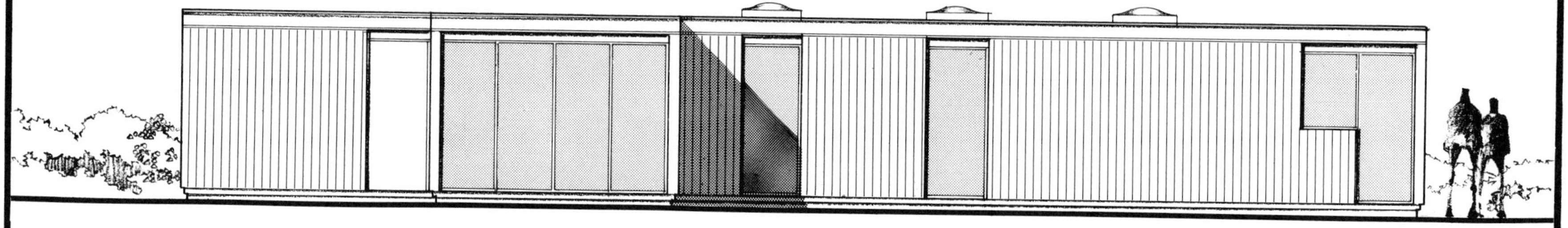

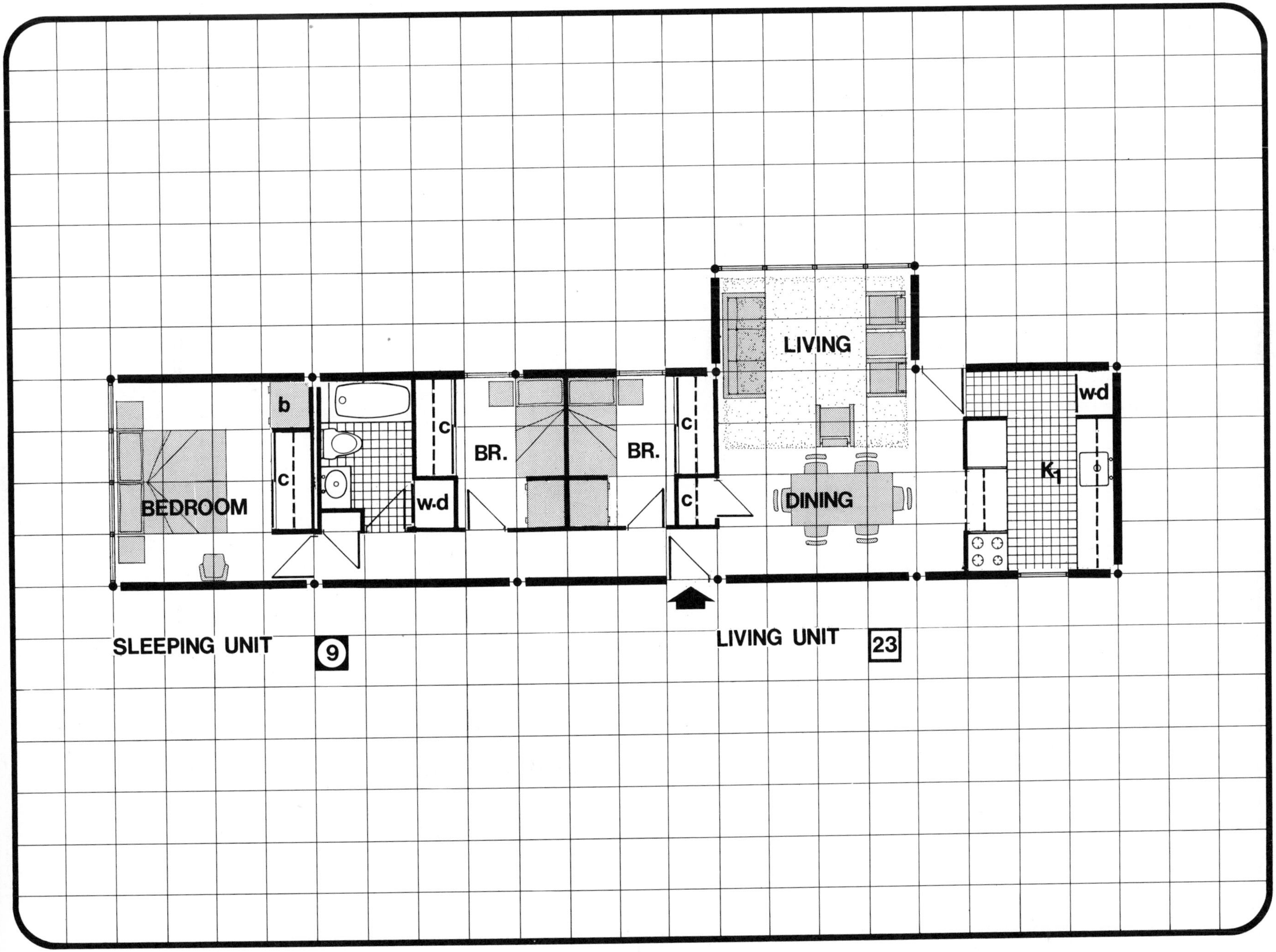

TOTAL UNITS

A 23 9

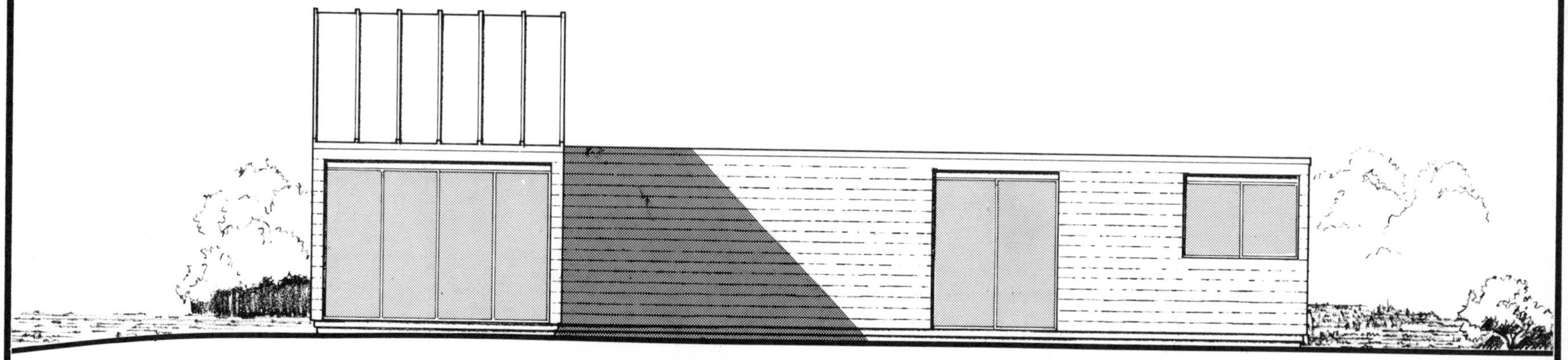

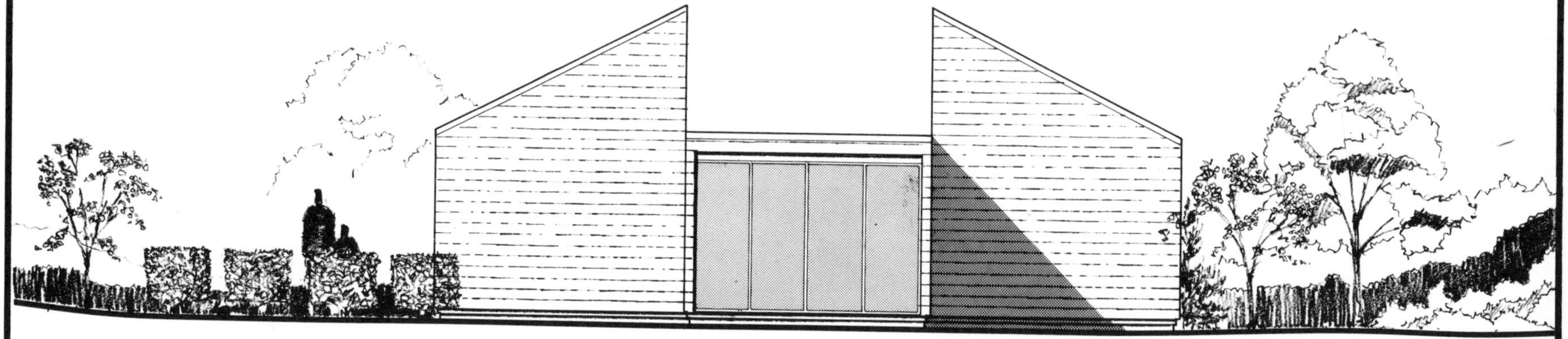

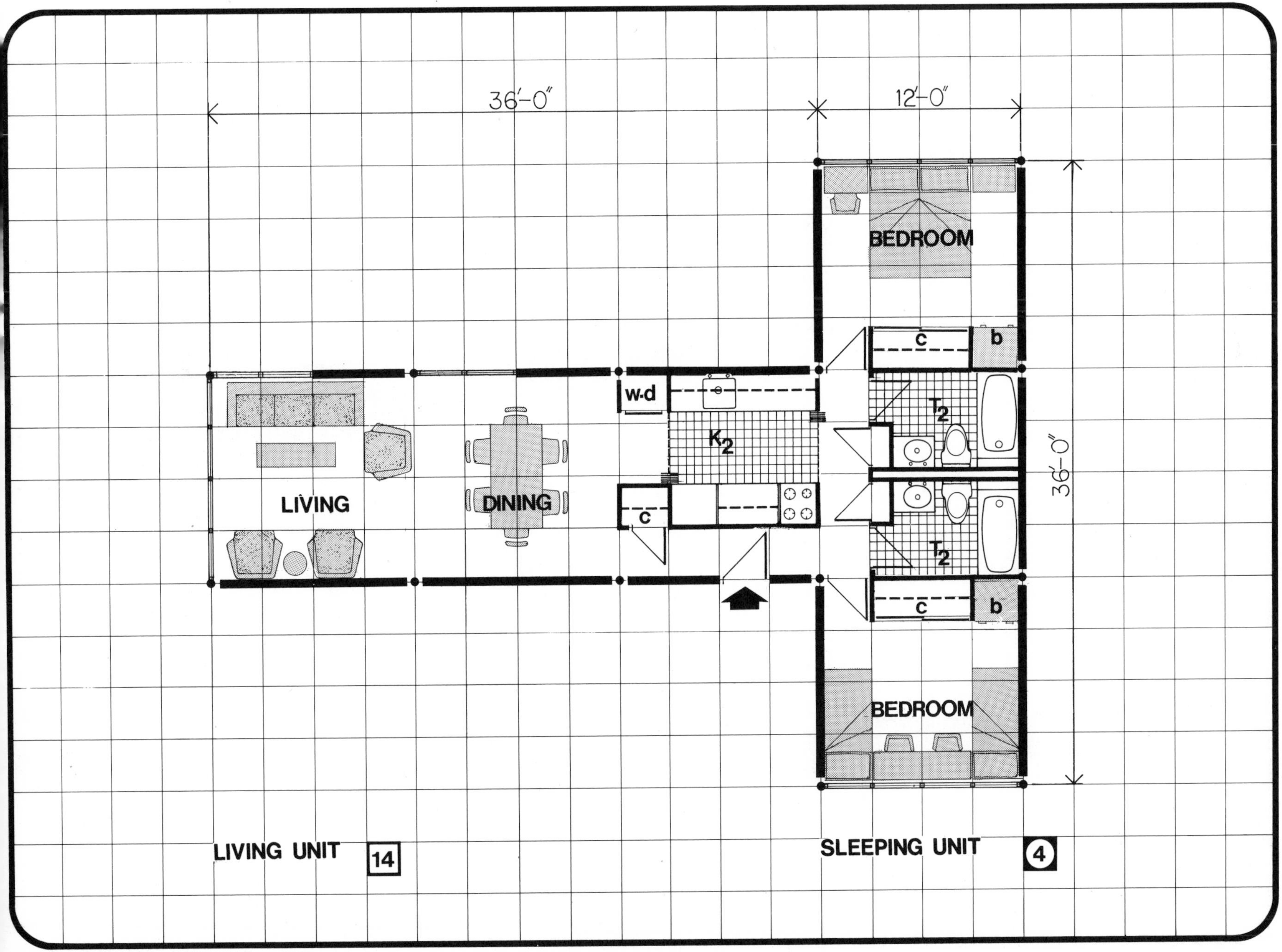

TOTAL UNITS

A 14 4

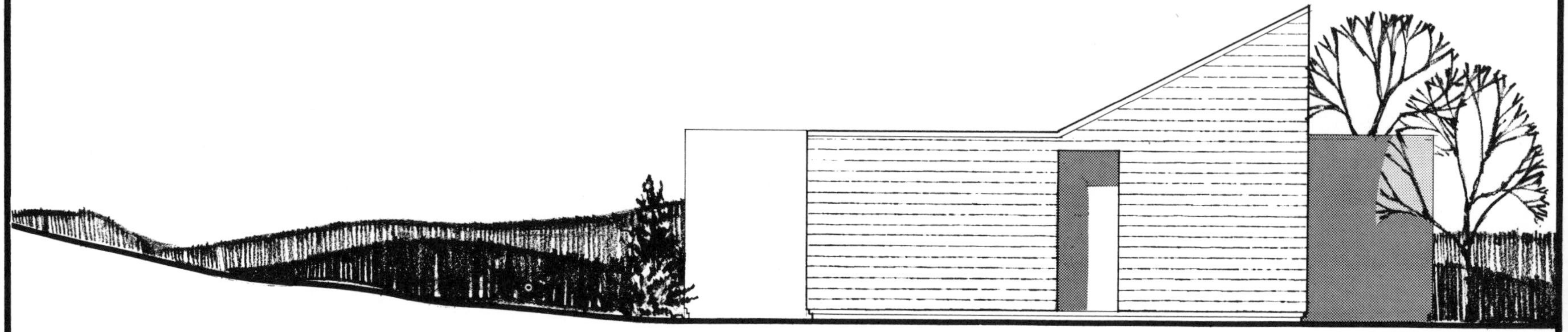

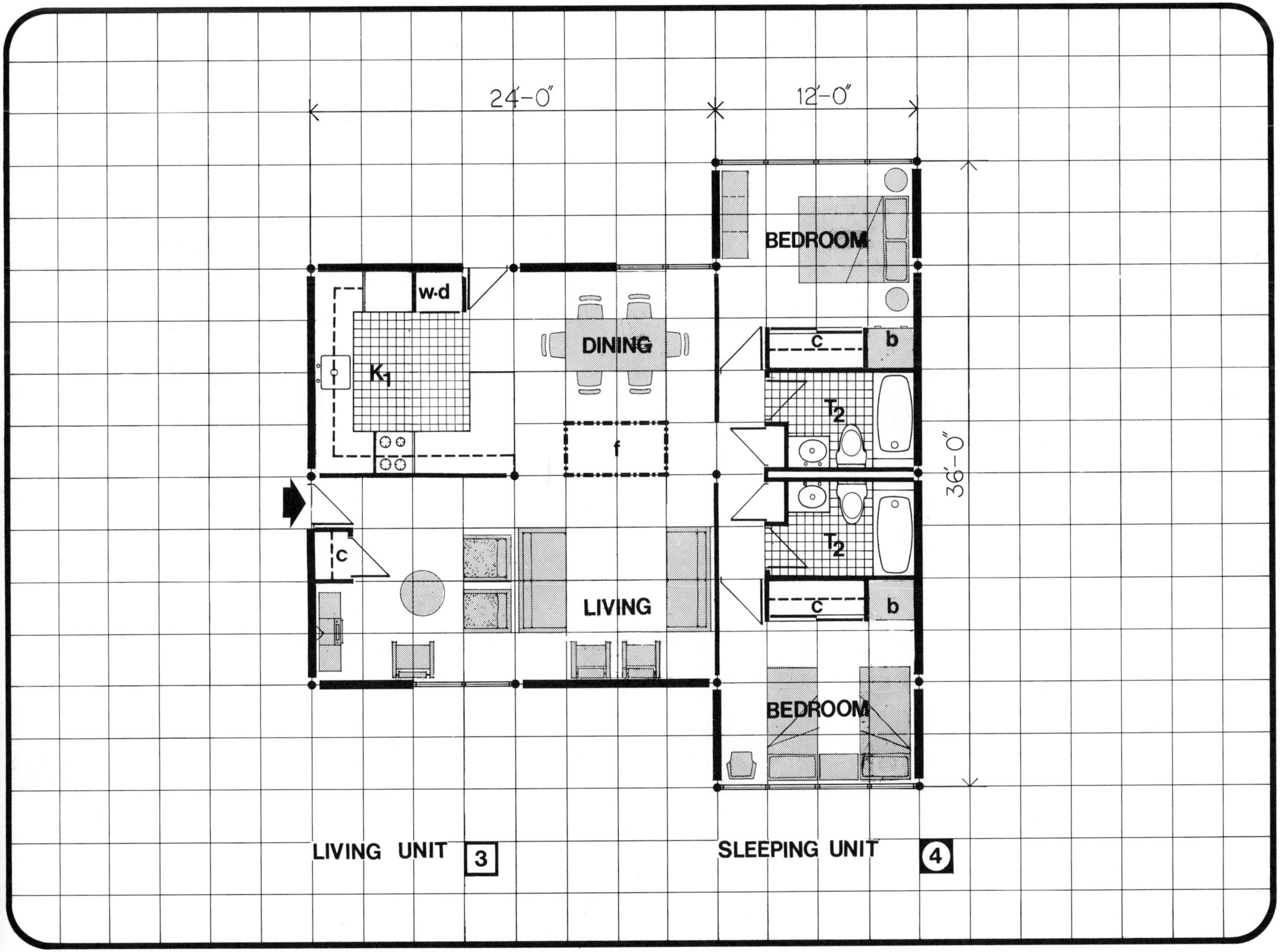
24'-0"
12'-0"
36'-0"
w-d
K_1
DINING
f
c
LIVING
BEDROOM
c
b
T_2
T_2
c
b
BEDROOM
LIVING UNIT 3
SLEEPING UNIT 4

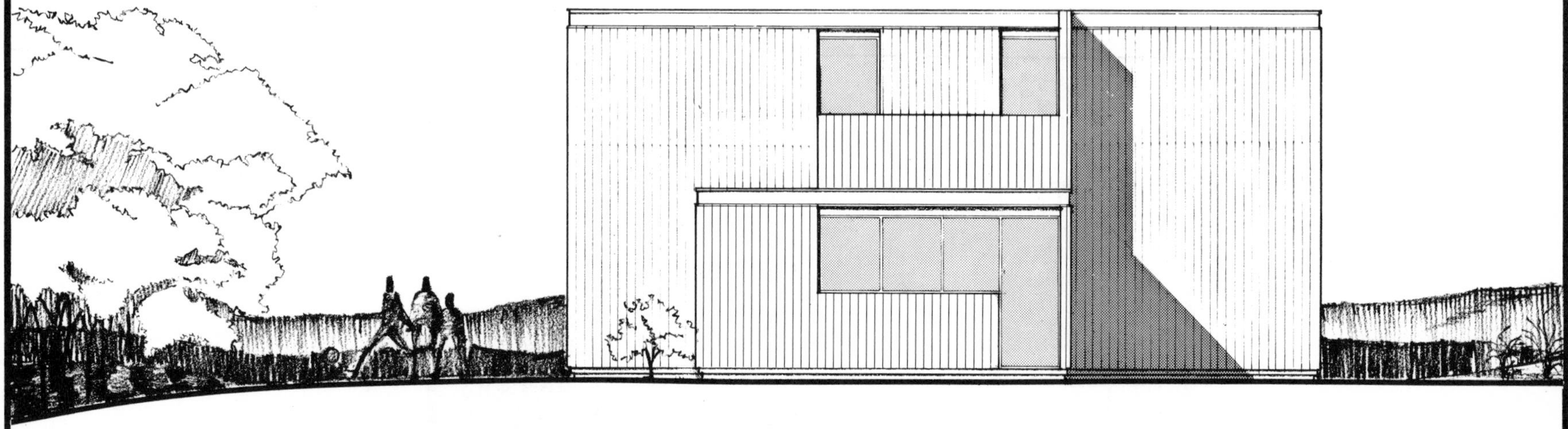

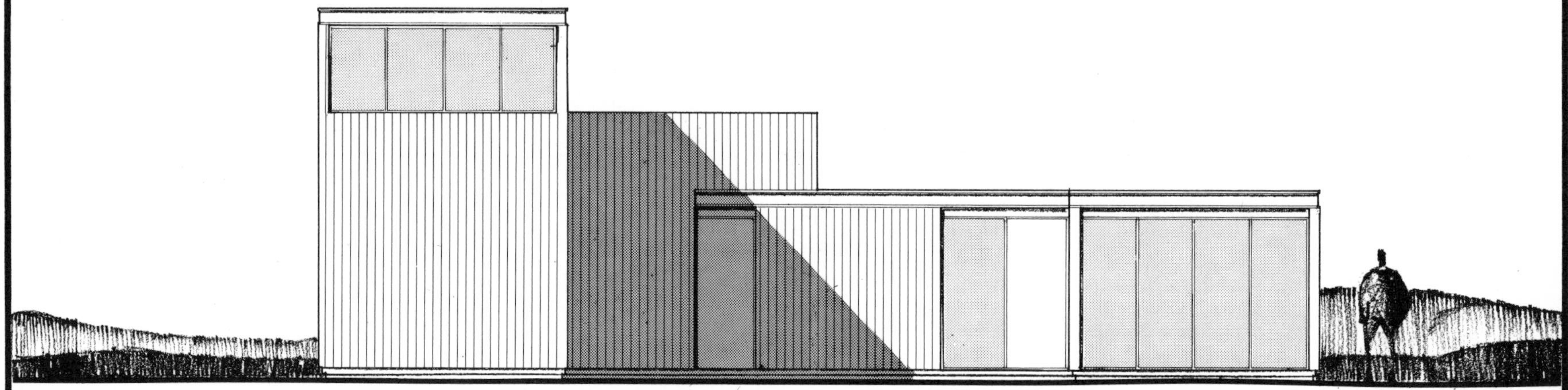

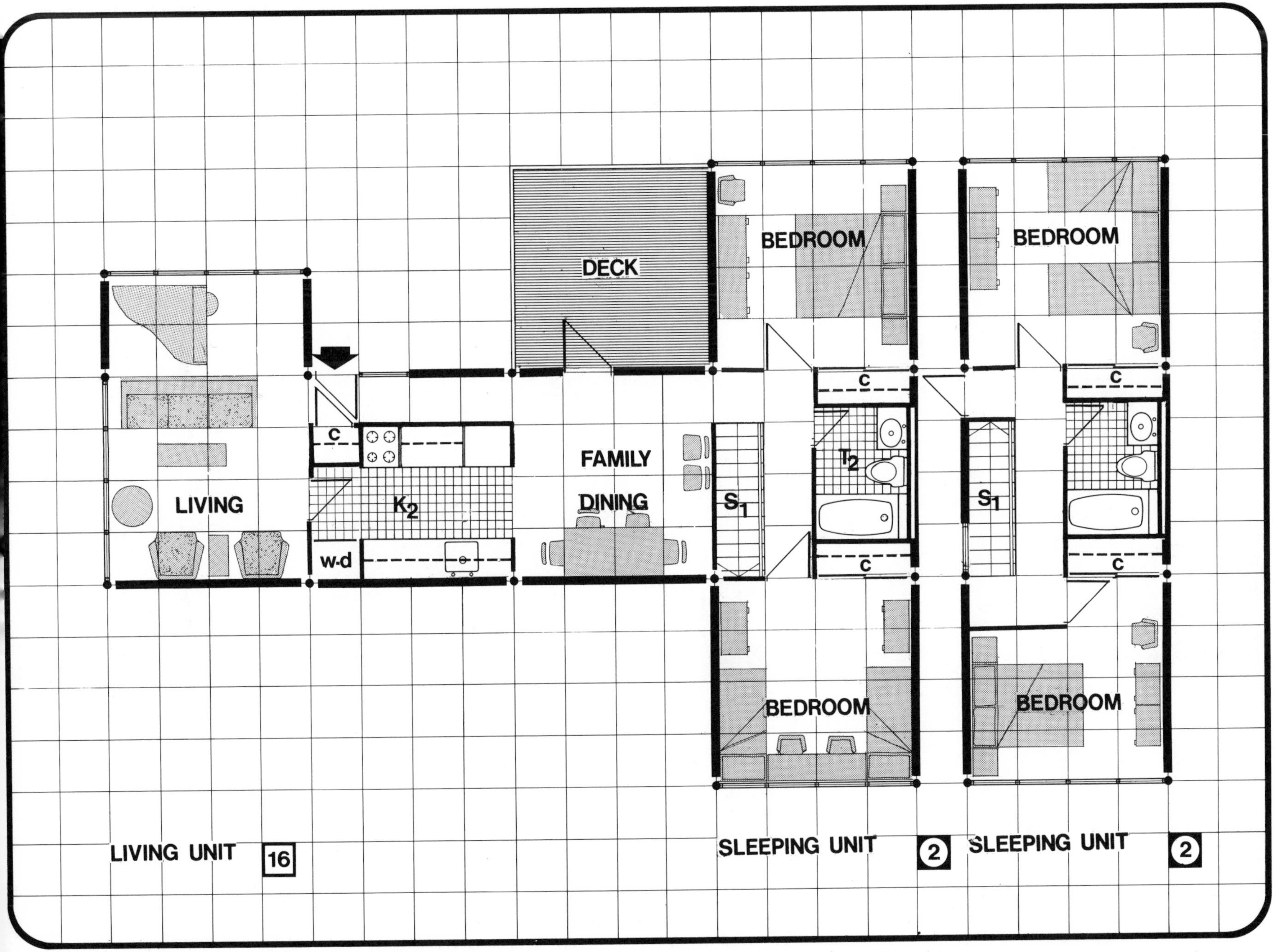

TOTAL UNITS

A 16 2 2

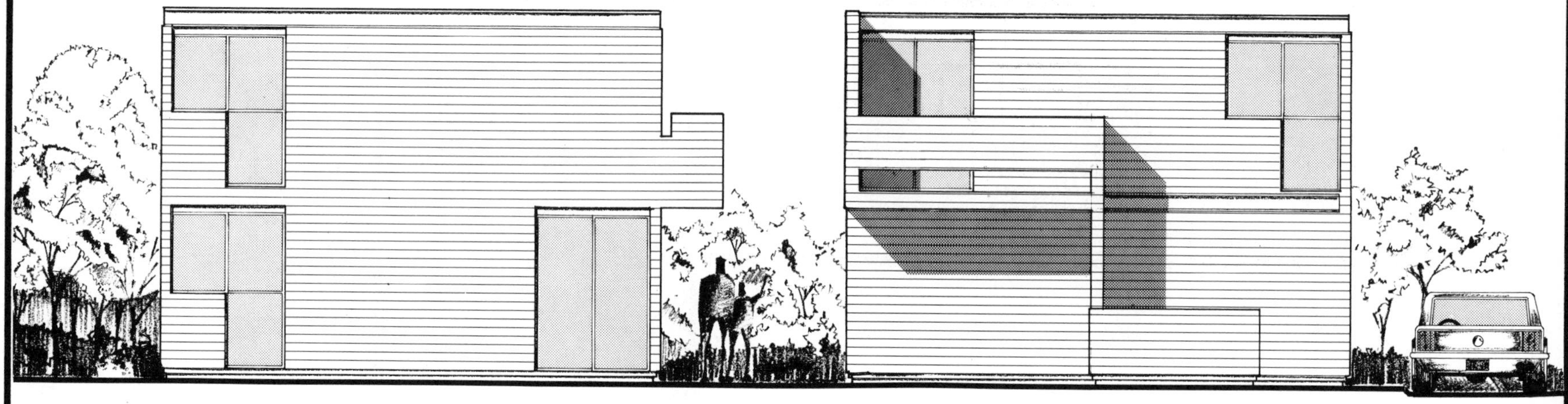

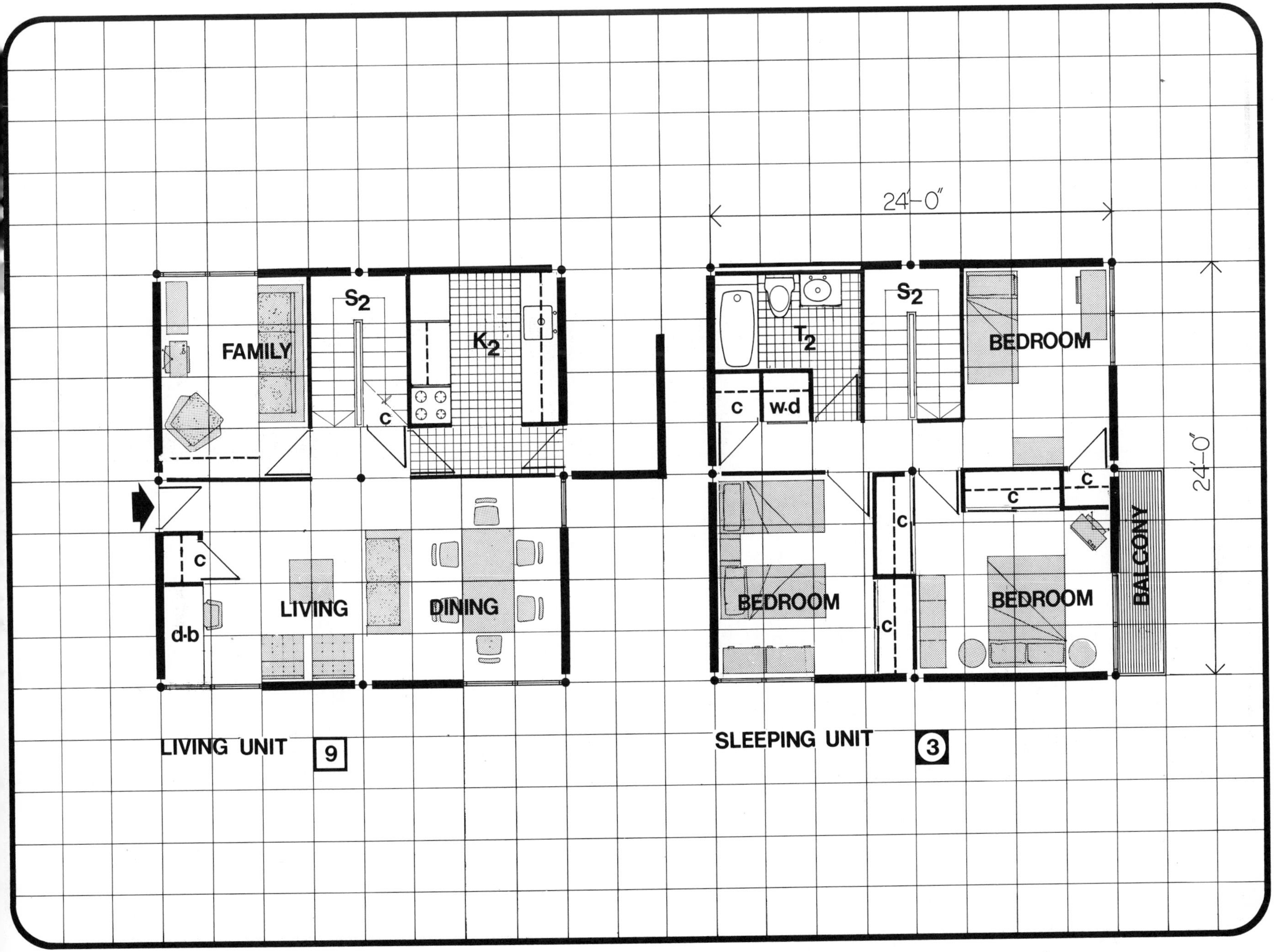

TOTAL UNITS

B 9 3

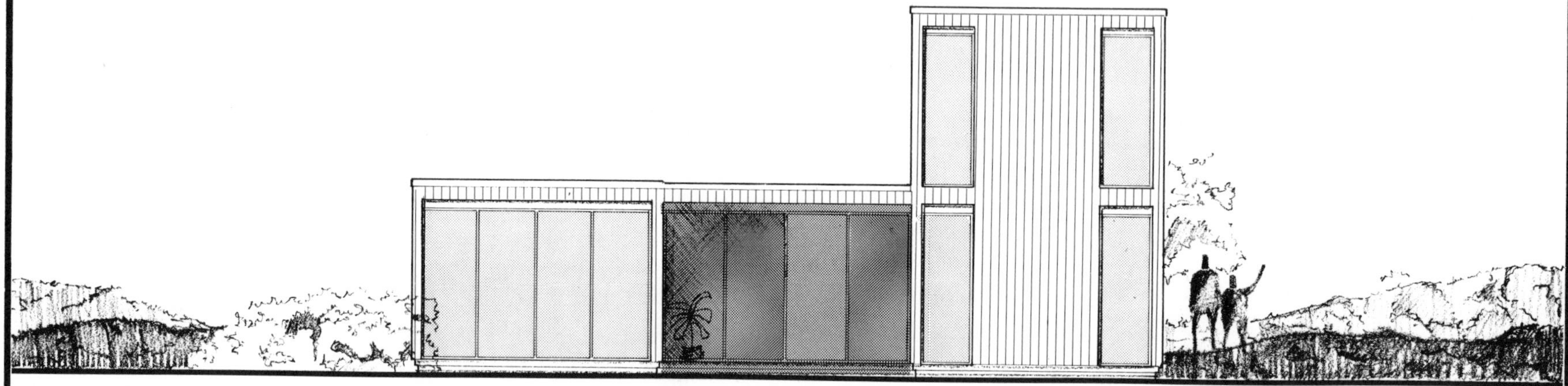

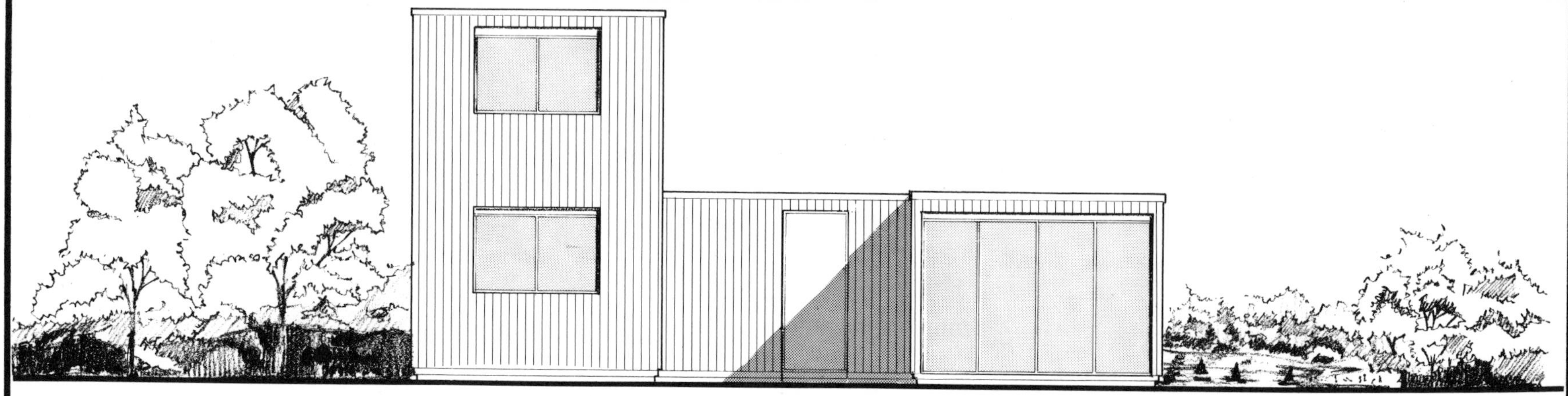

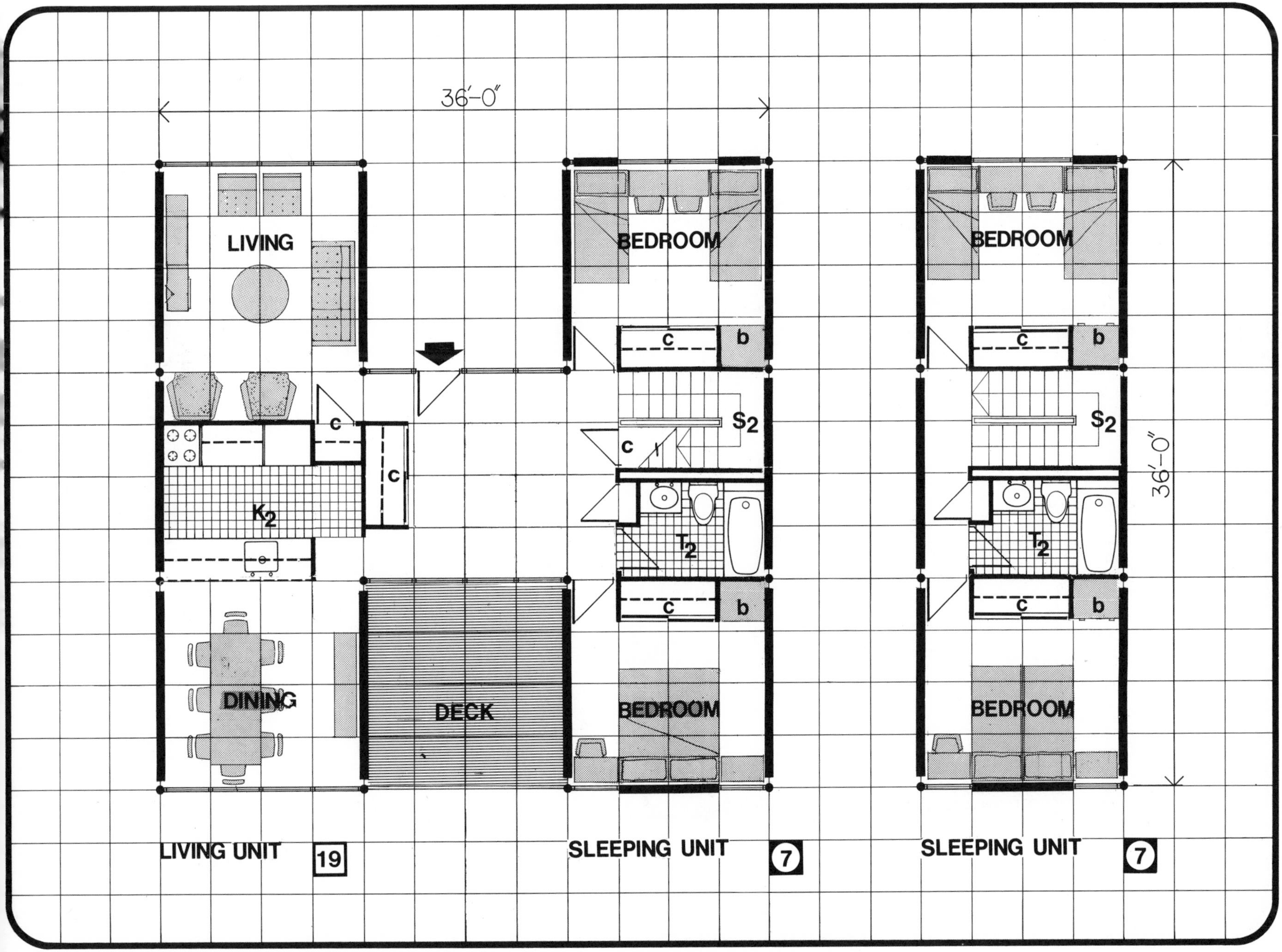
36'-0"
LIVING
K2
c
DINING
DECK
BEDROOM
c
b
S2
T2
BEDROOM
36'-0"
LIVING UNIT
19
SLEEPING UNIT
7
SLEEPING UNIT
7
TOTAL UNITS
A
19
7
7

ELEVATIONS

B 26

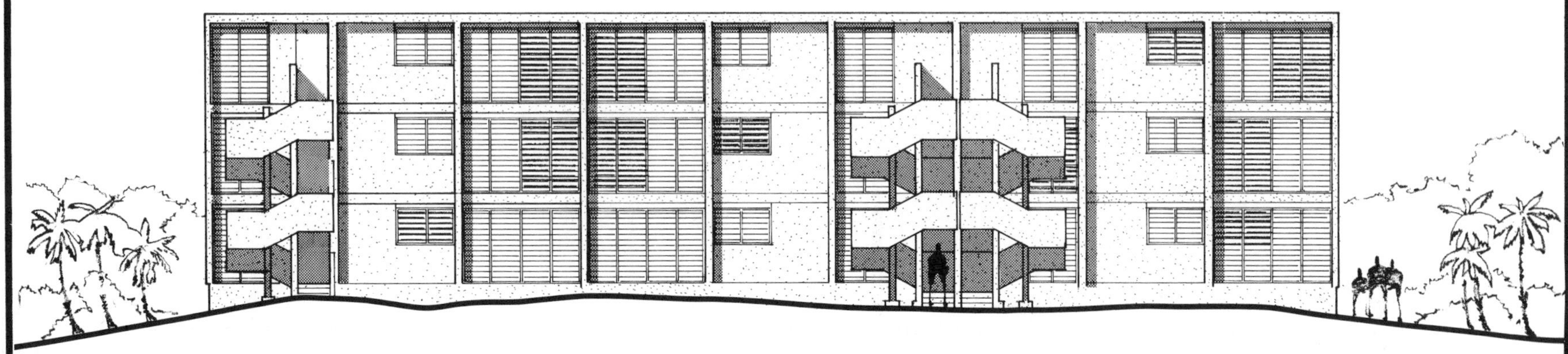

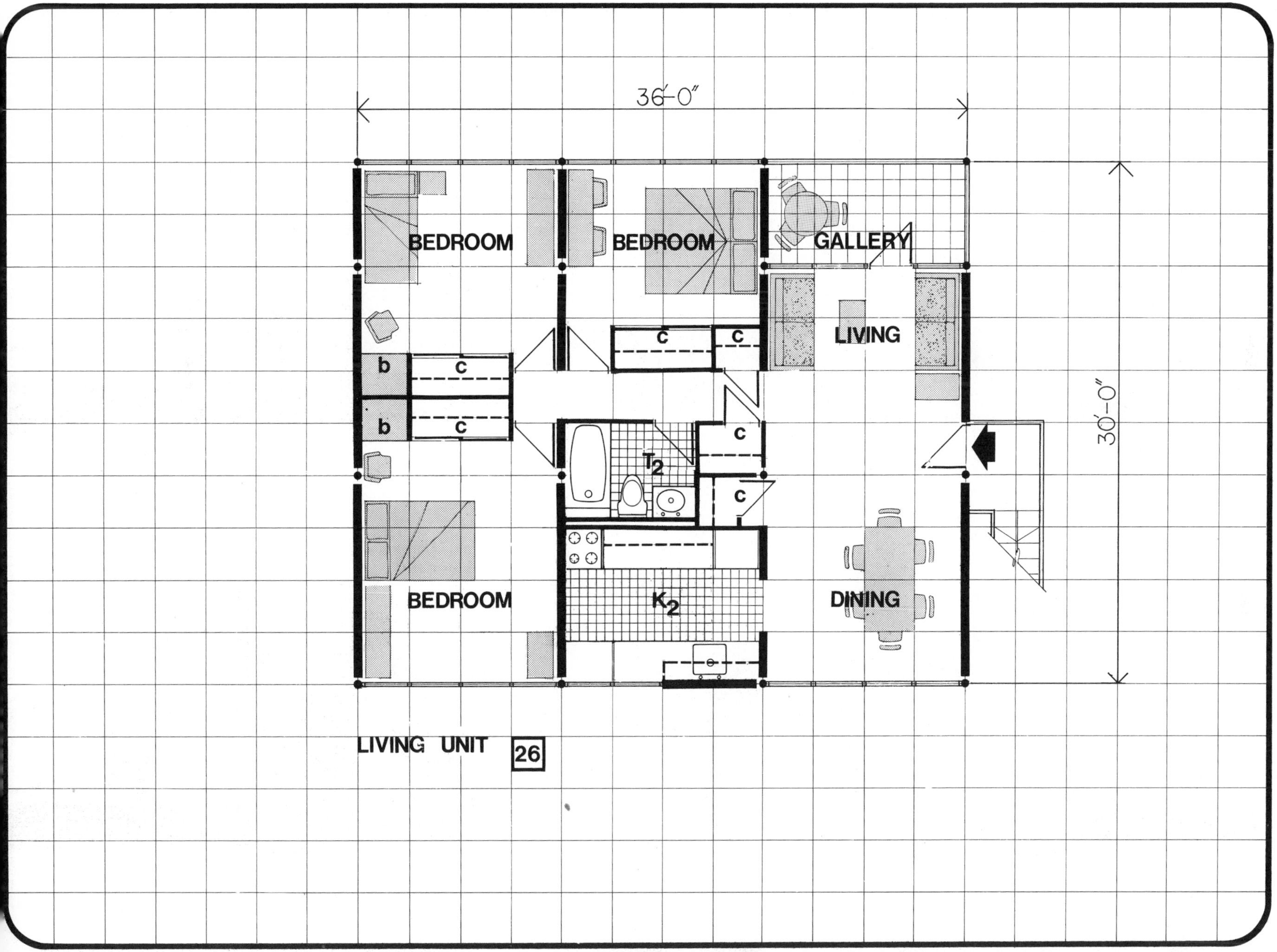
36'-0"
30'-0"
BEDROOM
BEDROOM
GALLERY
LIVING
b
b
c
c
c
c
c
c
T2
K2
BEDROOM
DINING
LIVING UNIT 26

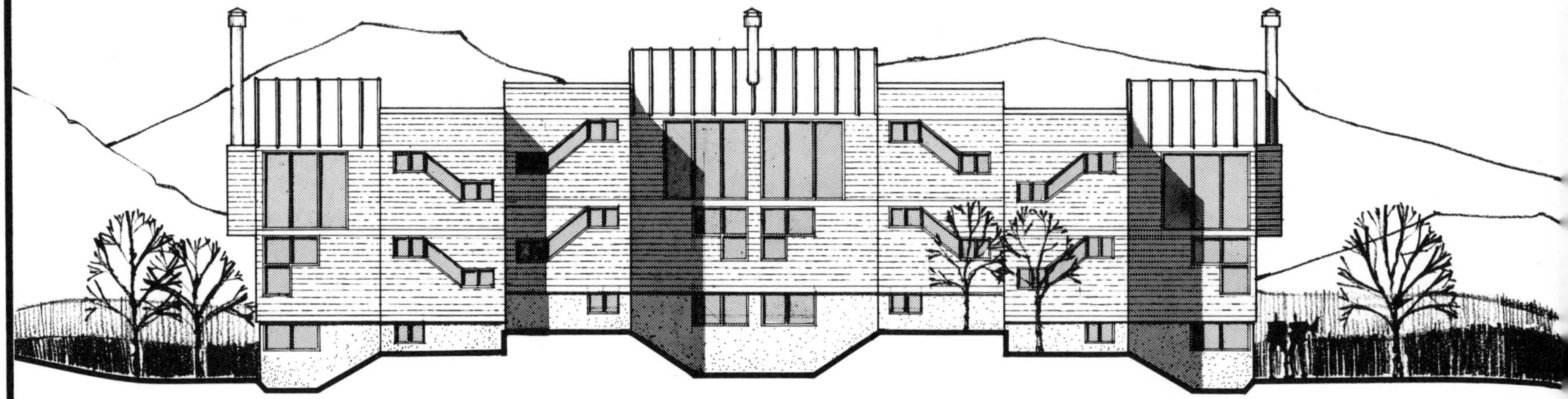

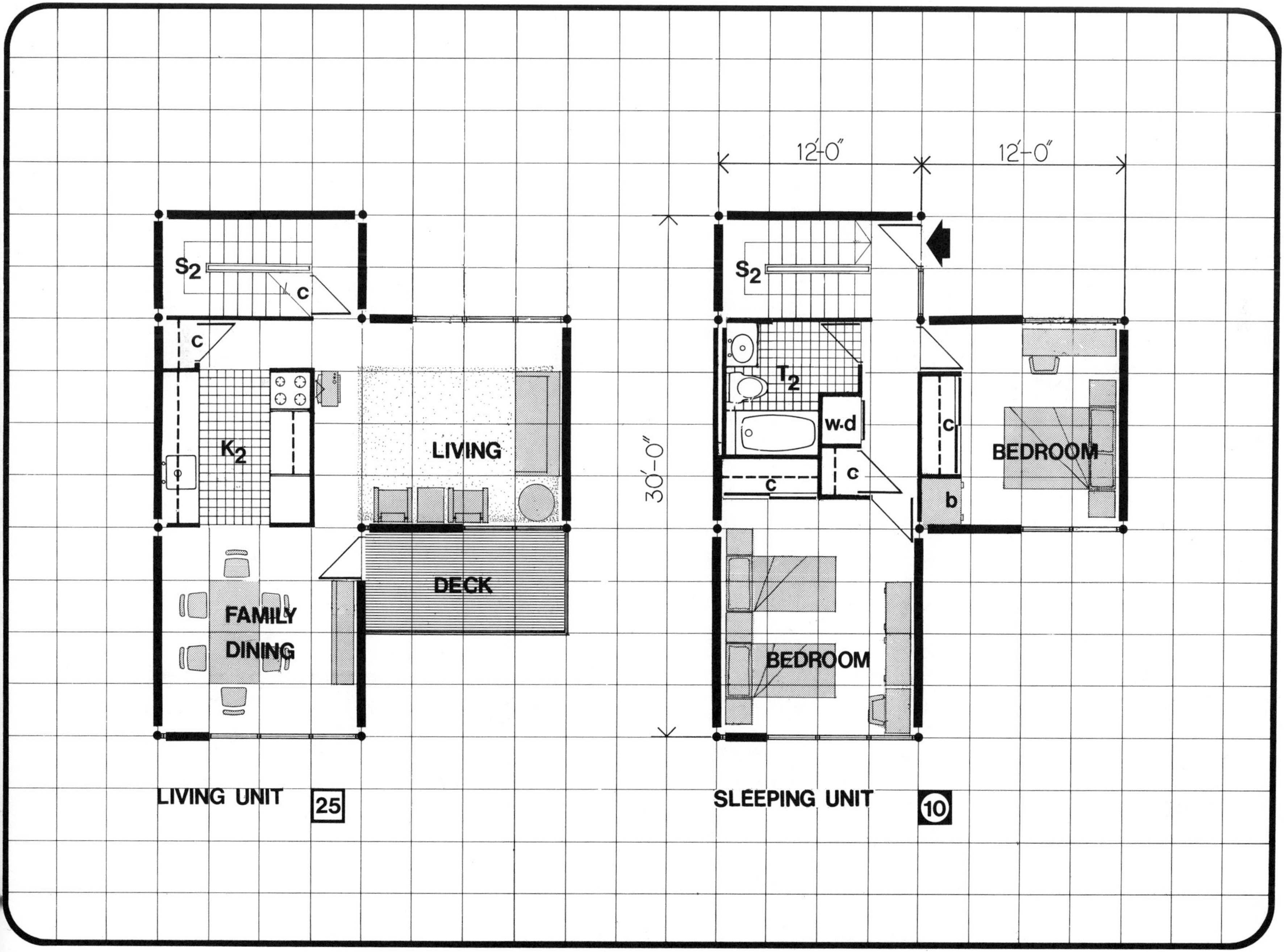
12'-0"
12'-0"
30'-0"
S2
c
c
K2
LIVING
DECK
FAMILY DINING
S2
T2
w-d
c
c
c
b
BEDROOM
BEDROOM
LIVING UNIT 25
SLEEPING UNIT 10

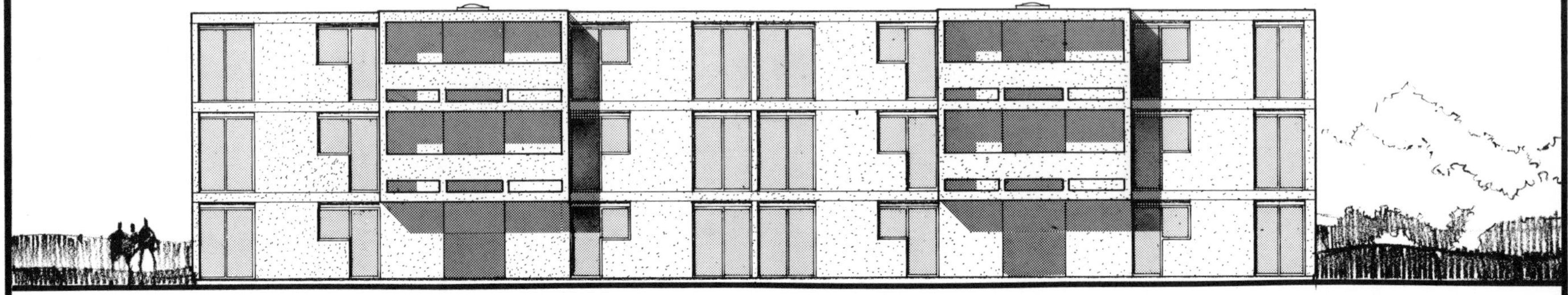

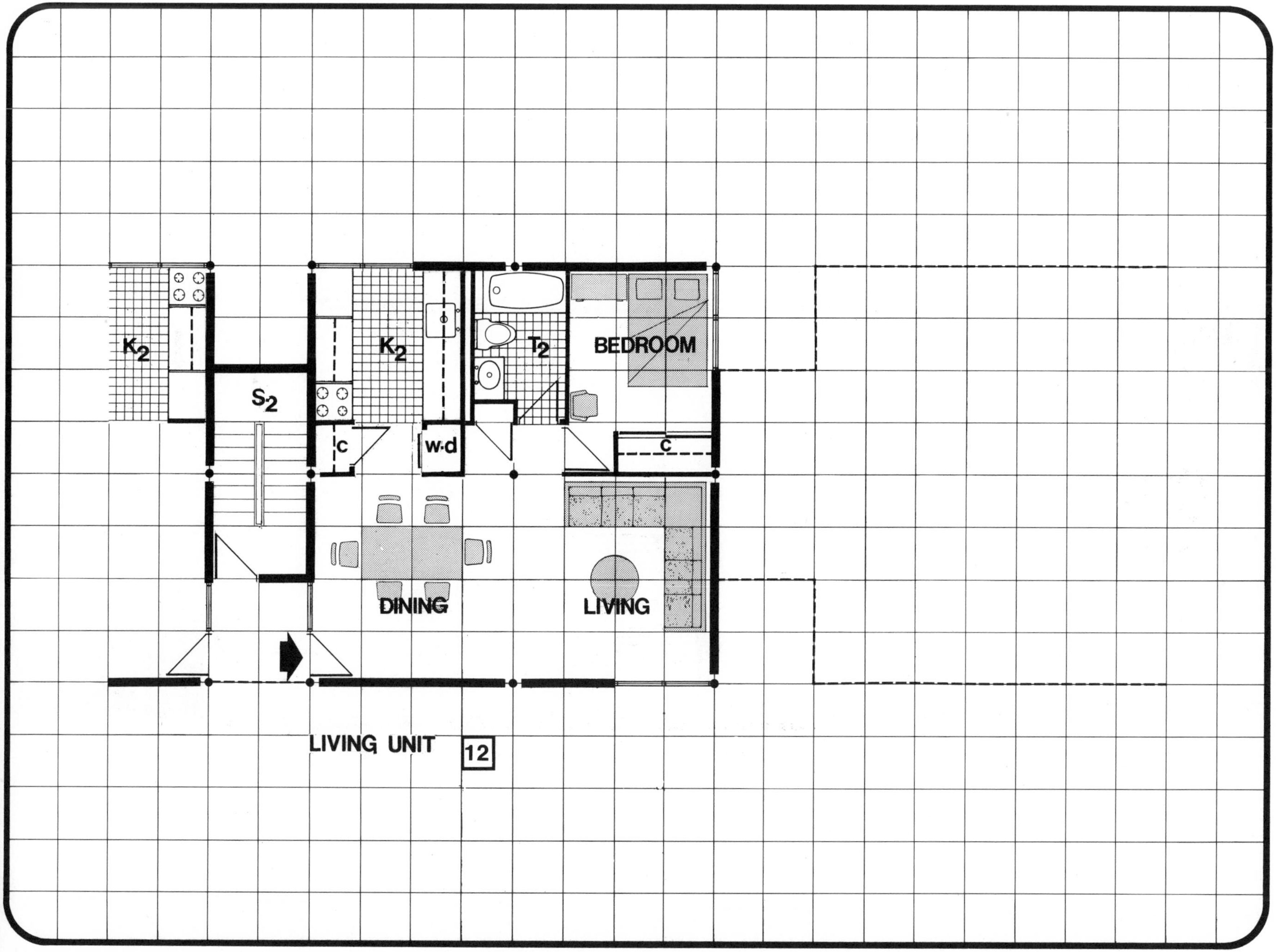

TOTAL UNITS

B 12

ELEVATIONS

A C 15

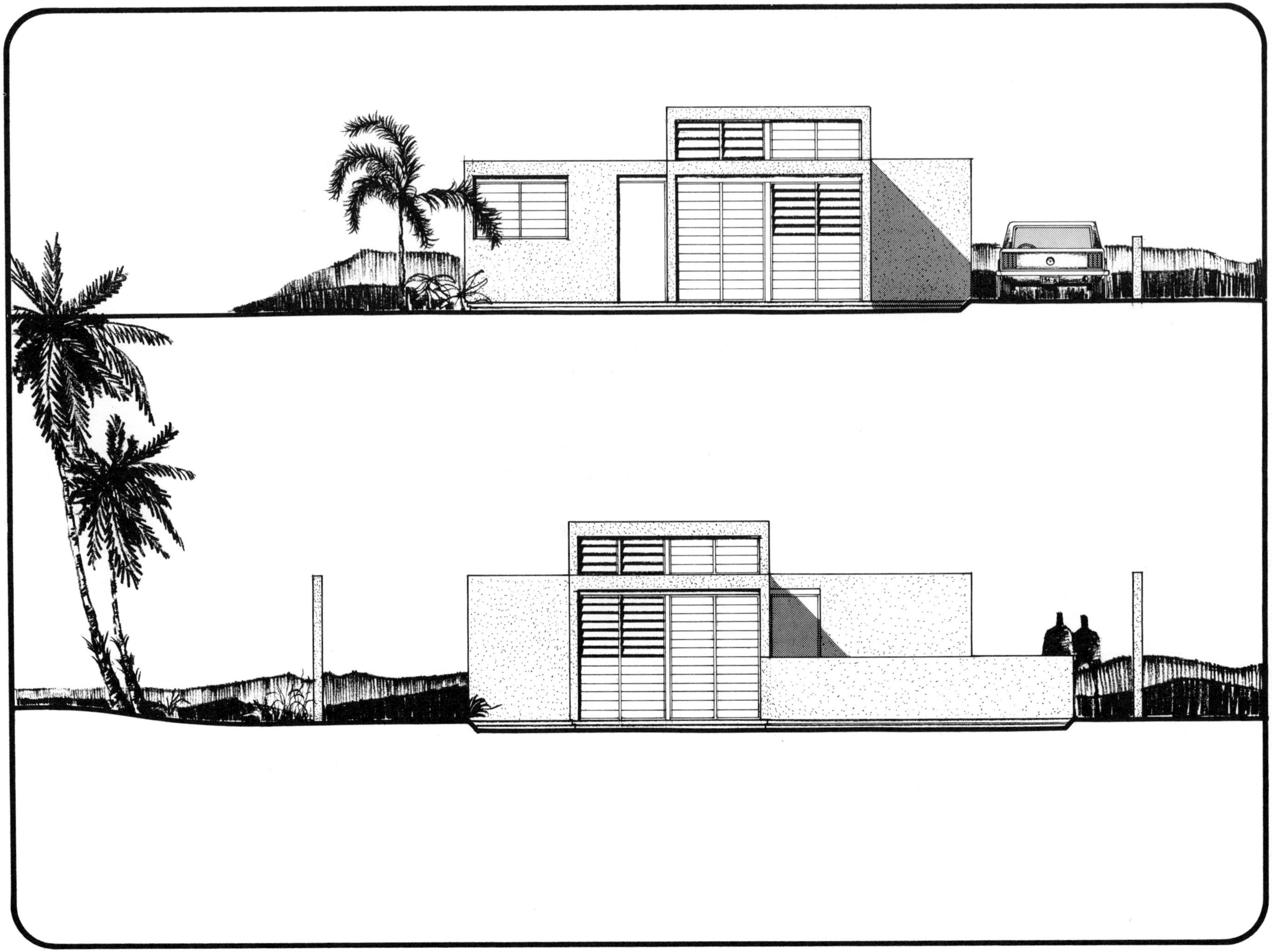

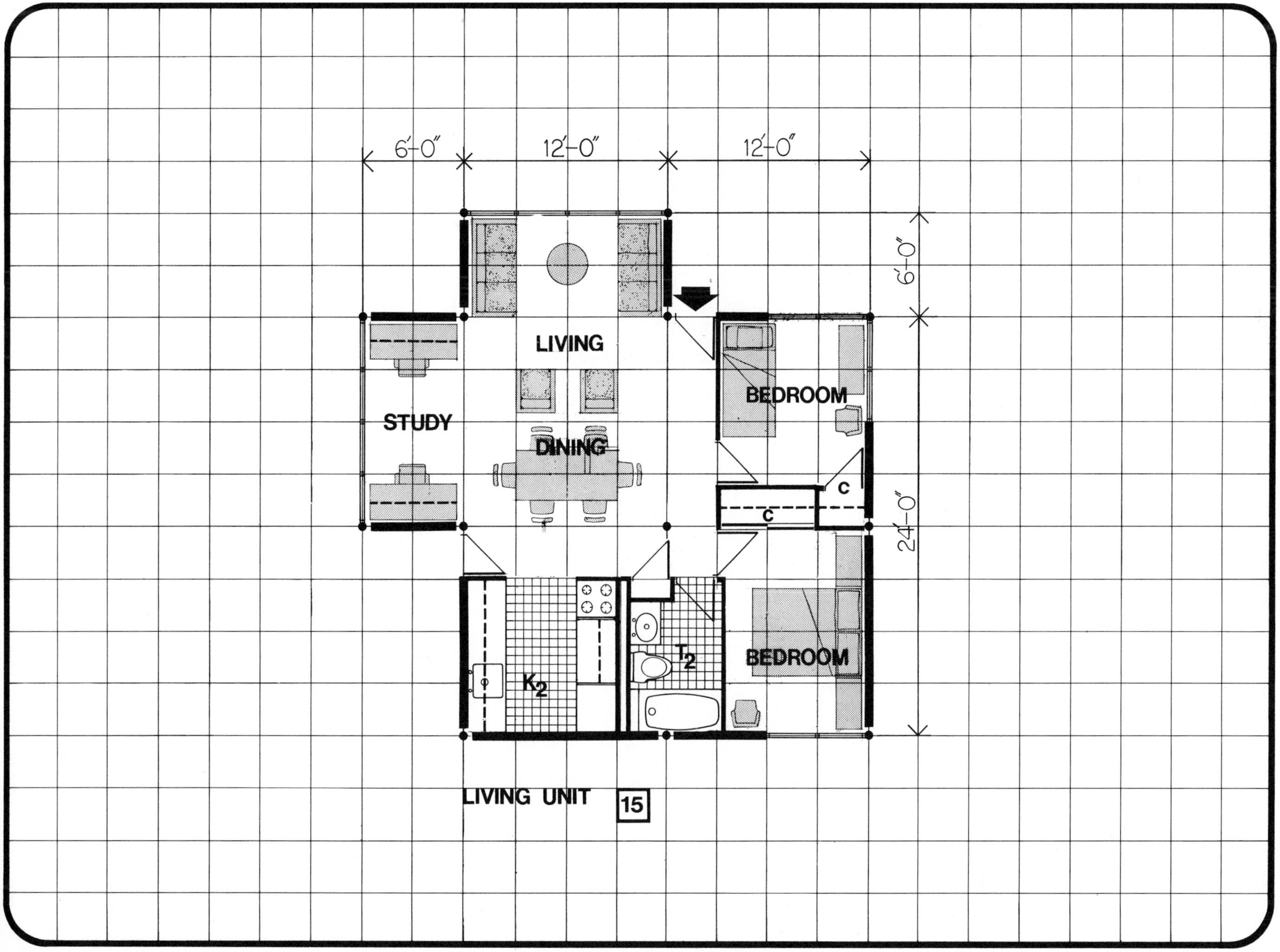
6'-0"
12'-0"
12'-0"
6'-0"
24'-0"
LIVING
STUDY
DINING
BEDROOM
C
C
K2
T2
BEDROOM
LIVING UNIT 15

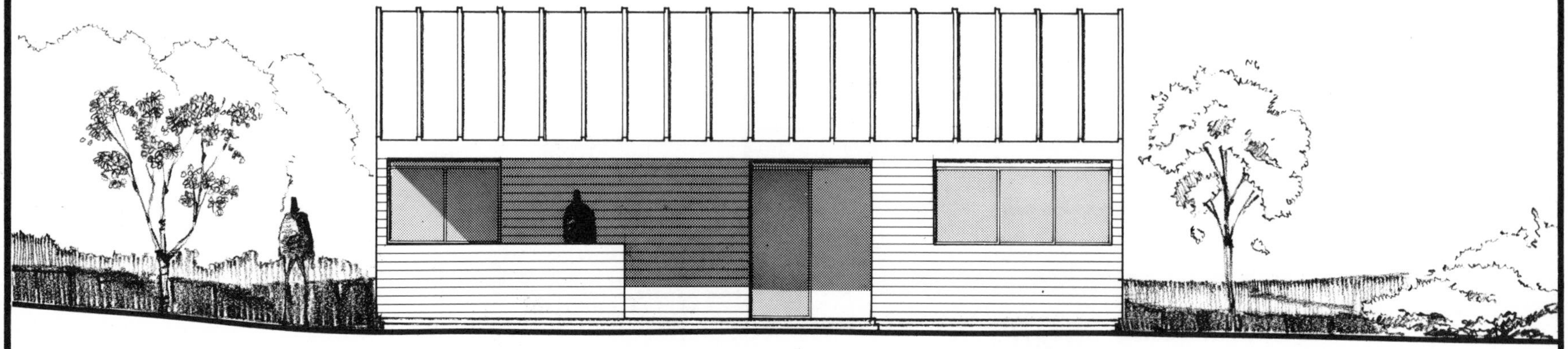

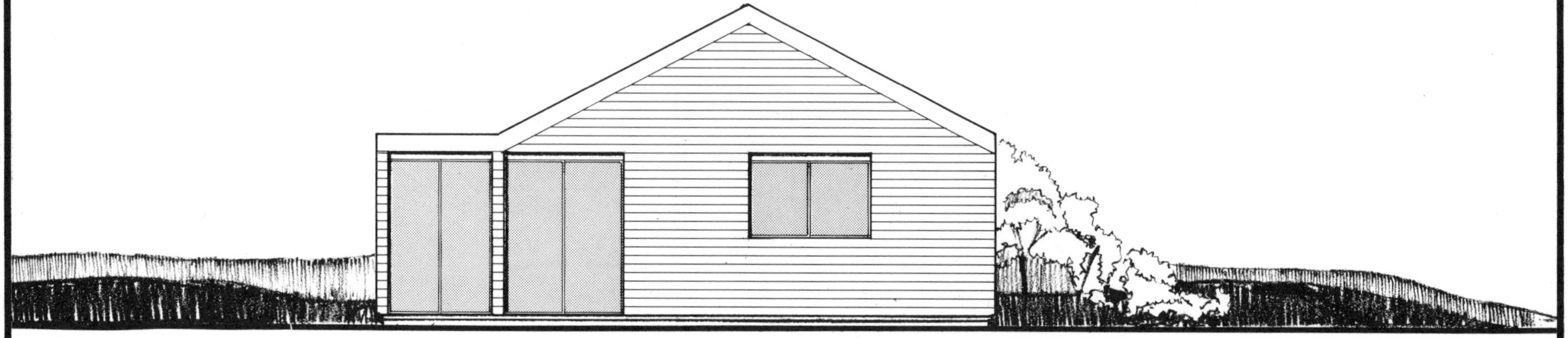

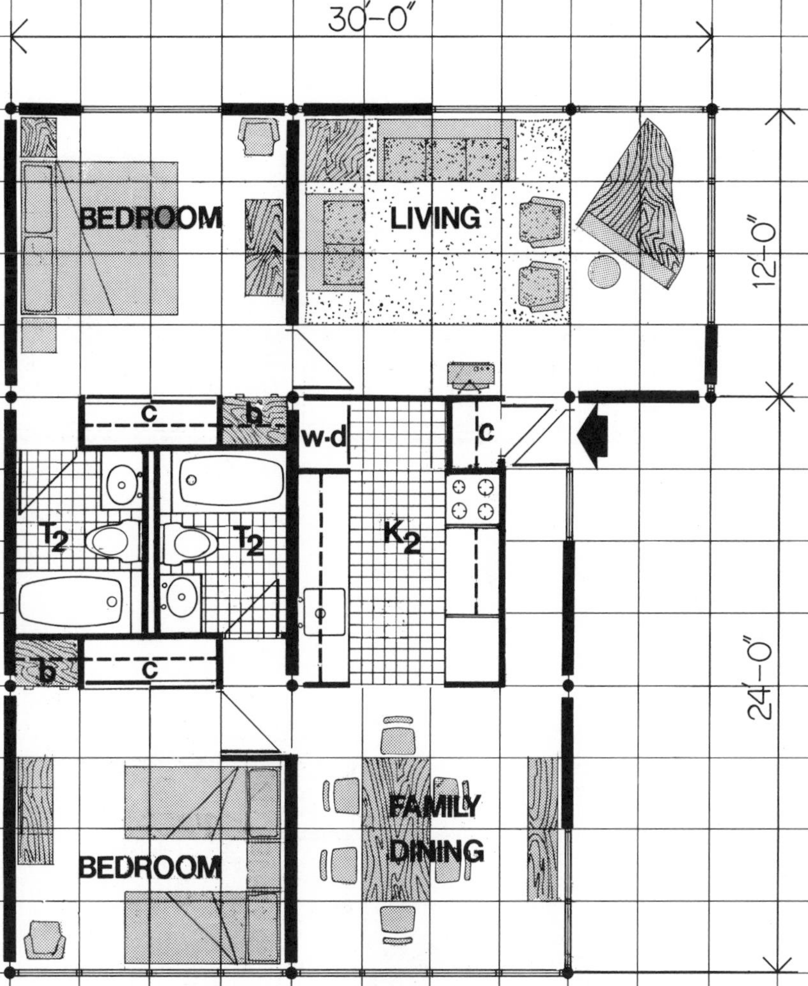

LIVING UNIT 5 SLEEPING UNIT 16

TOTAL UNITS

A B C 5 16

ELEVATIONS

A B C 3 6

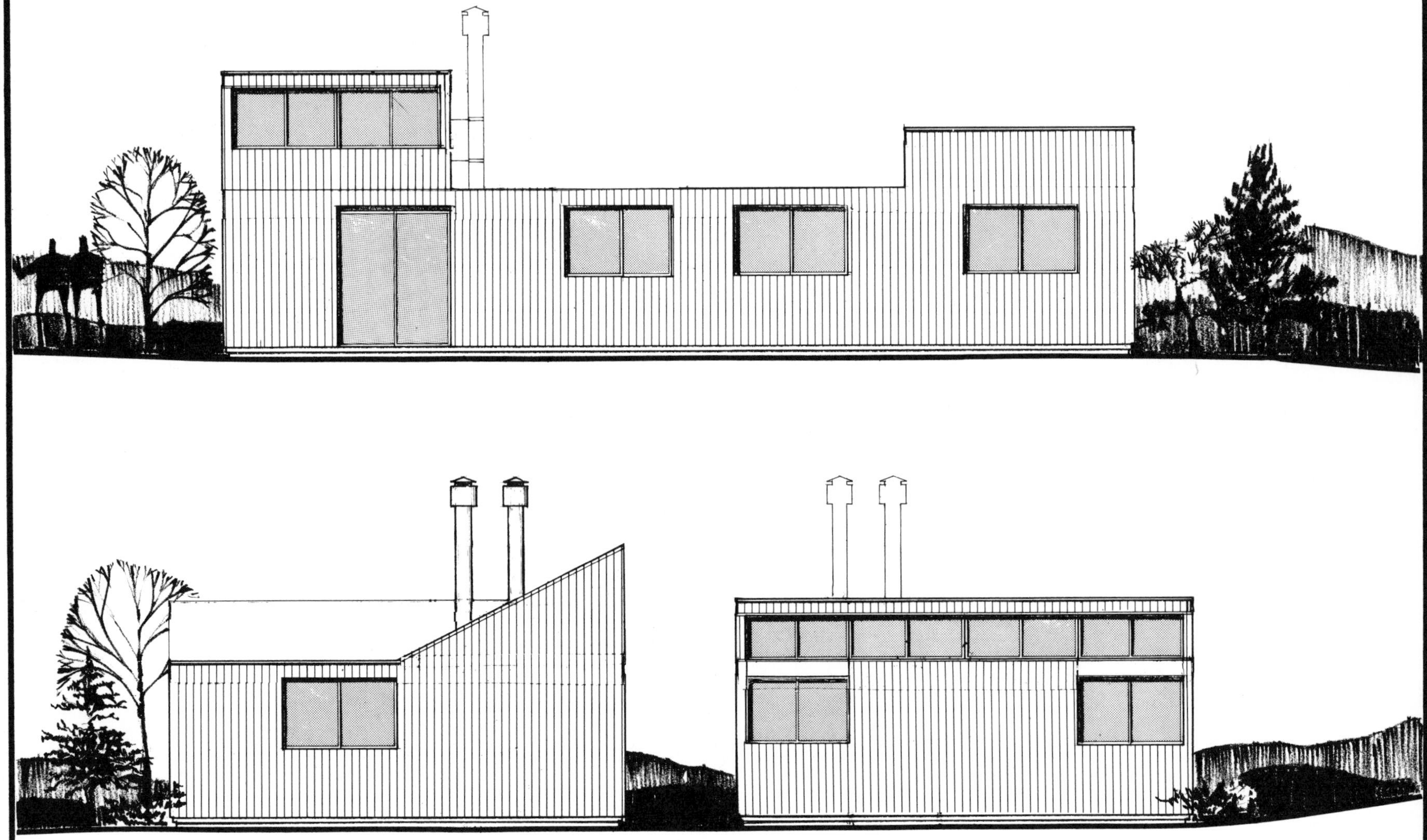

48'-0"
24'-0"
BEDROOM
BEDROOM
BEDROOM
DINING
LIVING
c
c
c
c
c
f
w·d
K_2
T_2
T_2
d·b
SLEEPING UNIT
6
LIVING UNIT
3

TOTAL UNITS

A B C 3 6

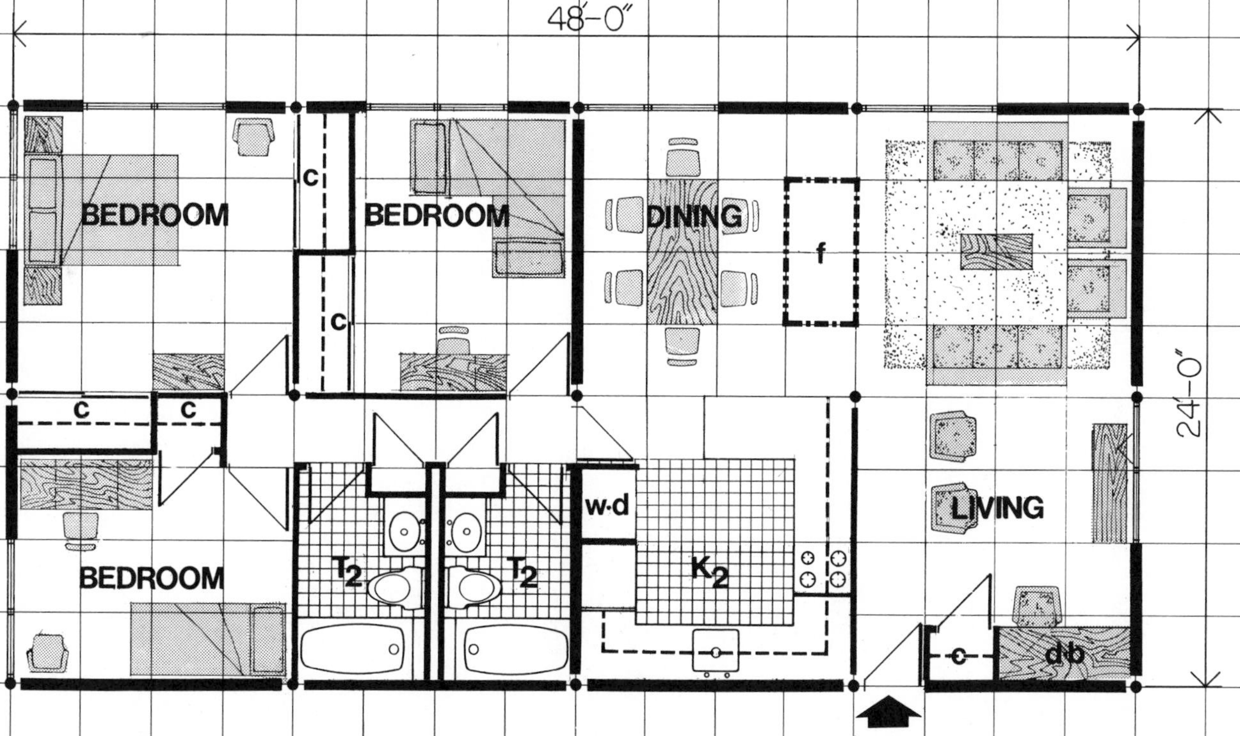

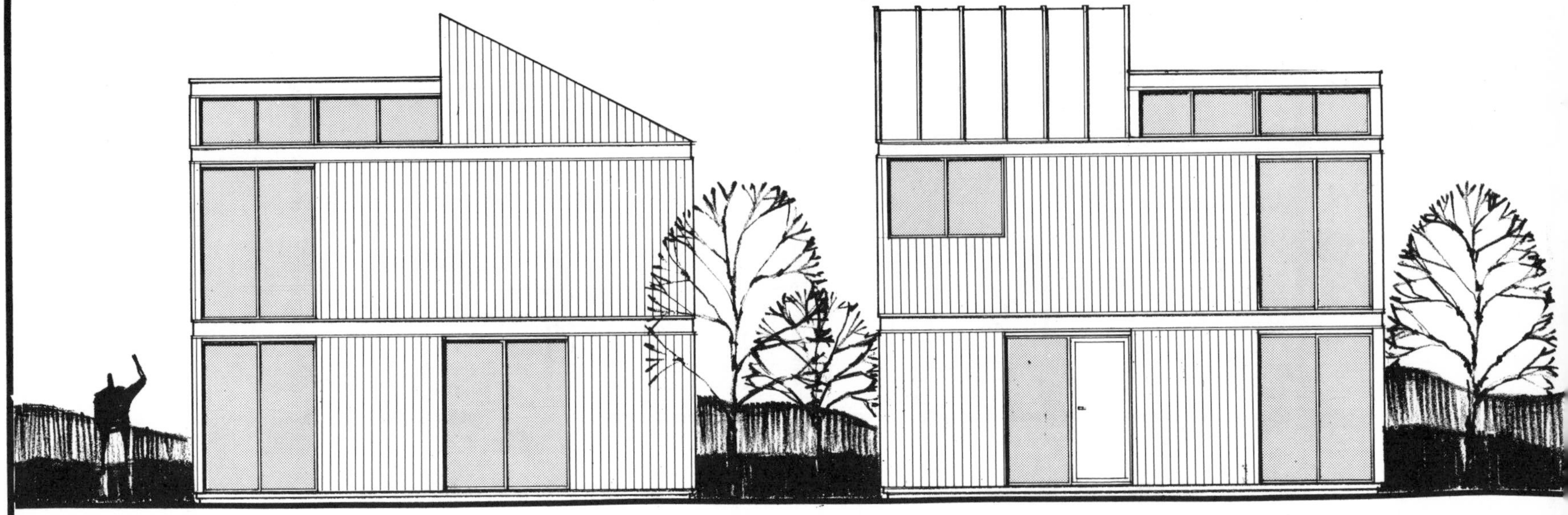

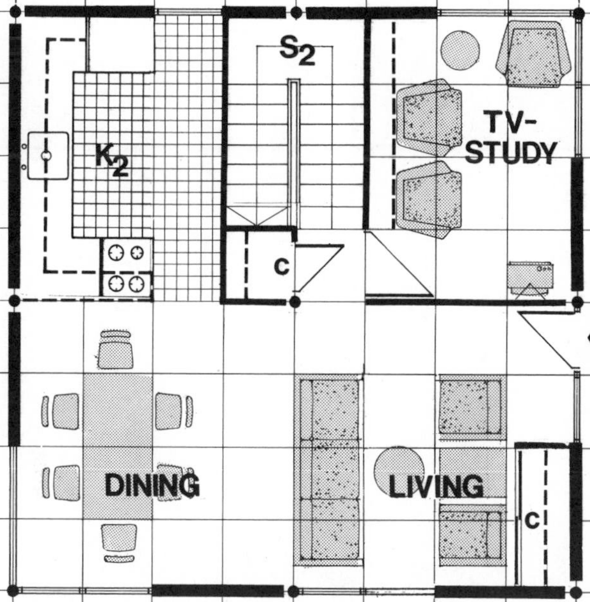

LIVING UNIT 6

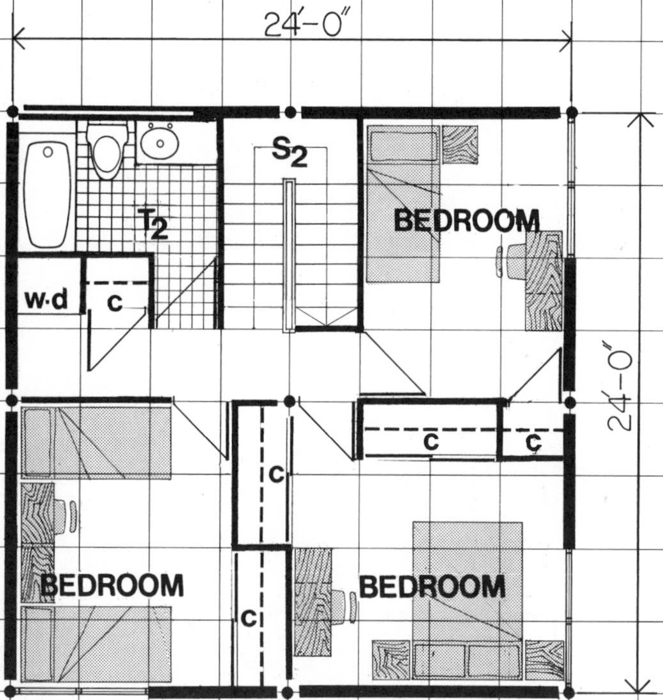

SLEEPING UNIT 3

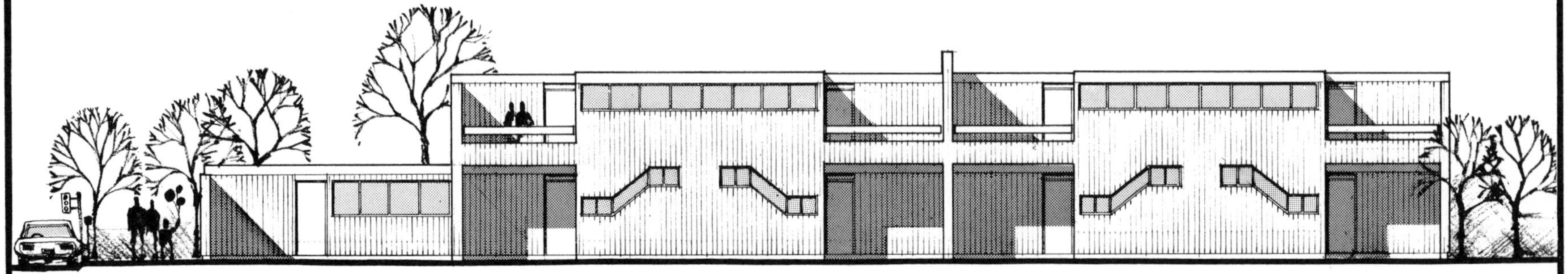

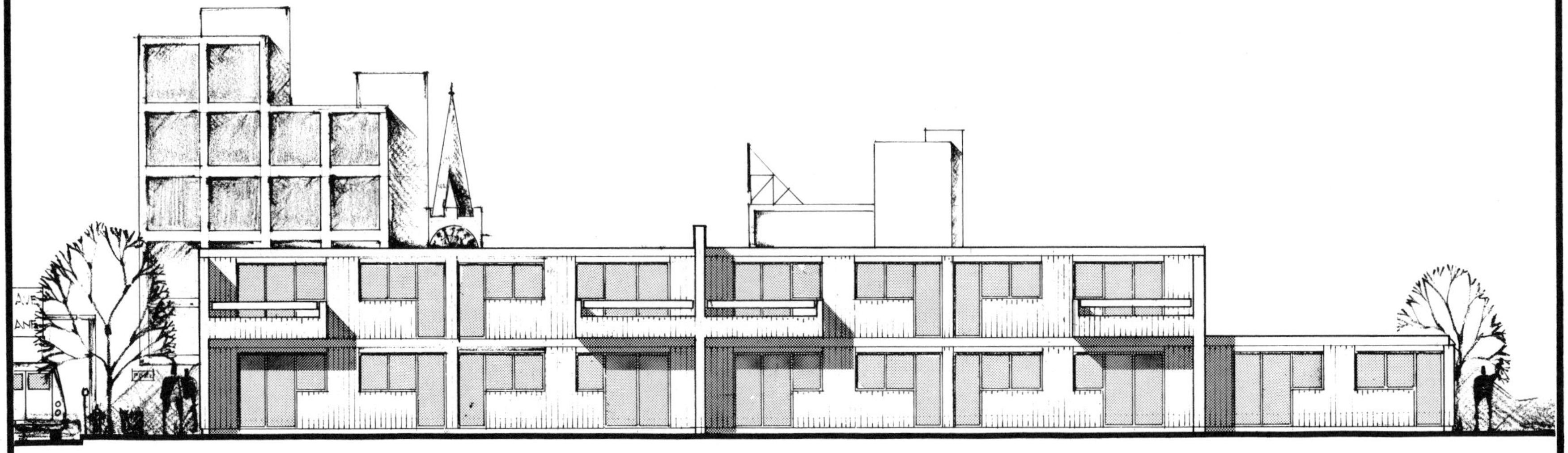

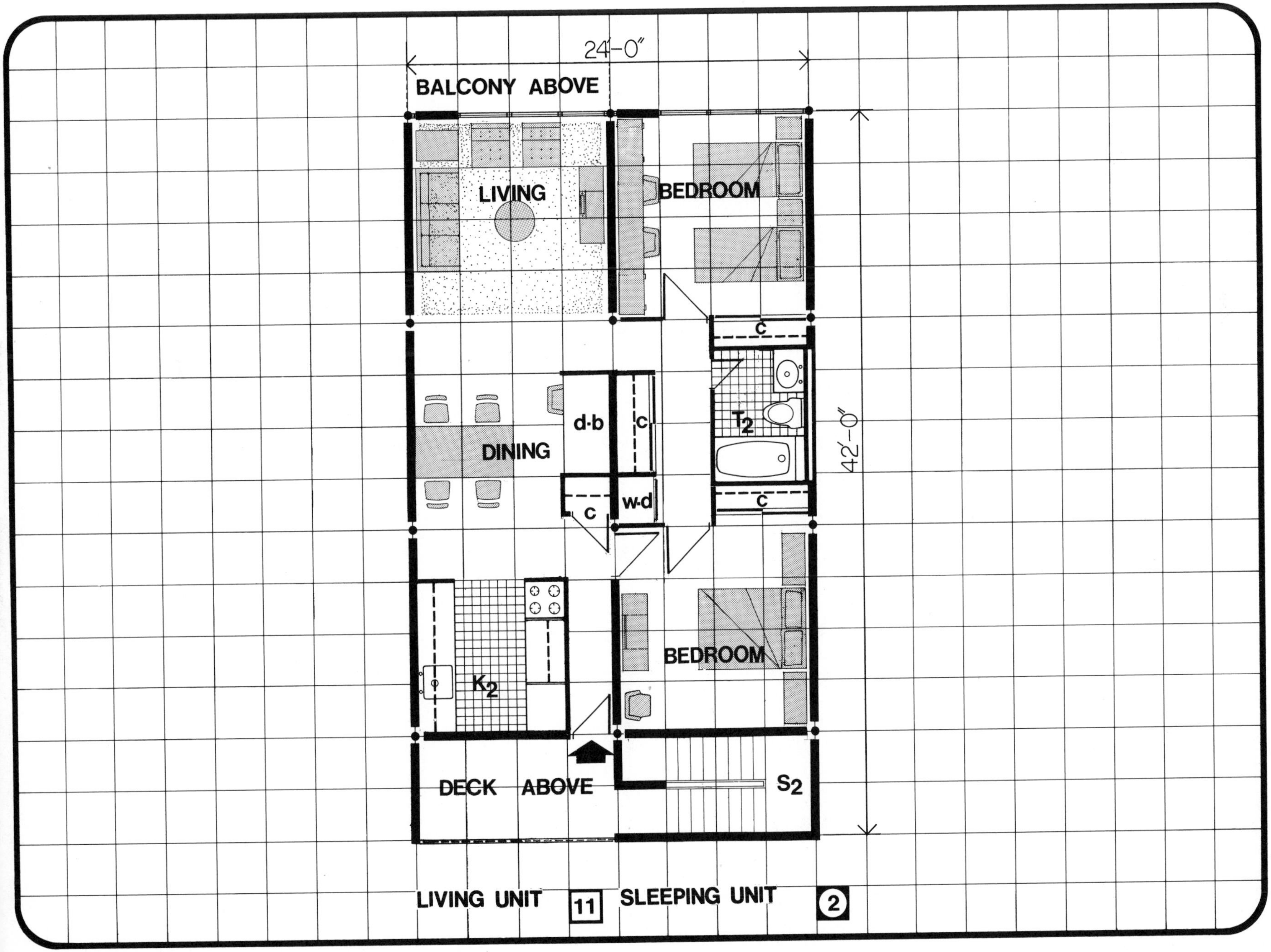
24'-0"
BALCONY ABOVE
LIVING
BEDROOM
c
d-b
c
T_2
42'-0"
DINING
c
w-d
c
K_2
BEDROOM
DECK ABOVE
S_2
LIVING UNIT 11
SLEEPING UNIT 2
TOTAL UNITS
B C 11 2

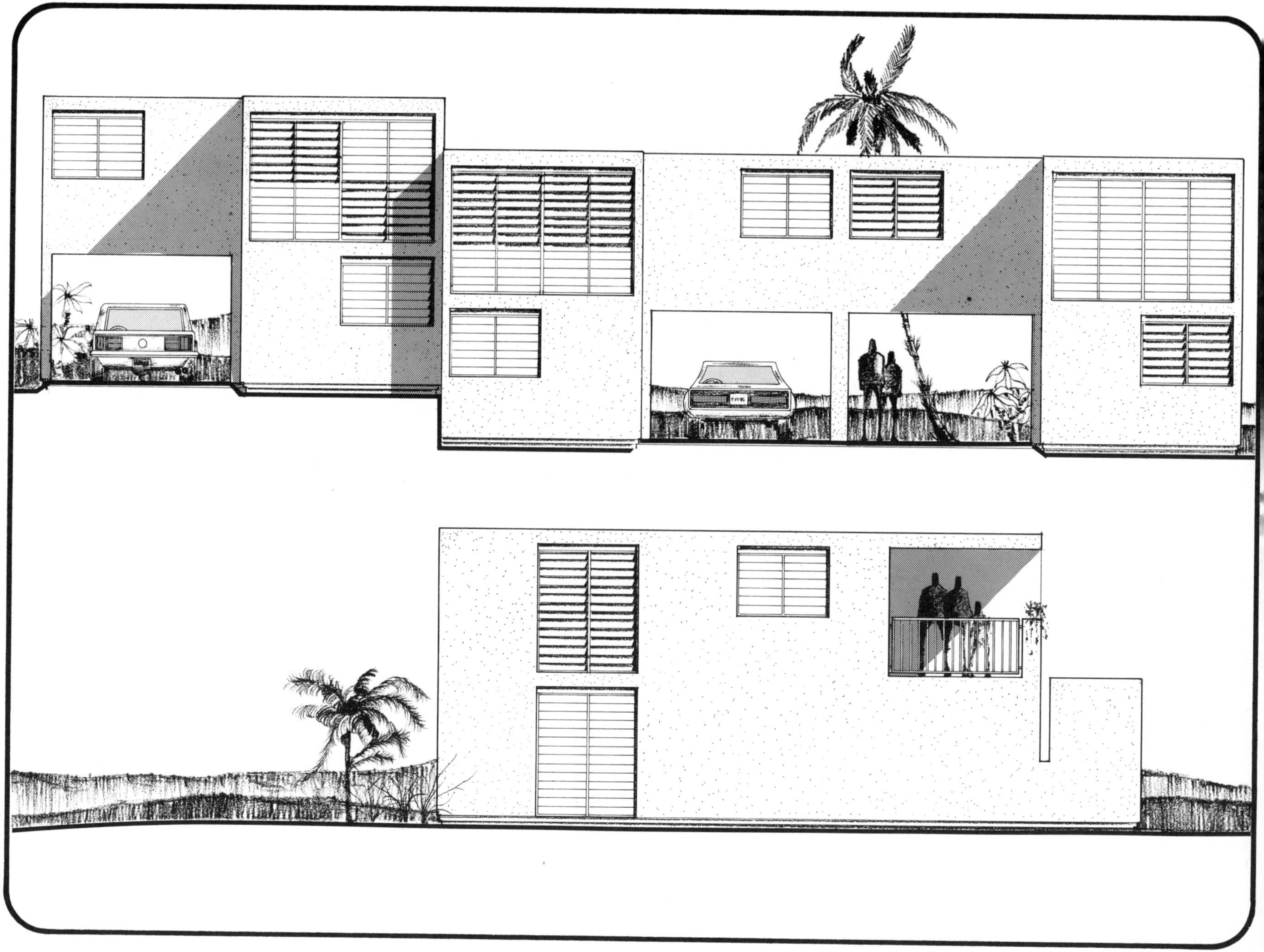

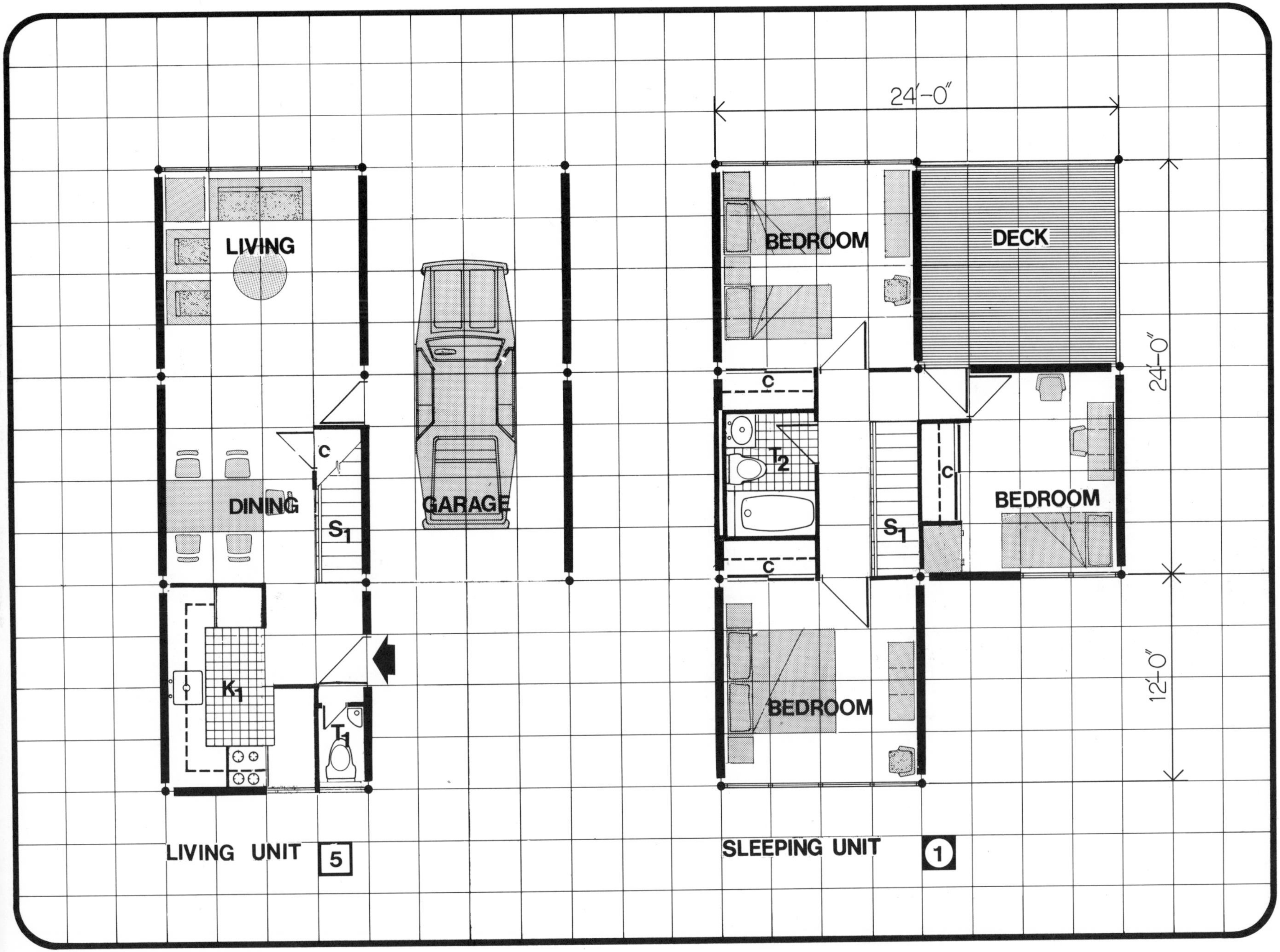
LIVING
DINING
C
S1
K1
T1
GARAGE
24'-0"
BEDROOM
DECK
C
T2
S1
C
BEDROOM
C
BEDROOM
24'-0"
12'-0"
LIVING UNIT 5
SLEEPING UNIT 1

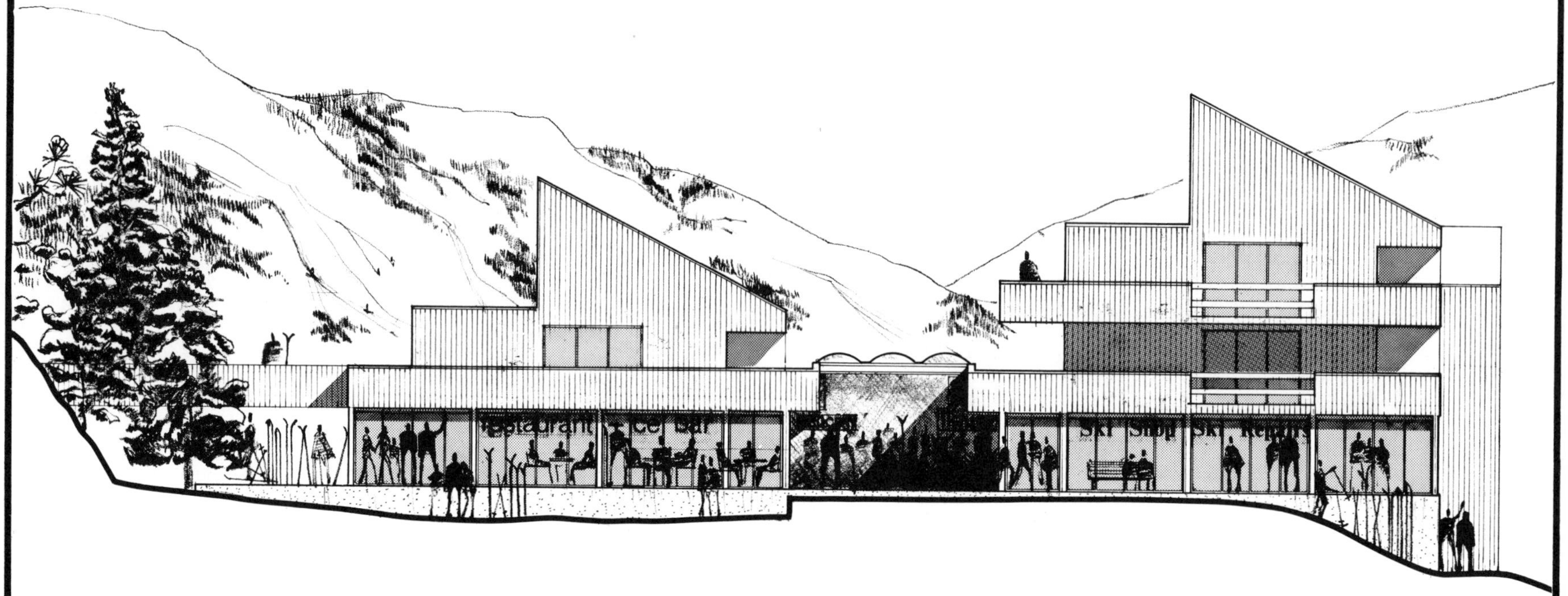

SLEEPING LOFT
C
THIRD & FOURTH LEVELS
BALCONY
LIVING
DINING
K3
T2
C
C
SECOND & THIRD LEVELS
TYPICAL COMMERCIAL BAY
FIRST LEVEL

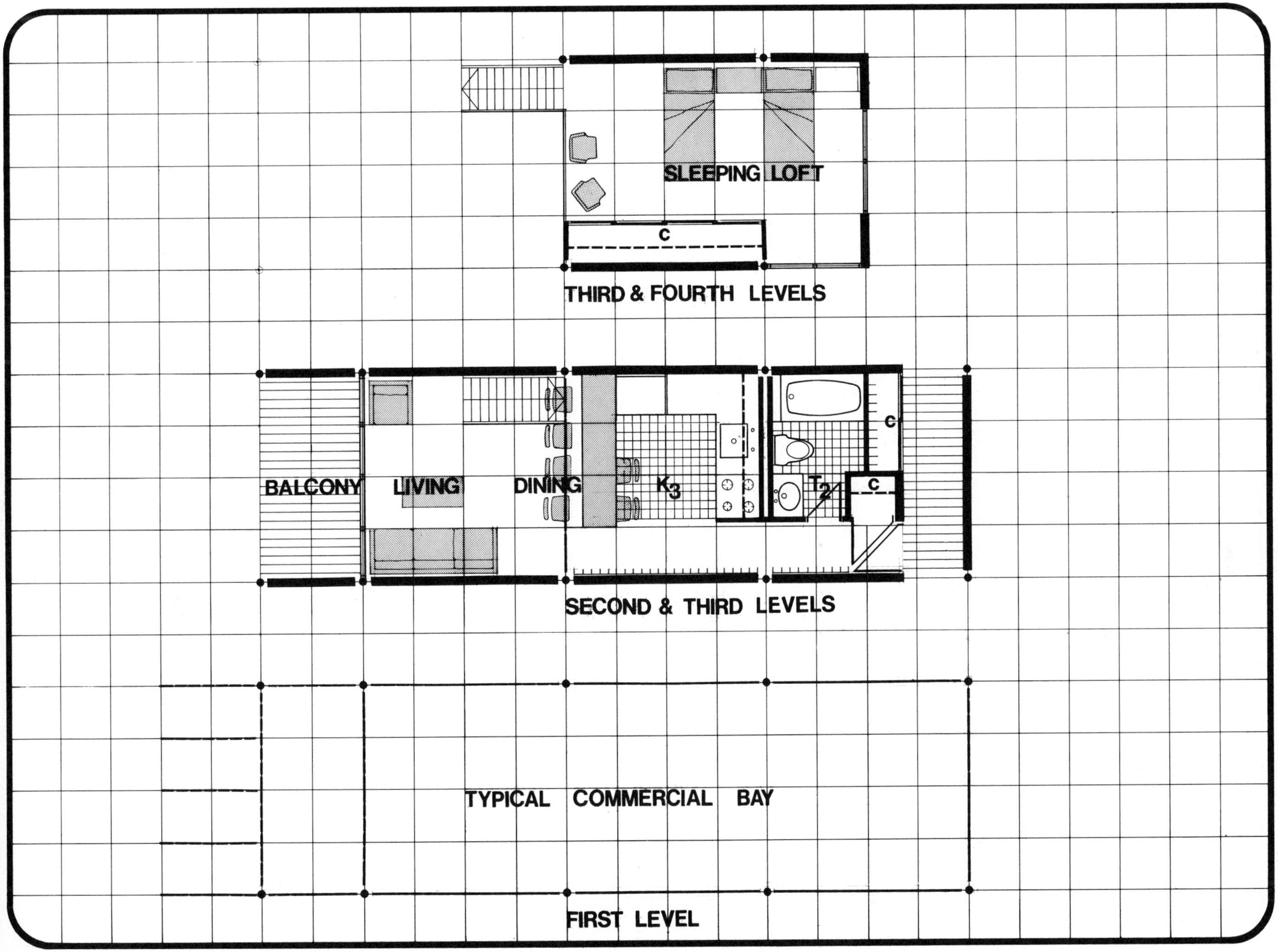

TOTAL UNITS
SKI VILLAGE UNIT

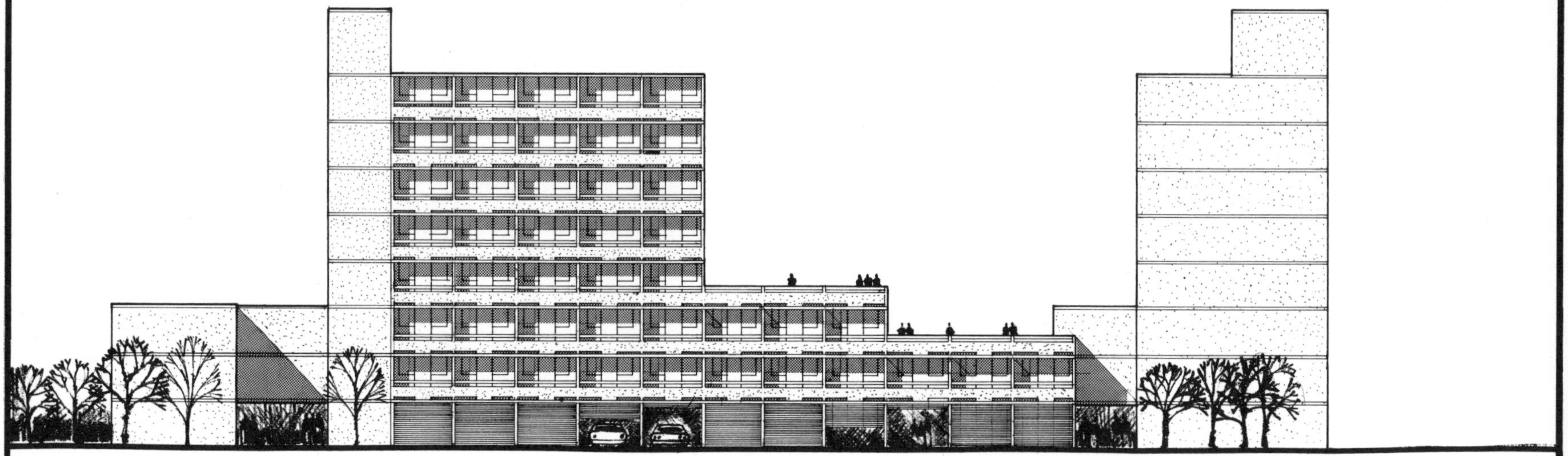

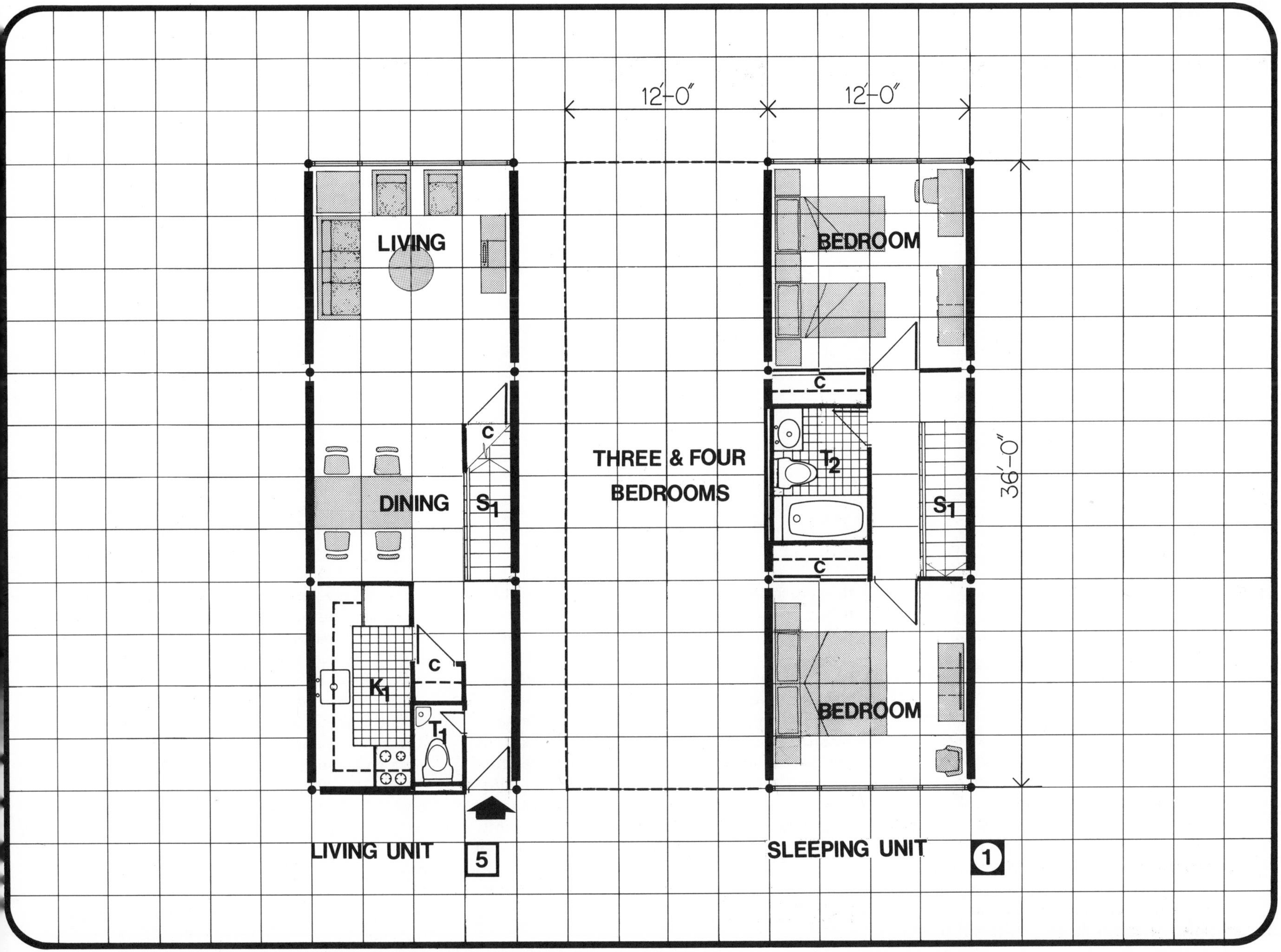

TOTAL UNITS

C 5 1

9 LAND PLANNING AND LOCAL ECOLOGY

THE NEED

"The new $12 million Playboy Club-Hotel is in the *rolling hills* of Southern Wisconsin, two miles east of *Lake Geneva.*

"Recreational facilities include a pair of championship *golf courses* . . . other summertime guests can *fish* and *sail* in a *private 25-acre lake,* ride horses on more than *20 miles* of *bridle paths,* or make use of a complete skeet and *trapshooting* range. For year-round swimmers there are indoor and outdoor pools—both equipped with bikinied Bunny lifeguards."

Today everyman is a playboy in that he wants the amenities described above in a recent promotion brochure. With these amenities, if one really thinks about it, it is the use of the land that is really being sold. Finally, it has become fashionable to realize the inherent goodness of the Earth. Books abound with titles such as *Design with Nature,*[1] *Townscape,*[2] and the like, and yet with the impending solution to the so-called "housing problem," i.e., industrialized building systems, no one talks about the large-scale environment which they will create or, rather, cause. Professor Burnham Kelly of Cornell University and the author of *The Prefabrication of Houses* in 1951 was recently awarded a Graham Foundation grant to search out some of the large-scale environmental problems caused by putting so much foreign matter on a small bit of earth. But this study is just beginning. Until now we have witnessed only two serious undertakings by systems builders along these lines, i.e., the affect of the site plan on user needs—one by the Stavebni Zavody Praha in Czechslovakia, in a research of environmental conditions in the satellite community, Invalidovna; the second, a small neighborhood study by the Wates Company in England of one of their systems-built housing estates.

1. Design with Nature, Ian McHarg, Natural History Press, New York, 1970.
2. Townscape, Gordon Cullen, The Arch'l Press, London, 1961.

Construction Techniques as Influence

Systems builders and developers still do not realize the importance of controlling the physical setting.

9-1
LAND PLANNING AND LOCAL ECOLOGY These have become of increasing concern and fashionable subjects. Nevertheless, environmental activist groups must fight to keep certain projects from imposing their form of life on the as yet untouched work of Mother Nature. *William Garnett, the Center for Advanced Visual Studies at M.I.T., 1953*

They seem to think that the human being, when added to a completed project, will act as the catalyst to make for a successful living environment. Sporadic and uncontrolled growth has taken place throughout this country by stick builders for many years, as evidenced by the vast acres of mansard-roofed garden apartments, Mediterranean shopping plazas, and so on. In some of the more progressive states, such as Vermont, legislation has already been enacted which controls growth and sets parameters for large-scale developments, but these controls are primarily ecological balance controls rather than a more sensitive interpretation of the human "eco." The systems builder must be more aware of the dangers of environmental destruction, since he holds a nuclear weapon relative to the hammer of the traditional builder.

9-2
INVALIDOVNA An experimental satellite community near Prague. Czechoslovakia, where large-scale environmental problems are put under a microscope. *Stavebni Zavody Praha*

9-3
ROWLATTS HILL, LEICESTER, ENGLAND This plan exhibits the concept of high-density housing with mixed land use including a tightly planned single-family residential community. A parking structure is located on the north side of the tower buildings, permitting more open space northeast of the high rise by concentrating urban development. *Concrete Ltd. and Stephen George, Architect*

One beneficial side effect of the requirements of systems building is the need to *concentrate* development. In England, the nature of subsidized systems building employed in the series of new towns ringing London might actually have precluded sprawl. But there was also a simultaneous awareness inherent in the concept of the new towns of the need to guide and concentrate urban development to avoid ruining the natural environment.

The nature of the housing industry in the U. S. has encouraged sprawl. Each private entrepreneur has generally built where land was most economical and available to him, and he has catered more to the individual desires of the American Dream—the one-quarter acre for everyman who could afford it—rather than planning for the needs of an evolving new form of community life which would eventually have to provide the less selfish forms of amenities such as service by mass transportation, educational systems, proximities to eliminate the complete dependency on the automobile, and response to ecological problems. Stick construction has not required any "economic radius to plant" and even the concept of rationalization of the building site for more efficient labor-man hours and use of equipment has not altered the "quarter acre" concept since its original post-war years.

It would seem that the only "efficiency" applied to the on-site construction practices of the developer-contractor of the 1940s and 1950s was the total razing of all on-site trees, rock outcroppings, etc. Only a slow realization of this substantial economic value of these natural features has altered the practice!

An entire chapter should be devoted to site planning, because it is here that the developer-con-

9-4
THE AMERICAN DREAM In a sense this has caused the urban sprawl and suburban nightmares which snake their way across our land. These illustrations clearly show low-density sprawl and ultrahigh-density development in the ancient settlement at Tell El Amarna. *Architectural Design, December 1968*

9-5
EVERY SITE Natural or man-made, is to some degree unique, and the general criteria one sets for site planning have to be flexible. *"Messages from Perugia," Bruno Zevi, Industrie Buitoni, Perugina, 1971*

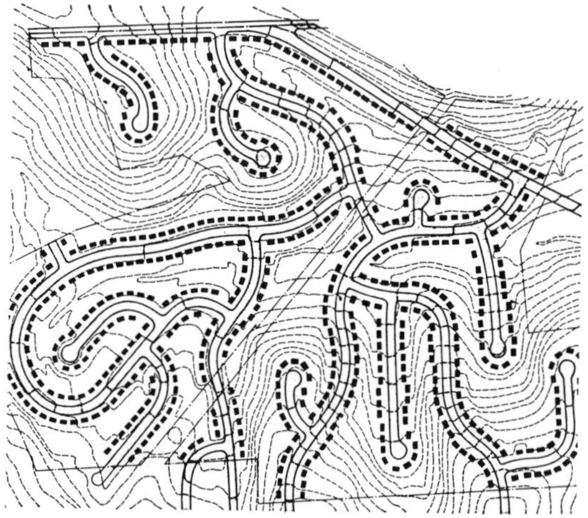

9-4

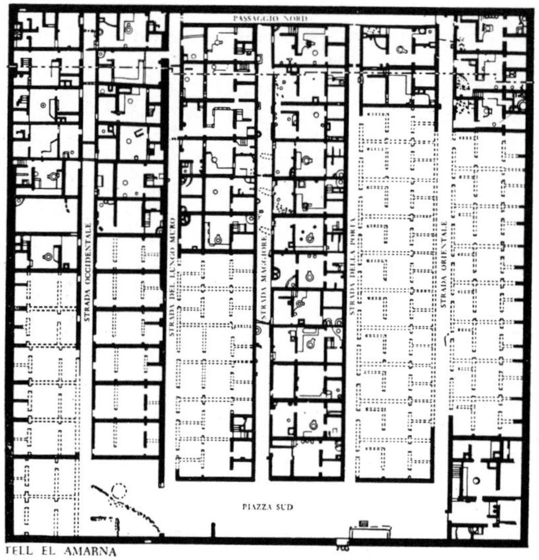

TELL EL AMARNA

9-5

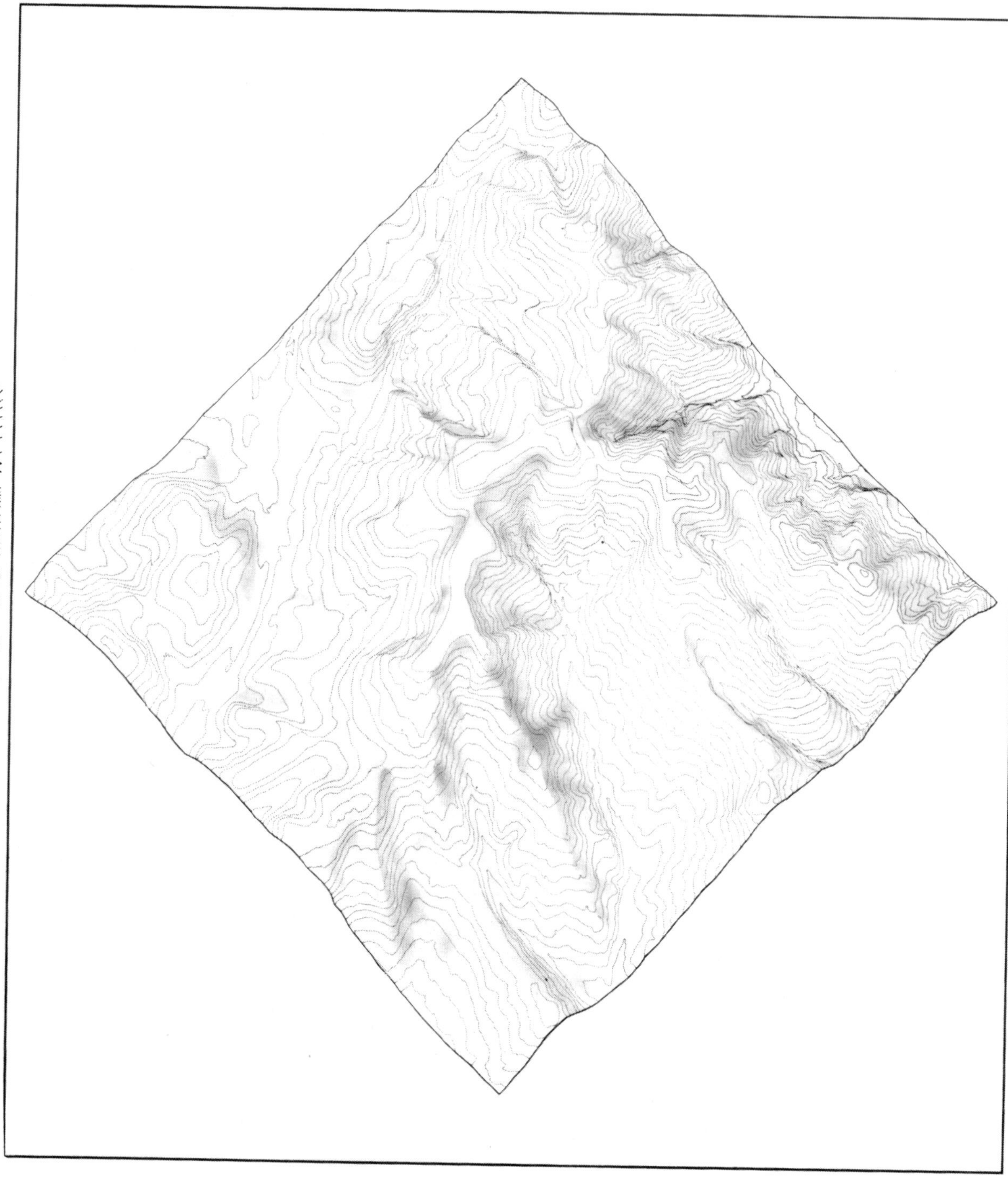

tractor has most often failed both himself and his customer. By answering an obvious demand for a self-centered life style, he has produced a "home" with inner amenities—but ignores its larger environmental contexts of neighborhood and site planning.

SITE PLANNING

Every site, natural or man-made, is to some degree unique, and the general criteria one sets in site planning have to be flexible. Understanding what standards and relationships are of significance takes a strong philosophical basis for decision-making. Objectives have to be clarified and the values of the developer, the user, and the designer must be harmonious in order to resolve *a sense of the place itself.*

Site plans are prepared firstly by analysis of the site itself, then establishing the goals of all parties involved. The technology will implement the specific plan. It is a constant play between the general and the specific.

PROGRAMMING

The project begins with a general idea of program and site type and works towards more specific identification of all building types and facilities, exterior space, pedestrian and vehicular circulation, and parking. Functional and quantity requirements are first roughly defined by the site size and context, population, expected market demands, mode of financing, etc., and expressed in terms of what specific types of buildings, exterior uses, and circulation patterns are to be program elements and the size and quantity of each that can be supported.

Possible program elements include:

Building types
- —Town house
- —Low-rise or walkup apartments
- —Medium- to high-rise multifamily units
- —Shopping and entertainment
- —Service facilities
- —Community meeting and work spaces
- —Day care
- —Education

Exterior spaces
- —Unit/family spaces—garden, patio, yard, balconies, driveways, entry paths, etc.
- —Community spaces—passive sitting, sunning, viewing, and public gathering, tot lots, playgrounds, fields, courts, park
- —Visual elements, plantings, sculpture, texture, land form, buffering

Circulation
- —Primary, secondary, tertiary walkways and access
  Common spaces, plazas, courts, intersections
- —Primary, secondary, tertiary roadways and access
- —Parking

Community
- —Social patterns and units
- —Social balance
- —Usage zones
- —Impact studies of usage alternatives
- —Phasing and financing determinants

The program statement includes not only specific types, sizes, and quantities but also outlines ideal functional relationships which should exist between these elements. These can often be expressed as schematic plans or a series of diagrams of ideal functional conditions. Program quantities *at this stage,* are generally set only by site size, market expectations, and financing specifics.

A sense of community must also be programmed in at an early stage and expressed in terms of social patterns and desired ethnic, age, and economic balance.

(a)

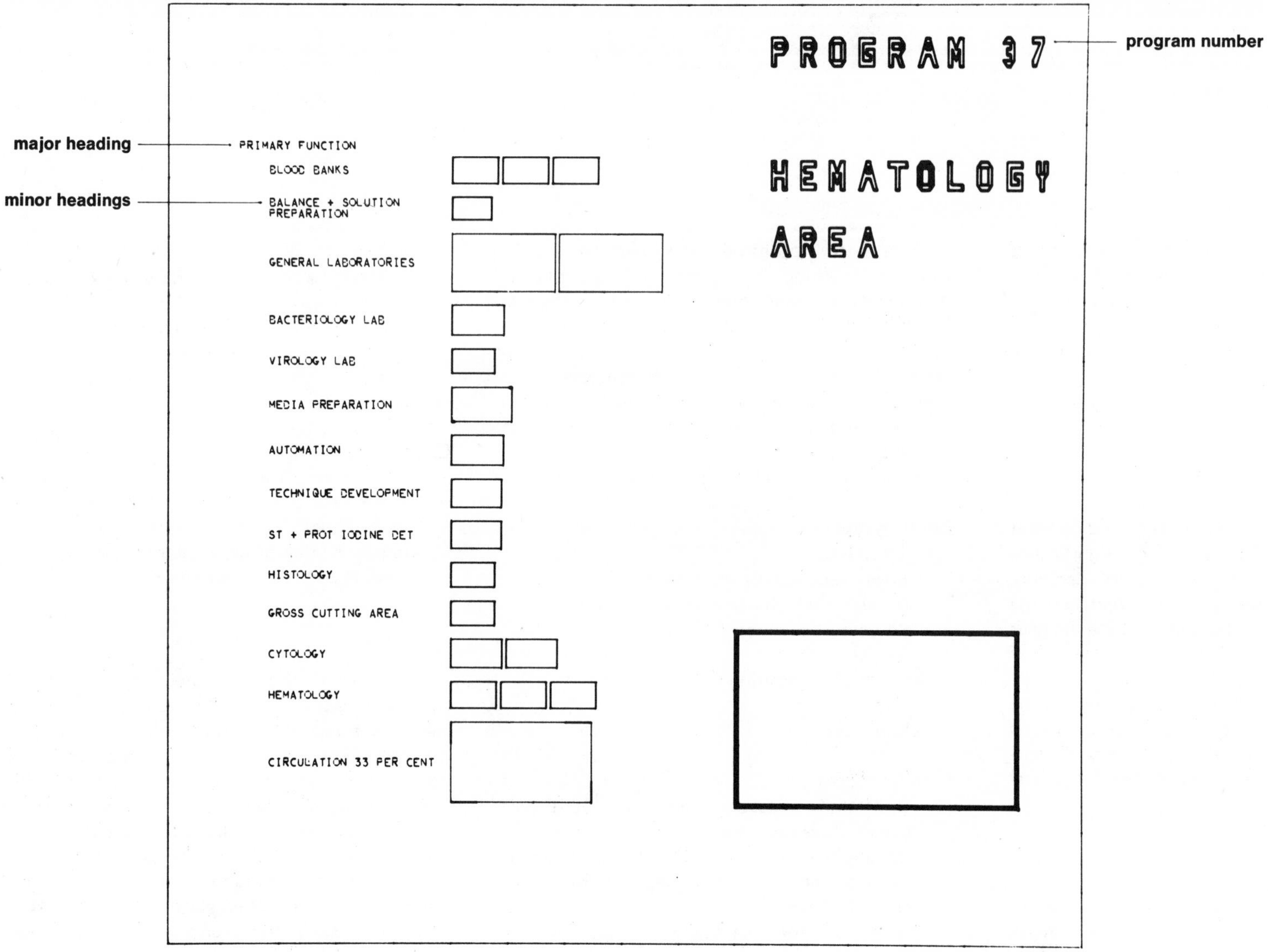

Comprograph 1 printout

9-6
PROGRAMMING Usually beginning with a general idea of the building requirements, these items become more specific, and identification of particular spaces and their relationships and juxtapositions develop. Computer planning of these functional and quantity requirements is a time-saving tool in architectural planning. *Design Systems, Inc., Cambridge, Massachusetts*

(b)

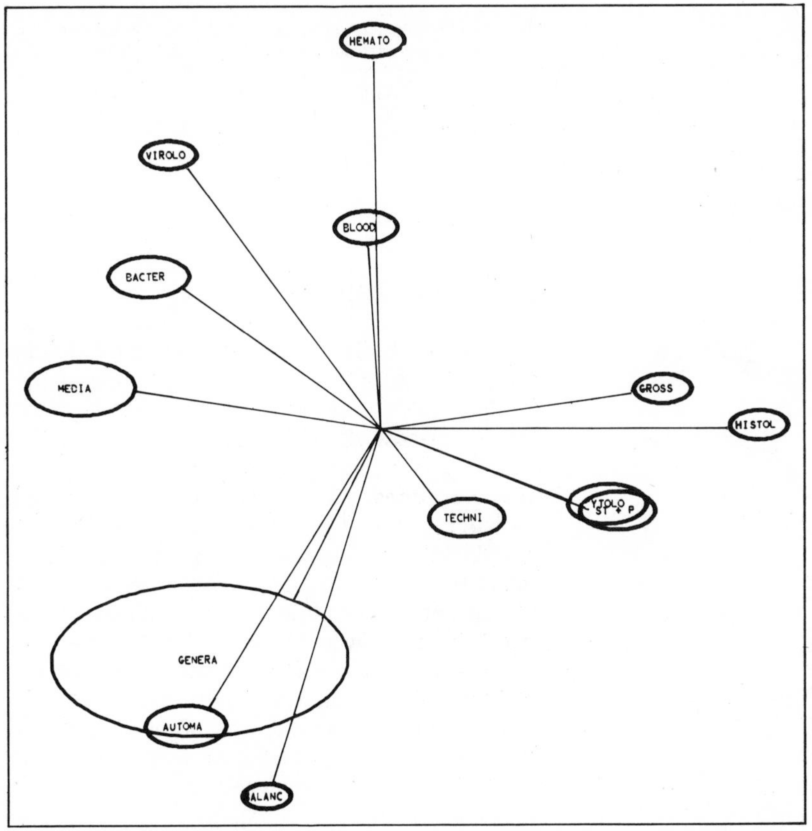

program 2A — radial scheme

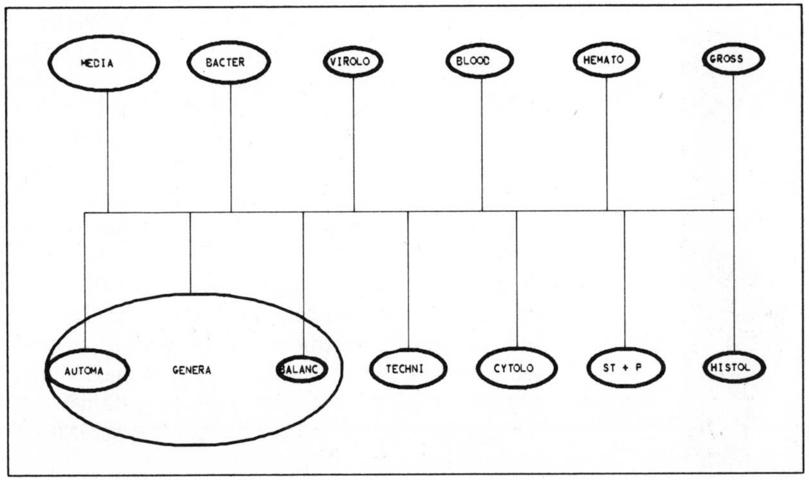

Comprograph 2 printout

program 2B — linear scheme

(c)

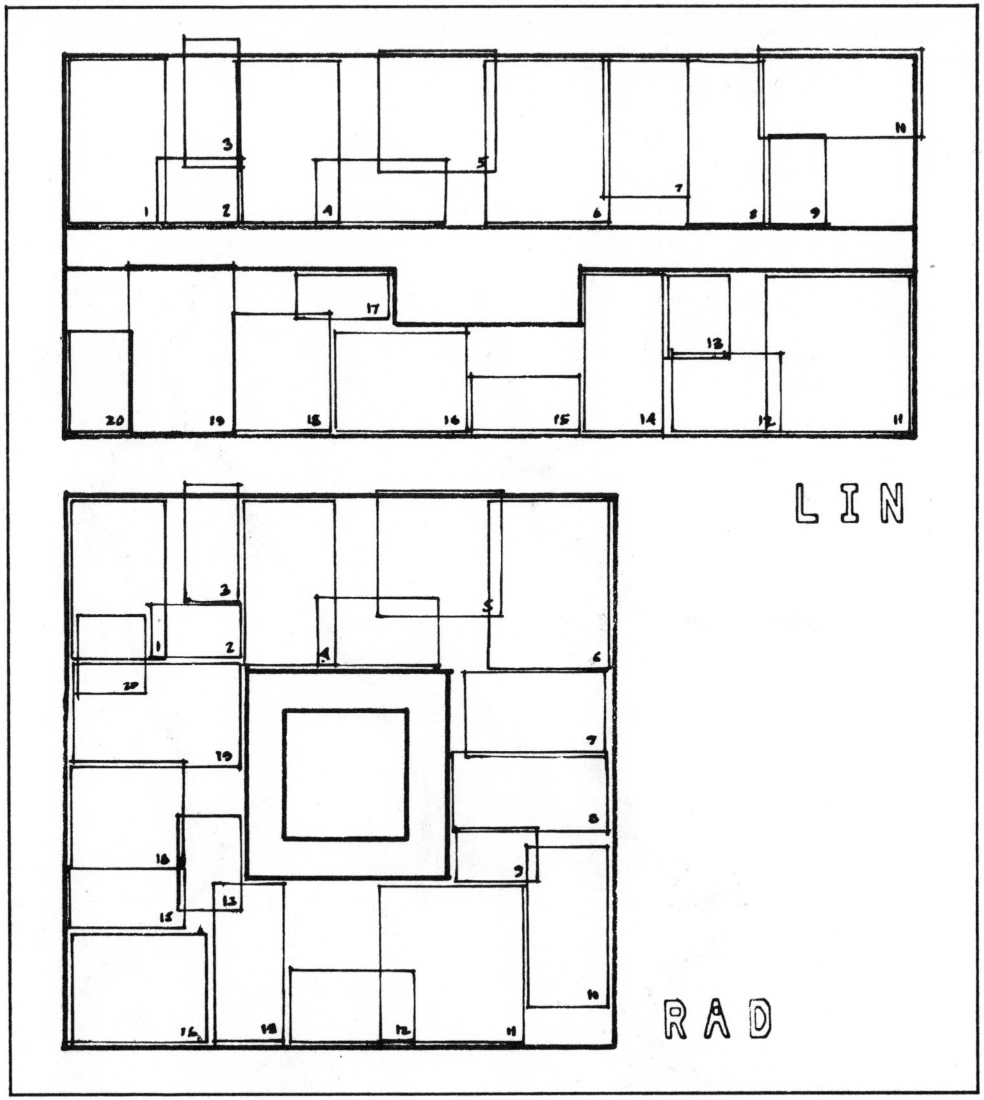

Comprograph 3 printout

9-7
SITE ANALYSIS This should be an unprejudiced evaluation of a site's specific topography, vegetation, geology, borings, utilities, access, and zoning—amenities, densities, micro-climate. *"Messages from Perugia" a figurative view of the geology and land sections, 15th century (anonymous)*

9-8
THE MASTER PLAN Acting as a visual goal, it is something which the lay person can use as a lofty ideal—a direction, a guide, a framework for development—the infrastructure network. *U.S. Department of Housing and Urban Development, "New Communities Project," executed by members of the Harvard Graduate School of Design in 1969*

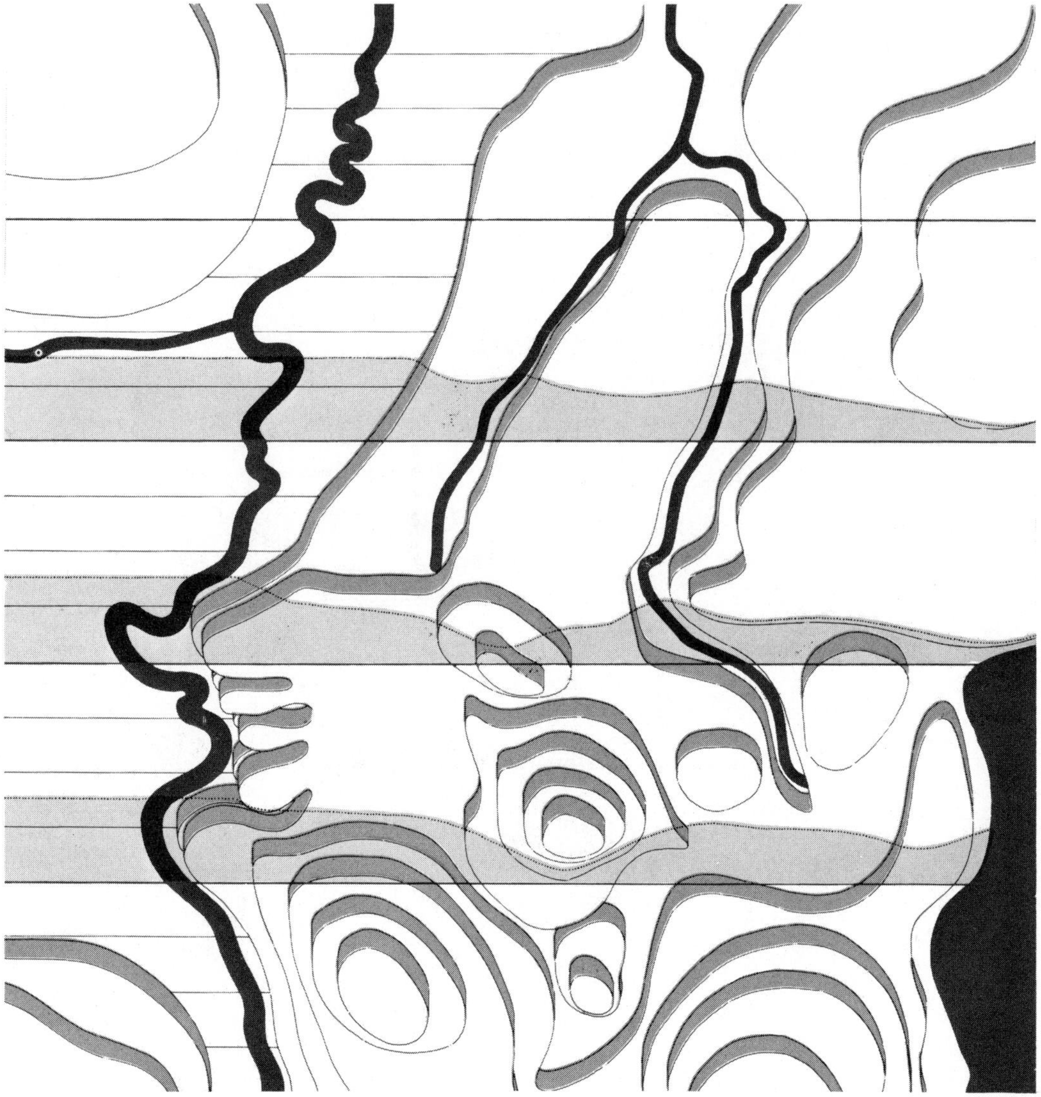

SITE ANALYSIS

Site analysis should be an unprejudiced evaluation of a site's specific topography, vegetation, geology, borings, utilities, access, and zoning, the neighborhood's adjacent densities and amenities, and the specific climate, sun, and wind orientation of the area. Locations are recorded and incompatible overlappings are given priorities. Site areas most suitable for building and natural features which should be preserved or developed are defined, and drainage or soil conditions to be avoided or utilized begin to indicate preferred locations for open space, parking, and major circulation. The neighborhood survey would also give direction as to aspects of local life style and an environmental character to be perpetuated or to be altered to better serve potential occupants.

THE MASTER PLAN

The master plan evolves when the generalized "ideal functional plan" is overlaid by the specific site conditions, and the specific program size and number requirements are worked out in terms of "optimal relationship diagrams." This constant interplay of the general and the specific is worked both ways, and the *site-related functional diagram* can then be interpreted for specific building type massing, open space, and circulation requirements, to provide a range of open spaces (private areas, tot lots, courts, recreational fields, planted areas) and to define the hierarchy of pedestrian and vehicular circulation and parking. The master plan sets the visual character and provides a framework for various life styles. It also sets the organization for future development.

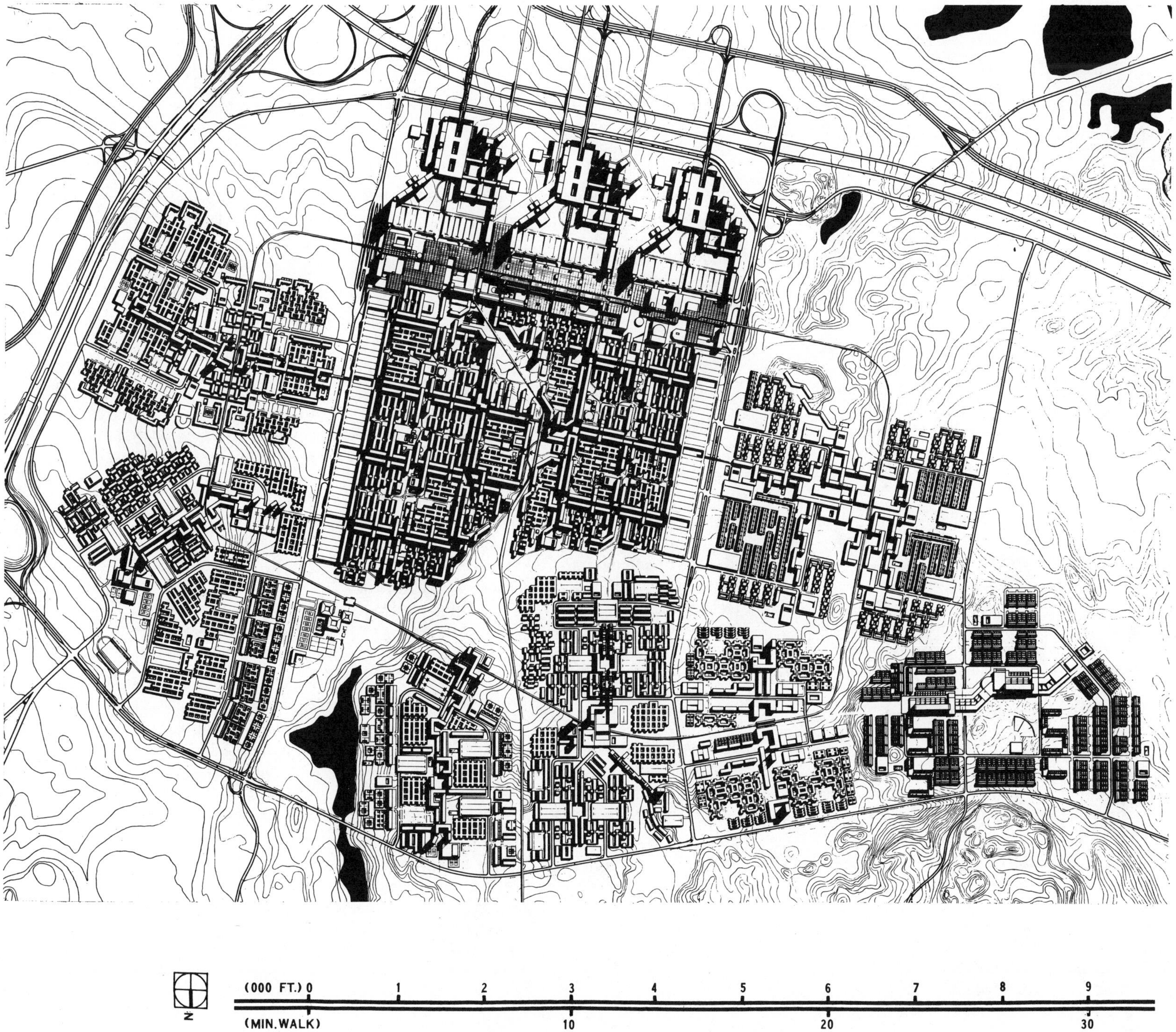
N
(000 FT.) 0 1 2 3 4 5 6 7 8 9
(MIN. WALK) 10 20 30

DESIGN DEVELOPMENT

The design development stage returns to the specific with detailed development of the program's elements oriented towards economic site treatments and optimizing the environmental quality.

Access
- —Public transportation, automobile and pedestrian routes
- —Time and distance from major origins
- —Approach and arrival (visual and psychological)

Circulation
- —Grid, linear, loop/radial
- —(Regional) (spine collector) (residential and service distribution)
- —Desire lines (destination—origin)

Parking:
- —Access from collectors or distribution
- —Ramp (straight, circular, switch back), sloped floor, elevator
- —Dimensional requirements
- —Underground, platform, street, lot, cul-de-sac
- —Buffers and screening

Spaces
- —Hierarchy: Private/public
passive/recreational
- —Pattern alternatives: ring, radial, open-ended, hierarchical

Climate and Orientation
- —Sun exposure and shadows
- —Desirable and undesirable views
- —Wind, rain, and snow conditions
- —Special orientation requirements (trade winds, cistern water collection, winter shadows, hurricane zone, etc.)

Land Form and Vegetation
- —Existing contour and watershed
- —Vegetation: canopy, understory, groundcover
- —Scale, direction, focus, accent, screening
- —Land forms, space dividers, and screen
- —Architectural vs. natural treatment

BUILDING TYPES AND MASSING

For the individual, the essential units of identity and association still remain the home, the street, the block. It is logical to adopt the pedestrian street and the cluster of dwellings as the fundamental items of identity and association in each residential community. The low-, medium-, and high-rise types of housing for which ECOLOGIC is appropriate have been studied in the context of clusters. The building types studied are:

1. *Row houses*—Groups of attached dwellings not exceeding three floors in height and with access to each unit at ground level.
2. *Walkups*—Groups of attached dwellings, mainly duplex in layout and up to 4–5 floors in height with the access to the upper-level units by way of a common access gallery.
3. *Elevator-served buildings*—These buildings take the form of continuous "spine buildings" and towers, or single or double high- and low-rise "blocks." In most cases the dwellings on the lower four floors would be designed as walkups, and elevators are generally designed on the "skip-stop" principle.
4. Combination cluster.

Once an appropriate cluster density for each of these is established, mixes—which would give both social and design variety—can be developed to suit the density required in a particular site and situation.

Row House Cluster

Dwelling Units per Acre—34 Net Project Density
Building Coverage—43.5% Average
Floor-Area Ratio—0.81 Average

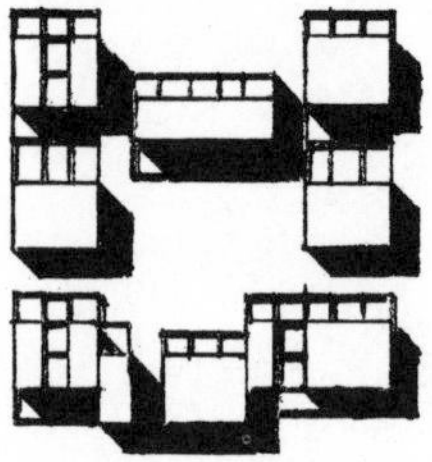

9-9

This type of cluster is particularly appropriate for ECOLOGIC I's and II's wood panel or steel systems. It is also a formation that can be included as part of a higher-density scheme using medium- and high-rise buildings in varying proportions. The ground-oriented row house unit is the preferred housing type for family living.

9-9—9-13
BUILDING TYPES, MASSING AND DENSITIES.
U.S. Department of Housing and Urban Development, "New Communities Project," executed by members of the Harvard Graduate School of Design in 1969

Walkup Cluster

Dwelling Units per Acre—69 Net Project Density
Building Coverage—49% Average
Floor-Area Ratio—1.63 Average

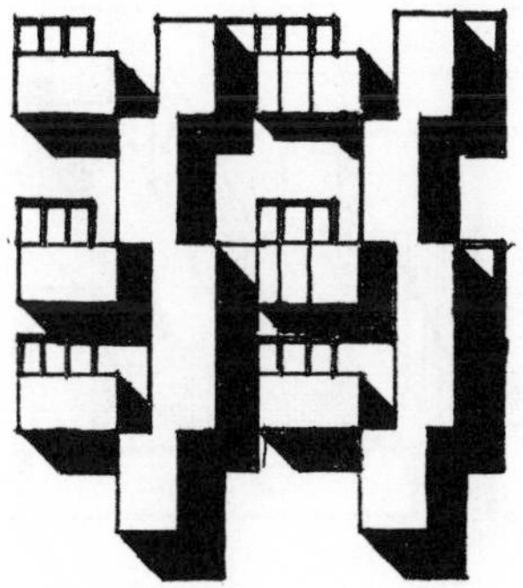

9-10

Each block can consist of one to six bedroom units or combination. Ground level units could have generous private courts while upper level units may gain private open space by deleting a unit or a bedroom for a deck, extending stairs to the roof level.

High-Rise and Infill

Dwelling Units per Acre—100–250 Net Project Density
Building Coverage—Varies

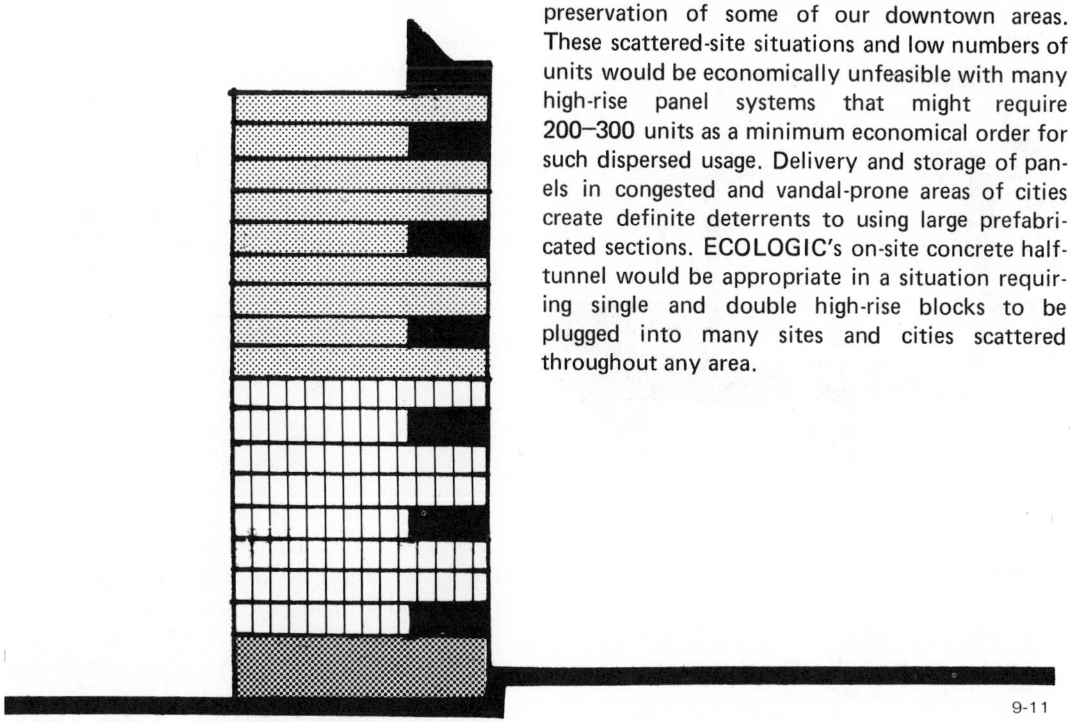

9-11

Unusual market demands exist in the area of high-rise infill. High-rise apartment buildings built on sites scattered throughout a city or even several cities are the type of housing needs that exist in this country and are a critical concept in the preservation of some of our downtown areas. These scattered-site situations and low numbers of units would be economically unfeasible with many high-rise panel systems that might require 200–300 units as a minimum economical order for such dispersed usage. Delivery and storage of panels in congested and vandal-prone areas of cities create definite deterrents to using large prefabricated sections. ECOLOGIC's on-site concrete half-tunnel would be appropriate in a situation requiring single and double high-rise blocks to be plugged into many sites and cities scattered throughout any area.

Combination Cluster—High- and Medium-Rise, Walkup, Row House

Dwelling Units per Acre—60–150 Net Project Density
Building Coverage—30% Average

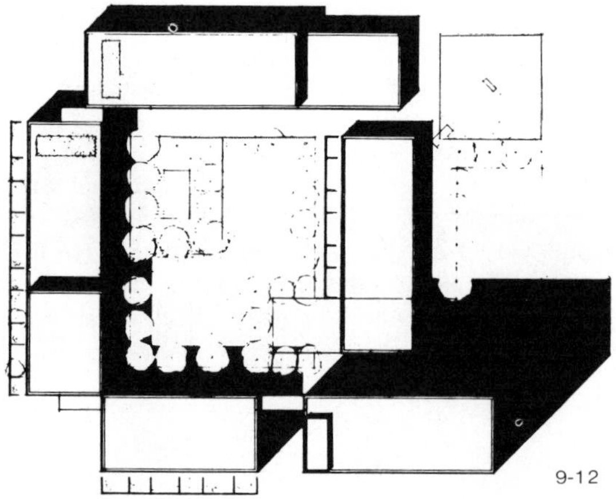

9-12

This type of cluster creates a more diverse environment. The full range of building types shows an appropriateness to the capacities of a combination of lightweight and high-rise systems as with the ECOLOGIC concept. The extra density of this type of arrangement allows a project to include communal recreational facilities (playfields, swimming pools, tennis courts) which would not be economical with less dense clusters.

This type of cluster plan also integrates easily into both urban and more suburban locations. The low 3–4 story buildings would be designed adjacent to the low walkups of an existing neighborhood and then step up toward the high-rise buildings. Being adjacent to high-rise structures also makes 5–6 story structures an economic possibility, by joint use of the elevator core. Otherwise, 4 story walkup limitations and elevator costs make these in-between heights unfeasible. This cluster also makes a suitable prototype project as it can illustrate the range of building types from row houses and walkups to high-rise apartments.

Spine Cluster

Dwelling Units per Acre—150–250 Net Project Density and Up, If Required
Building Coverage—Varies

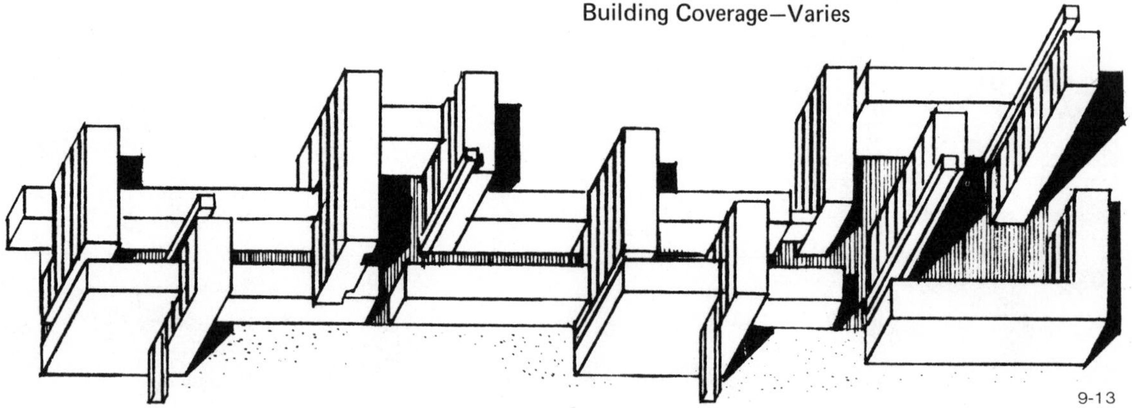

9-13

High-Rise Spine Cluster (Mixed Land Use) A totally high-rise building cluster would require, because of the density implied, a substantial amount of ancillary commercial and recreational facilities.

Variety in building height, form, and usage provides a rich environment. Because of this, the massing of such a complex might take the form of high-rise along a spine.

This type of project is applicable for use in core sections of densely populated metropolitan areas, as a format for air rights projects or as a typical model for a new town center. There are areas where land is very expensive and desirable or where a high density is desired to serve an important transportation node and establish an activity center. It is almost certain that a project of this scale would be, or at least should be, a mixed land usage scheme.

New Community Developments

Gross Project Density—Approximately 40 DU/Acre

Net Project Densities—From 60 to 200 DU/Acre

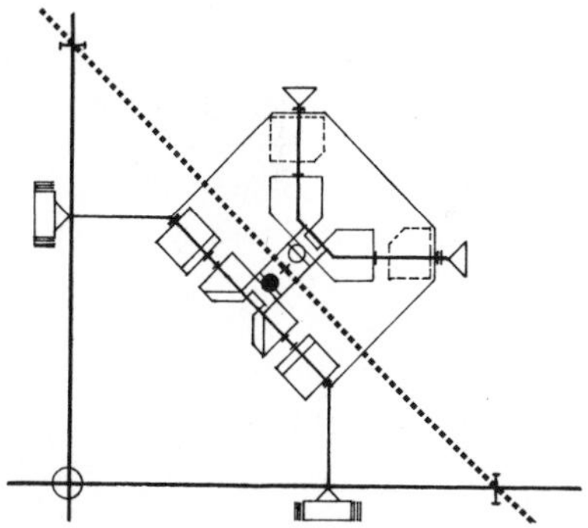

9-14

SYSTEM ECOLOGIC could be used to build the initial "seed units" of new town and new town-in-town developments—continuing in the following years to build units on a schedule set by the impetus of the town itself, not by the demands of a factory. A developing new town, or new town in-town does not usually maintain a constant rate of growth and therefore might have difficulties in supporting a factory.

It is often the nature of "new town" type developments to be located in areas which are at the present remote or even in areas of wilderness in order to develop the area. These might be inaccessbile mountain or desert wildernesses or previously underdeveloped areas that now have the potential to boom, such as areas of Vermont and Alaska. Not being tied to a central plant or limited by extreme transportation requirements, would indicate an on-site system.

9-14—9-15
NEW COMMUNITY—TOWN CENTER Systems building is fully utilized as an appendage to the utilities and circulation infrastructures. Note housing on lower portion "steps" back to open itself to the southern sun while the slab buildings on the northern side of pedestrian mall area are designed with an east-west orientation and a double-loaded corridor.
Baldwin, Cutler, Heder, Stephens: Masters Theses, Harvard University 1967

9-15

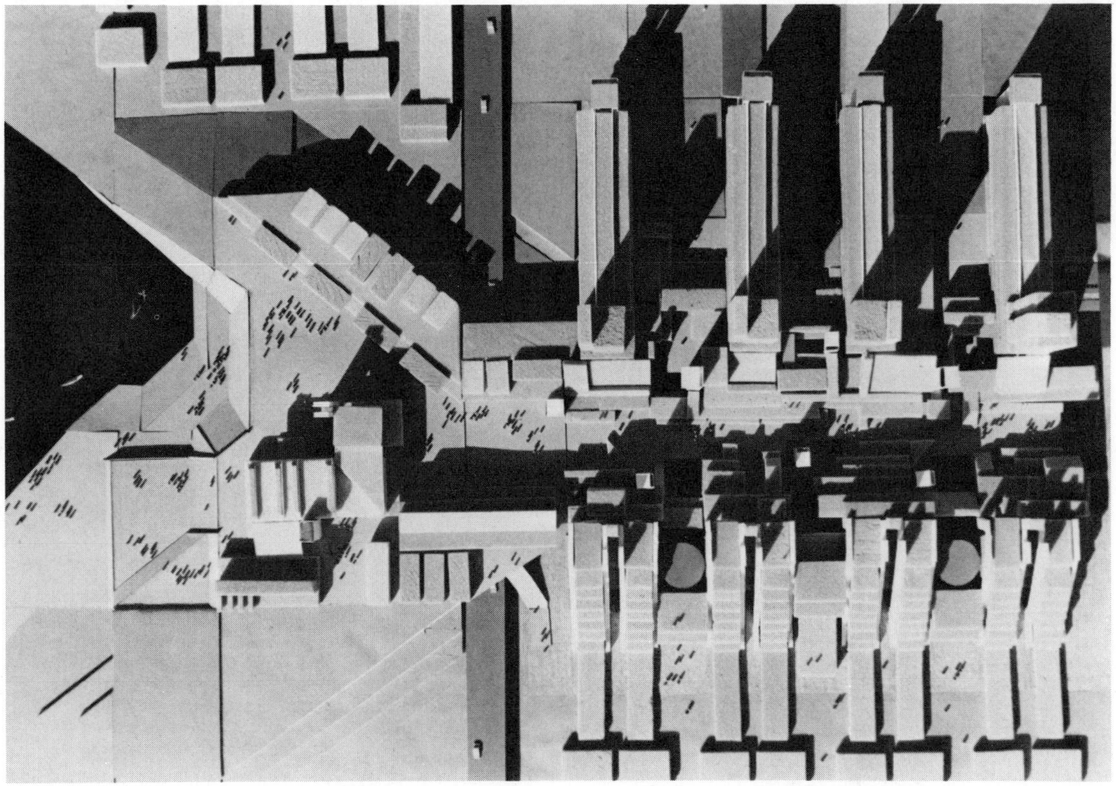

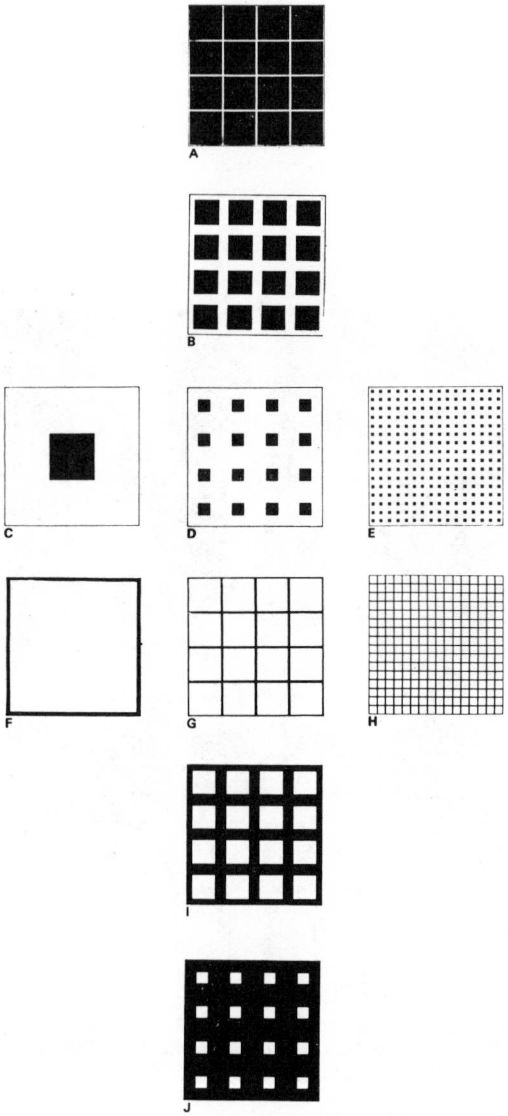

9-16
DENSITIES, LAND USE, AND GEOMETRICAL ORDER
The proportional use of urban land and its resultant effects are illustrated. Figure C shows 10% land coverage distributed in a concentrated nuclear form, E in a dispersed nuclear pattern, however, F—H show the same amount of urban land distributed in a linear manner. A shows 90% land coverage, B shows 50% coverage, while I and J shows the inverted scheme of linear coverage. *Lionel March, RIBA conference, July 1967, Brighton, England. Land Use and Built Form Studies, University of Cambridge.*

DENSITIES

Higher densities can provide an extension of choice and a desirable increase in the complexity of social patterns.

In 1960, there was a density of 4,200 persons per square mile in cities over 100,000; by 1980, it is projected that there will be 3,800 persons per square mile. This indicates an increasing rate of land use and, adding the population growth, it shows us that in order to conserve our natural resources, we must consider higher densities. During the last 10 years, we developed approximately 5 million new acres and, in the next decade, we will probably develop 11 million acres.[3]

There is an even greater tendency to sprawl in the face of the population explosion, which can be controlled only by innovative land-use concepts, such as Planned Unit Development. Developers must build "communities," not housing.

More economical use of the land available does not mean wholesale increases in the scale of all projects, yet almost all projects built in this country do not properly use the land and are at very low densities. Increasing the density, however, can substantially reduce overall costs of infrastructure, utilities, roadways, services, etc. More importantly PUD offers the housing designer the opportunity to do repetitive planning with variety and flexibility.

In single-family housing, 13 units per acre is considered a modest density, allowing 3,000 square feet per each unit and backyard. Densities of up to 60 persons per acre (about 20 dwelling units per acre) are perfectly practical with two-

3. NAHB Chief Economist, Michael Sumichast.

story townhousing. Densities of 90 persons per acre or 30 dwelling units per acre can be done in 2–3 story housing. High-rise need not be utilized until about 140 persons per acre are reached.

Higher densities can provide an extension of choice and a desirable increase in the complexity of social patterns. A richer matrix of social relationships results when a sensitive use zoning accompanies an increase in density. Through concentration of pedestrian movement and local "social buildings," for instance, we can increase the possibility and frequency of accidental or intentional social encounters. Higher densities do not force sociability and must be controlled so that privacy within the home is not threatened, but they increase the choice of available jobs, shops, restaurants, bars, schools, churches, recreational facilities, and acquaintances that also form the habitat. Concentration also offers the only proven alternative to the automobile by increasing services and amenities within walking distance or by providing a "critical mass" of riders to support mass transportation.

LAND USE PLANNING

Tighter communities must be built—not scattered housing. With greater densities must come environmental factors and amenities such as (a) variety of building types, (b) alternatives and mixtures of dwelling unit size, plan, and economic type, (c) the full extent of mixed land use amenities supportable.

If we are going to produce a tremendous volume of new housing, we must insure that the amenities are a proportional part of the total complex. Mixed land usage and variations of building and unit types are really the only available means of nurturing the well-rounded living community.

Obviously, the economies of size provide the yardstick for determining the extent of external (outside unit) amenities. For example, a *combination cluster* of high- and low-rise in a project with densities of about 60–90 dwelling units per acre net project density might support several convenience shops, offices, private garages, meeting rooms, laundromats, nurseries/day care, consulting rooms, a luncheoneete, and a few hotel-type units to be rented out to visitors or residents of the project who need the "guest room" space. Site planning would include a paved game table area for older residents as well as tot lots and ample spaces for outdoor social encounter.

A larger-scale project such as a *high-rise spine scheme* with densities of 200 dwelling units per acre over many acres might be incorporated in a new town in-town, large renewal section of a city, or as a format for the core of a new town. In a situation like this, intensive implementation of mixed land use concepts would be appropriate. This core type could include (as well as the smaller-scale facilities already listed) hotels, motels, offices, shops, dormitories and classrooms, hospital patient care units, housing for the elderly, etc. All of these could be system-built and the whole project can be built on a platform over a multilevel parking garage or an air-rights site. A long-span podium could also incorporate other open space facilities such as theaters and department stores.

Building systems can be flexible enough so that dormitory dwellings, housing for the elderly, hospital patient care units (convalescent, psychiatric, nursing homes, etc.), classroom facilities and even penal institutions can be system-built, but it is worth noting that large panel systems would require, generally, retooling for each building type, and frame systems would have to provide fire walls and acoustic barriers. *On-site, techniques, perhaps, offer the greatest applicability to this type of multiple use.*

Any system or "stable of systems" should be able to adapt to a wide range of housing types, ranging from 2 to 3 stories to over 30 stories in height. This variety allows for the development of large family housing units in combination with medium- and high-rise types of housing. It is possible to develop family units close to the ground while using the high-rise towers for occupants with older or no children. A variety of building and unit type and considered massing can provide a total cross-section of ages, family sizes, economic backgrounds, by eliminating ghetto-izing families by unit size or elderly by provision of small high-rise units only.

SITE TOPOGRAPHY

Generally, there are no particular problems in adapting most systems to sloping sites other than those found with traditional work. Steps in plan between dwellings should be on the module used and steps between dwellings should be story height. Where possible, traveling tower cranes should run along the contours.

Extreme topographic conditions can be adapted to by building on platforms. Some other usual methods dealing with and profiting by sloping sites are shown in the sketches below.

Contour variations, rather than restricting, actually offer the possibilities for a wider range of design solutions. For instance, a sloped site might

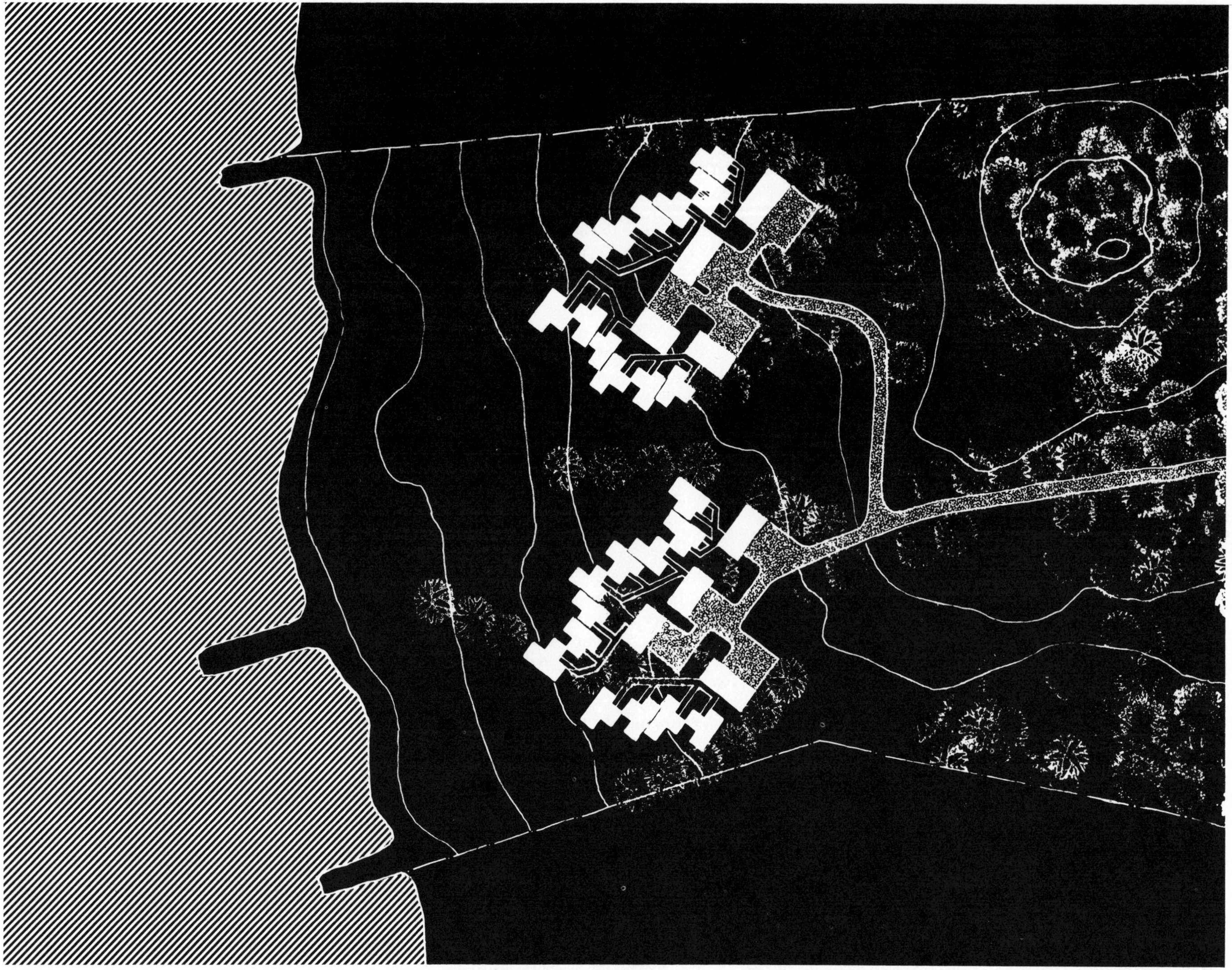

9-17
ORGANIZATION OF SITE ELEMENTS When done in a diagrammatic manner, this permits the individual units each to have a view, to "hide" the parking, to cluster the utilities, to retain open space, to gain privacy, and to heighten the total environmental experience. *ECODESIGN*

make possible a 5 or 6 story walkup by using the slope to enter at the third level—eliminating the need for an elevator while maintaining the walkup limit of three floors.

UTILITIES

The impact of housing on utility systems will depend on the location of housing sites, the number of units developed on the sites, and the capacity of existing utility systems. On smaller urban "in-fill" sites, the capacity of existing systems will probably be sufficient to serve any new housing. The problem with the provision of adequate utilities will arise on larger sites in outlying regions. In more developed areas, local communities will be capable of providing adequate water, electricity, gas and telephone service to the site. With new town developments, etc., utilities would be a major consideration, especially in regard to sewer systems. It is quite possible that many larger sites, as well as new towns, may be located in communities which do not presently have sewer systems. There are diminishing alternatives available for resolving the problem of sewage disposal in these site situations. Natural on-site disposal systems, when allowed, require careful analysis of existing soil conditions on the site and in the immediate surrounding area. If the soil conditions are not suitable for on-site disposal, it will be necessary for the developer to install an artificial sewage disposal system which would result in an increase in the per-unit cost of the housing.

The design and construction of utility systems can be undertaken in a much more efficient manner when the housing is developed as a part of a Planned Unit Development (PUD). Since the entire

9-18
BUILDINGS PARALLEL TO SLOPE—
BUILDINGS PERPENDICULAR TO SLOPE.
Professor Waclaw Zalewski and students in the Department of Architecture, M.I.T.

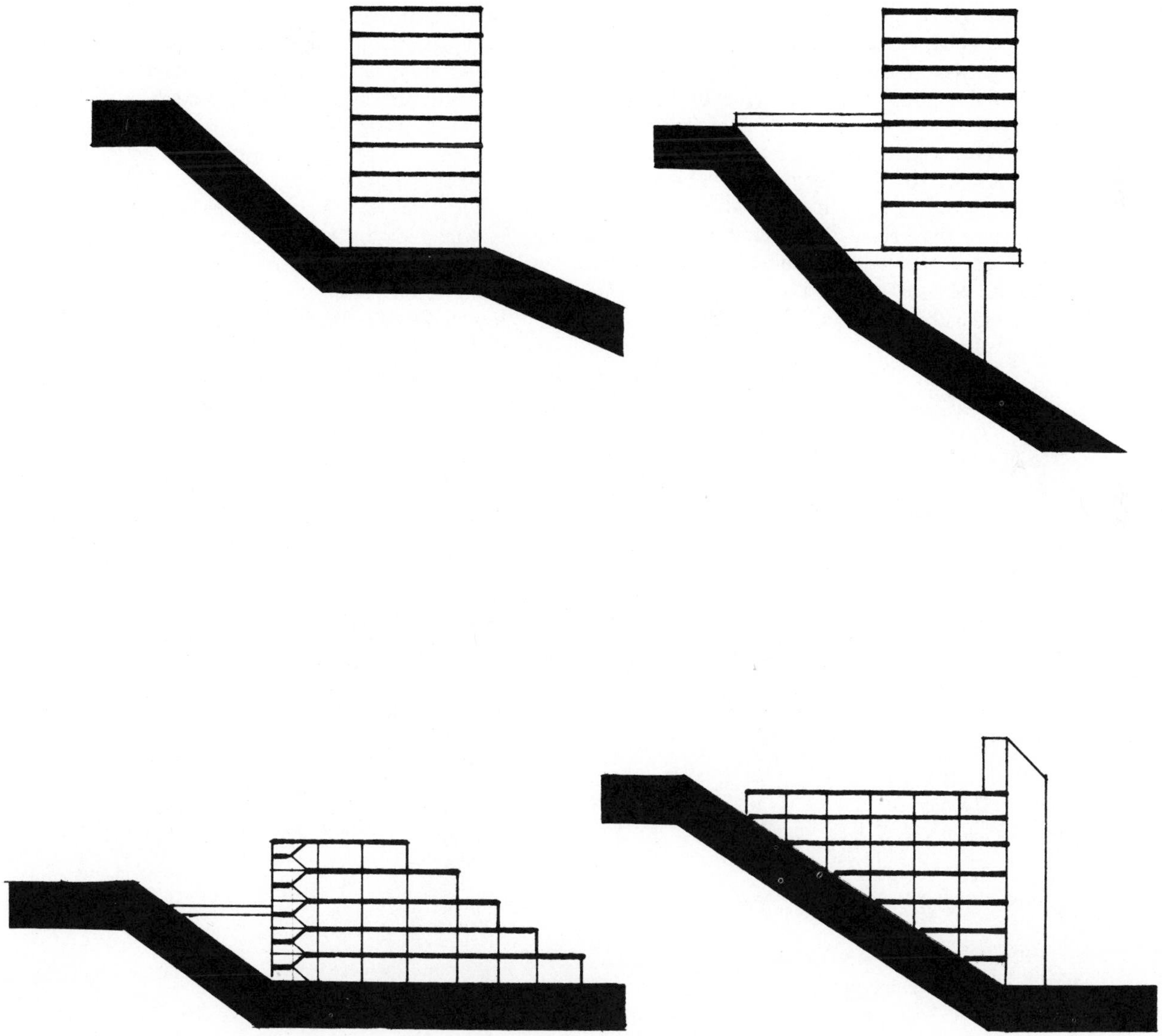

development is designed at one time, it is possible for the community to design utility systems to much closer tolerances than normally allowed. Presently, communities must design utility systems for the eventual possibility of saturation development. This frequently means that the systems are overdesigned for the development which actually occurs. Since more economical utility systems result when they are designed to serve the specific needs of a Planned Unit Development, it follows that the cost of these utilities would then bear a more realistic relationship to the ultimate cost of the development.

SUMMARY

The ingenious American who provides most housing in the U. S.—the small scale developer/contractor who wants to rationalize his operations and offer "the systems building approach" without taking the risks involved in paving the way with a totally new concept for building in the U. S.—this is one element to whom this book is directed and for whom we have outlined the transitional system: ECOLOGIC I, II, and III. The essential difference between SYSTEM ECOLOGIC and other systems, or so-called systems, is that it is a design method and a construction method, a kit of planned spaces and components which can be composed in an almost infinite series of dwelling units and building types.

ECOLOGIC is a design tool as well as an actual building system ready for immediate utilization under the totally diverse and dispersed range of conditions that occur throughout the U. S. housing market. This is the reason for the three totally different construction technologies which we have developed for the transitional system. ECOLOGIC can be used as a framework for setting up on-site or factory fabrication techniques and for creating a new full *integrated industrialized building system* of all aspects of building techniques.

In any case, herein is a chance to rationalize the design and construction process and quicken the schedule while fully satisfying the user's needs. The building industry has not yet reached the point where any building system is anything but transitional!

9-19
ECOLOGIC Most suitable for use as a high-, medium-, or low-rise composite building type project.
This project includes several types of condominium units combined with commercial spaces as the first phase core of a mountain ski village at Sugarloaf/USA, Maine. *ECODESIGN*

BIBLIOGRAPHY

"Automatic Production of Lightweight Panels," *Industrialized Building Systems and Components,* London, October 1967.

Bemis, Albert Farwell, *Rational Design,* Volume III, *The Evolving House,* M.I.T. Press, Cambridge, Massachusetts, 1933.

Blachere, G., *Savoir Batir, Habitabilité, Durabilité, Economie Des Bâtiments,* Editions Eyrolles, Paris, 1966.

Bloc, Andre, editor, "Habitat," *L'Architecture d'Aujourd'hui, February–March 1967, Paris, 1967.*

Carriero, Joseph, et al., *The New Building Block,* A Report on the Factory-Produced Dwelling Module, Research Report No. 8, Cornell University, Ithaca, New York, 1968.

Cars In Housing/1; Cars in Housing/2;Residential Areas: Higher Densities; Service Cores in High Flats, Ministry of Housing and Local Government, London, 1962.

Cazaly and Huggins, *Canadian Prestressed Concrete Institute Handbook,* Ontario.

Corner, Eric, *Modular Primer,* Modular Society Ltd., London, 1963.

Cutler, Laurence S., *Industrialized Building Systems* (tape and cassette), McGraw-Hill, Hightstown, New Jersey, 1970.

Deeson, A. F. L., editor, *Comprehensive Industrialized Building Systems Annual—1967 (The),* Product Journals Ltd., Kent, England, 1967.

Deeson, A. F. L., editor, *Comprehensive Industrialized Building Systems Annual—1970 (The),* Morgan-Grampion, London, 1970.

Diamant, R. M. E., *Industrialized Buildings,* Volumes I, II, III, *The Architect and Building News,* Iliffe Books Ltd., London, 1968.

Dietz, A. G. H., "Future Potential of Building Systems," American Society of Civil Engineers Reprint 741, Pittsburgh, 1968.

Dietz, A. G. H., and Laurence S. Cutler, *Industrialized Building Systems for Housing,* M.I.T. Press, Cambridge, Massachusetts, 1971.

Effect of Repetition on Building Operations and Processes On-Site, United Nations, New York, 1965.

Europrefab Systems Handbook, Housing, Interbuild Prefabrication Publications, Ltd., London, 1969.

Flats and Houses 1958: Design and Economy, Her Majesty's Stationery Office, Henry Brooke, Ministry of Housing and Local Government, London, 1958.

General Plans, The National Building Agency, London.

Guy, R. B., et al., *The State of the Art of Prefabrication in the Construction Industry,* Battelle Memorial Institute, 1967.

Harris, Walter D., and Hans A. Hosse, *Vivienda en el Peru (La),* Union Panamericana, Secretaria General Organization de los Estados Americanos, Washington, D. C., 1963.

Harris, Walter D., and Hans A. Hosse, colaboradores, *Vivienda en Honduras (La),* Union Panamericana Secretaria General Organization de los Estados Americanos, Washington, D. C., 1964.

Harrison, H. W., *Performance Specifications for Building Components,* Ministry of Public Building and Works, Watford, England, 1969.

Hoffman, Hubert, *Row Houses and Cluster Houses: an International Survey,* Praeger, New York, 1957.

Hosken, Fran P. The Functions of Cities, Schenkman Publishing Co., Cambridge, Mass., 1973.

Housing Construction, Novosti Press Agency Publishing House, Moscow, 1967.

Housing Primer: Low and Medium Rise Housing, A. and P. Smithson, editors, Architectural Design, London, 1967.

Industrialization of Building (The), R.I.B.A. Building Study Team, R.I.B.A. Publishing Office, London.

Industrialization Forum, "Communicating Industrialization: two courses and a symposium," Montreal, January 1970.

Industrialized Building and the Structural Engineer, The Institution of Structural Engineers, London, 1966.

Industrialized Housing and the Architect, R.I.B.A. and National Building Agency, London, 1967.

Industrialized Housing for London County Council at Morris-Walk, Western House, Western Avenue, London W 5.

Institute of Building Types Design, *Building Industrialization Technical Design Typification in Hungary,* Budapest, 1969.

Interbuild, International Building News, Prefabrication Publications, Ltd., London (all back issues).

International Council for Building Research, Studies, and Documentation—CIB, editors, *Towards Industrialized Building,* 3rd CIB Congress, Copenhagen, Elsevier Publishing Company, Amsterdam, 1966.

Kelly, Burnham, *The Prefabrication of Houses,* M.I.T. Press, Cambridge, Massachusetts; John Wiley & Sons, Inc., New York, 1951.

Kjeldsen, Marius, *Adoption of Modular Coordination in Denmark (The),* Ministry of Housing, Copenhagen, 1966.

Kjeldsen, Marius, and W. R., Simonsen, *Industrialized Building in Denmark,* 3rd CIB Congress, 1965, published in collaboration with "Byggeindustrien" and A. Jesperson & Sons Foundation.

Kubicek, R., Ing., *Konstruktivy 35 Let,* Prague, 1964.

Marchand, P. Eugene, *Report of the Canadian Technical Mission of Prefabricated Concrete Components in Industrialized Building in Europe, September 2–22, 1966,* Department of Industry, Ottawa, 1967.

Martin, Bruce, editor, *The Coordination of Dimensions for Building,* R. I. B. A., London, 1965.

Metric House Shells, The National Building Agency, London.

Meyer-Bohe, Walter, editor, *Vorfertigung—Atlas der Systeme,* Vulkan-Verlag Dr. W. Classen, Essen, 1967.

Mobile Home Minimum Body and Frame Design and Construction Standards, c. 1967, Mobile Homes Manufacturing Association.

Modular Coordination in Building, United Nations, New York, 1968.

Modular Coordination of Low-Cost Housing, United Nations, New York, 1966.

Nordiske Komite for Bygningbestemmelser, NKB, 1965, *Modular ABC,* Reprt #4, Stockholm.

Notter, George, comp., *Comparative Housing Study,* Cambridge, Massachusetts, Harvard Graduate School of Design, 1958.

Pratt Institute, *Cost Reduction Methods for High-Rise Apartments,* Brooklyn, New York, 1967.

Private Housing in London—People and Environment in Three Wates Housing Schemes, Shankland-Cox, London, 1968.

Progressive Assembly Construction Methods, Stavebni Zavody Praha, Prague, 1968.

Ramsey, Charles G., and Harold R. Sleeper, *Architectural Graphic Standards,* John Wiley & Sons, Inc., New York, 5th Edition, 1956.

Rowland, Norman, *Reston Low Income Housing Demonstration Program,* a report on factory produced multi-family housing utilizing light gauge steel modules, Washington, D. C., 1969.

Schmid, Thomas, and Carlo Testa, *Systems Building: An International Survey of Methods,* Pall Mall Press, London, 1969.

SCSD Project, California (School Construction Systems Development), Ministry of Public Building and Works, London, 1965.

Sebestyen, Gyula, *Large Panel Buildings,* Akademiai Kiado, Hungarian Academy of Sciences, Budapest, 1967.

Symposium on High Flats: Part I, R.I.B.A., London, 1955.

System Building 2, Interbuild, Manchester Square, London.

Thomas, Mark Hartland, *Modular Design of Low-Cost Housing,* United Nations, New York, 1966.

U. K. Ministry of Housing and Local Government, *Co-ordination of Components in Housing,* metric dimensional framework, London, H.M.S.O., 1968.

U. K. Ministry of Housing and Local Government, *Industrialized Building,* London, H.M.S.O., 1965.

U. S. S. R., *Industrialized Techniques in Housing,* Moscow, 1963.

U. S. Department of Housing and Urban Development, *Bibliography on Housing Building and Planning,* Washington, D. C., 1969.

U. S. Department of Housing and Urban Development, *In-Cities Experimental Housing Research and Development Project,* phase 1, composite report, volumes I–IV, Washington, D. C., 1969.

U. S. Department of Housing and Urban Development, *Industrialized Building,* a comparative analysis of European experience, Washington, D. C., 1968.

U. S. Department of Housing and Urban Development, *Operation BREAKTHROUGH: Mass Produced and Industrialized Housing,* Washington, D. C., 1970.

U. S. National Commission on Urban Problems, *Building the American City,* Report to the Congress and to the President of the United States, Washington, D. C., Government Printing Office, 1968.

GLOSSARY

Assembly: An arrangement of building elements to make a whole.

A. S. T. M.: American Society of Testing Materials.

Basic Module: (*m*:System Ecologic) A module with the size of 4 inches or 10 millimeters used to coordinate the dimensions.

Building Process: The process which embraces every stage from the conception to the total satisfaction of a building requirement.

Closed System: A building system having only internalized interchangeability of its subsystems.

Closure: The cladding or facing material or system used for exterior walls and openings.

Component: An industrial product which is manufactured in a factory. Although it is an independent unit, it is intended to interface with other building materials.

Dead Load: The weight of all permanent construction including walls, floors, partitions, and service equipment.

Dimension: A distance between two points, lines, or planes.

Eco: The habitat or environment that influences the mode of life or the course of development of a community.

Ecology: A branch of science concerned with the interrelationship of organisms and their environment.

Forming: The casting of concrete in molds whether on-site or in a factory.

Geometry: The aspect related to the spatial and dimensional characteristics of the building design.

Grid Line: A line in a reference grid (See *Planning Grid*).

H. M. S. O.: Her Majesty's Stationery Office (U. K.)

H. U. D.: Department of Housing and Urban Development (U. S. A.).

Industrialized Building System: The total integration of all subsystems and components into an overall building process fully utilizing mass production and assembly techniques.

Interface: The point of contact or blending of two objects.

Joint: The space between and the meeting of two or more building elements.

Live Load: The weight superimposed by the use and occupancy of the structure, not including wind loads or dead loads.

Mobile Home: A portable structure built on a chassis and designed to be hauled to a site and used with or without a permanent foundation as a dwelling unit, when connected to site utilities.

Modular Coordination: Coordination achieved through a dimensional discipline based on the basic module.

Modular Housing: A dwelling unit fabricated of two or more modular units and erected on a foundation.

Module: A convenient size which is used as an increment or coefficient in repetitive planning or production.

Module Unit: A building block, which is self-contained and factory fabricated.

Motor Home: A temporary dwelling, which is portable and is used for travel recreation and is self-propelled.

Open System: A building system having an externalized interchangeability of its subsystems, components, or building elements.

Parker Morris Standard: A set of guidelines used as a basis for local authority and town council housing in the U. K.

Part: A portion of a component having its own identity.

Planning Grid: (9*m*: SYSTEM ECOLOGIC) A reference grid used for the preparation of plans and elevations of buildings based on 3′-0″.

Prefabrication: The fabrication of building elements before they reach the building site.

R. F. P.: Request for Proposal.

Sectional: A dwelling unit of two or more mobile homes put on a permanent foundation and joined to make a single-family residence.

Structural Module: (12*m*: SYSTEM ECOLOGIC) Horizontal location of load-bearing elements and components related to any multiple of *m* but based on 4′-0″.

Subsystem: Identifiably complete, physically integrated, dimensionally coordinated, series of parts which function together and perform a complete building requirement, i.e., structural, mechanical, flooring.

System: Integrated components, materials and/or building elements, performing in aggregate some specified function.

Tolerance: An allowance for the lack of accuracy which must be accepted for the positioning of a component on site.

Trailer: A vehicular, portable structure built on a chassis designed as a temporary dwelling.

Wall Types: Loadbearing exterior, loadbearing interior, non-loadbearing exterior, non-loadbearing interior.

Wind Load: The lateral or vertical pressure or uplift forces due to wind blowing in any direction.

APPENDIX

A SELECT LIST OF SYSTEMS BUILDING MANUFACTURERS

I. Monolithic (Boxes)

Conbox A/S
Filstedvej 12
Aalborg, Denmark

Uniment
Conrad
249 East 32nd Street
New York, New York 10016

USSR Monolithic Blocks
Moscow, U. S. S. R.
(Reference: "Housing Construction" Novosti Press Agency Publishing House Moscow, U. S. S. R.)

W System
9 Avenue D'Orsay
Paris 7e, France

Zachry
H. B. Zachry General Contractors
P. O. Box 21130
San Antonio, Texas 78221

A B Patio
Vattravagen 8B
Uttran, Sweden

Calder Homes
Biddick Lane
Washington, County Durham, England

Canister Housing
17 Compton Terrace, Candbury
London, N. 1, England

U. W. S. Tri-Ten Container
United States Steel Corporation
525 William Penn Place
Pittsburgh, Pennsylvania

Jones and Laughlin Steel Corporation
3 Gateway Center
Pittsburgh, Pennsylvania 15230

Lunnaville
Byggnads AB
Sweden

Magnolia Homes Manufacturing Coproration
P. O. Box 230
Vicksburg, Mississippi 39180

Mobile Homes Manufacturers Association
20 North Wacker Drive
Chicago, Illinois

Mucklow Plan, Ltd.
Waterfall Lane
Old Hill, Staffs., England

Portakadin Ltd.
Ouseacres, Boroughbridge Road
York, England

Ski-Units
Stephenson Developments (Huddersfield) Ltd.
Grosvenor Works
Leeds Road
Huddersfield, Yorks., England

Suspended Structures Inc.
47 Kearney Street
San Francisco, California 94108

II. Total Systems (Panels)

Bakelite, Ltd.
12 Grosvenor Gardens
London, S. W. 1, England

Behlen Manufacturing Company
P. O. Box 569
Columbus, Nebraska 68601

B. M. B.
N. V. Nederlandsch Bouwsyndicat
Binckhorstlaan 309
's-Gravenhage en Amsterdam
Holland

BR Plastics Building
British Railroad Board
Plastics Development Unit
Central Design and Development Establishment
London Road
Derby, United Kingdom

ECODESIGN, INC.
180 Franklin St.
Cambridge, Mass. 02139

G. H. Burgess and Company, Ltd.
(Burgess Skid Unit)
Jel House, Staines Road
Hounslow, Middlesex, England

Calverley (Industrialized Buildings) Ltd.
Evington Valley Road
Leicester, England

Carlton Contractors Ltd.
Carlton House
Ashley Road
Epson, Surrey, England

Cauvet Construction
Building and Construction Company, Ltd.
Penarth Road
Cardiff, Glamorgan, England

Coignet
Miller, Buckley & Coignet, Ltd.
Trident House, Station Road
Hayes, Middlesex, England

Cosmos
3 Child's Street
London, S. W. 5, England

A. B. Bostadsforskning
Norrlandsgatan 7
Stockholm, Sweden

Foulquier System
12 Rue Jacquemont
Paris, 17e, France

Fram Precast Concrete Ltd.
Daston Road, Wythenshawe
Manchester 22, England

Frameform
James Riley and Partner, Ltd.
31 Fulham Palace Road
London, W. 6, England

Gregory (Drury)
Gregory Housing Ltd.
21 Farncombe Road
Worthing, Sussex, England

Gypsolith
Spc. T. Iotti et Fils
Saint-Nom-La-Breteche, France

Hebel Gasbeton GMBH
8080 Emmering-Furstenfeldbruck
Postfach 10, Germany

Hosgaard & Schultz A/Z
Betonelement-Fabrikkerne
Vasekaer 9, Herlev
Denmark

I. B. S. E. (Pascal Method)
19, Rue Augereau
38 Grenoble, France

Kenkast Buildings Ltd.
Astley, Manchester, England

Koppers Stanley
Koppers Company, Inc.
Pittsburgh, Pennsylvania 15219

Larsen & Neilsen
Copenhagen, Denmark
(Factory—Ølstykke)

Linex
Platenfabriek Linex
Prins Hendrikstraat 6, Koewacht
O. Zeeuws-Vlaanderen, Nederland N. V.

System Locarn Ltd.
24 York Road
Hitchin, Hertfordshire, England

Loc-Pac
Monsanto Chemical Company
Building Products Department
St. Louis, Missouri

Marley Concrete Ltd.
Peasmarch
Guilford, Surrey, England

Misawa Home
I. Banshucho
Shinjuku-ku
Tokyo, Japan

Modus
The Amey Chivers Housing Company, Ltd.
26a Ock Street
Abingdon, Berkshire, England

Ocees Rab
Ocees Components and Structures Ltd.
49/54 Knightsbridge Court
Sloane Street
London, S. W. 1, England

Preton
Stahlton AG, Abteilung Vbk
Reisbachstrasse 57
8034 Zurich, Switzerland

Puutalo
Mannerheimtie 9B
W. Llewelly & Sons Ltd.
16/20 South Street
Eastborne, Sussex, England

Reema Construction Ltd.
Milford Manor
Salisbury, Wilts., England

SB 2
Wale-Sindall Developments Ltd.
Babraham Road, Sawton
Cambridge, England

SF 1
Department of Architecture and Civic Design
Greater London Council
County Hall
London, S. E. 1, England

Skarne System
International AB
Seavagen 153-155 Fack
Stockholm 23, Sweden

Spacemaker
Shepherd Building Group Ltd.
New Lane
Huntington, York, England

Star Manufacturing Company
Woods Corporation
Box 94910
Oklahoma City, Oklahoma 73109

Sundh Ltd.
31 Sun Street, Finsbury Square
Southhall, Middlesex, England

Tegelement
Tegel Industries Centralknotor AB
Drottningsgatan 99
Stockholm VA, Sweden

Tersons
Tersons Ltd.
Dollis Park
London, N. 3, England

Thermagard Mark III
Gardiner Building System Ltd.
G. P. O. Box 140 Borad Plain
Bristol 2, England

T08B
Stavebni Zavody Praha
Revlocni 7
Praha, Czechoslovakia

Tracoba
Gilbert-Ash Ltd.
Newcombe House, 45 Notting Hill Gate
London, W. 11, England

Urethane Structures
27 East 62nd Street
New York, New York 10021

United Steel Homes
Charlestown Road
New Albany, Indiana 47150

VAM System
Intervan N. V.
Generaal Spoorlaan 489
Rijswijk 2H, Holland

Wates Ltd.
1260 London Road
Norbury
London, S. W. 16, England

Wilma II
Wilma Aannemingsmaatschappoj NV
26 Van Berlostraat
Weert, Holland

III. Structural Systems (Frames)

Albee Industries
931 Summit Street
Niles, Ohio 44446

Aluminum Structures Company
2514 Laurel
Wilmette, Illionis 60091

Butler Manufacturing Company
13th and Eastern Avenues
Kansas City, Missouri

C.L.A.S.P.
Brockhouse Steel Structures Ltd.
Birmingham Road
West Bromwich, United Kingdom

Conder
Winchester, Hampshire, England

Filon
12333 South Van Ness Avenue
Hawthorne 90250
Los Angeles, California

G80 Panelwall/G80 Wingframe
G. 80 Developments
Hemstead House
Marlowes, Hemel Hempstead
Hertfordshire, England

Gangnail Truss Systems
Automated Buildings Components Inc.
7525 Northwest 37th Avenue
Miami, Florida

Guildway Ltd.
Portsmouth Road
Jarrow, County Durham, England

Steelcor Building System
Inland Steel Products Company
4101 West Burnham Street
P. O. Box 393
Milwaukee, Wisconsin

IPI
Via G.de Grassi, 14
Milano, Italy

J. E. Lesser (Homes) Ltd.
Jel House, Staines Road
Houslow, Middlesex, England

Lowton-Cubitt Housing
Cubitts Construction Systems Ltd.
1 Queen Anne's Gate
London, S. W. 1, England

5M
Ministry of Housing and Local Government
Research and Development Group
Caxton House, Tothill Street
London, S. W. 1, England

National Building Frame Manufacturers
Dickens House, 15 Tooks Court
London, E. C. 4, England

Owen Kleine Structures Ltd.
P. O. Box 10
Darlaston, Wednesbury, Staffs., England

Porte des Lilas (Self Lift)
Enterprise Quilery
8 & 10 Avenue du 4 Septembre
Saint-Maur-Des-Fosses, France

Public Building Frame
Ministry of Public Building and Works
London, S. W. 1, England

Officine Saira SpA
Via Marconi, 4
37069 Villafranca
Verona, Italy

SCSD (Educational Facilities Laboratories)
477 Madison Avenue
New York, New York 10022

Scola
Hampshire County Council
The Castle
Winchester, England

SEAC II
Essex County Council
County Hall
Chelmford, Essex, England

Simmcast
W. J. Simms Sons and Cooke Ltd.
Haydn Road
Sherwood, Nottingham, England

Star Manufacturing Company
Woods Corporation
Box 94910
Oklahoma City, Oklahoma 73109

Structurapid "Brevetti Gaburri"
Piazza Partigiani, 29
17021 Alassio (Savona), Italy

System T
British Railroad Eastern Region
Tinsley, Sheffield, England

T & N House
Turner & Newall Ltd.
Asbestos House, 77/79 Fountain Street
Manchester, 2, England

Technosider SpA
Rome, Italy

Trusteel Corporation Ltd.
Gate House, The High
Harlow, Essex, England

Vic Hallam Housing Mk 3
Vic Hallam Ltd.
Valley Works
Langley Mill, Notts., England

IV. Special Construction Techniques (On-Site)

Aerojet-General Corporation (Filament Winding)
1100 West Hollivale Street
Azosa, California 91702

Airfloor Company of California, Inc.
13729 East Rosecrans Avenue
Santa Fe Springs, California 90670

Kjeldsen System
Royal Academy School of Architecture
Kongens Nytorv 1
Copenhagen, Denmark

Lift Slab
British Lift Slab Ltd.
Lynton House, Lynton Square
Birmingham, 22b, England

Mowlem
John Mowlem & Company Ltd.
Westgate House, Ealing Road
Brentford, Middlesex, England

Parkwall
Sir Lindsay Parkinson & Company, Ltd.
88 Upper Richmond Road
London, S. W. 15, England

Storiform/Sectra/Easiform
John Laing Construction Ltd.
Page Street, Mill Hill
London, N. W. 7, England

Prometo
William Thornton & Sons Ltd.
38 Wellington Road
Liverpool 8, England

Smith-Emery Company
781 East Washington Boulevard
Los Angeles, California

Sheel Unit System
Oscar Singer, Ing. Arch., FRIBA
10 Dorland Court, 106 West Hill
London, S. W. 15, England

Stacksack
J. L Brett & Company
1918 South 77 Sunshine Strip
Harlingen, Texas 78550

Sunley Allbetong
Hambrook House, Brixton Hill
London, S. W. 2, England

Tracoba No. 4
IBS Industrialized Building Systems Inc.
110 East 59th Street
New York, New York 10022

V. Components and Performance Specifications

Air-Tech Industries, Inc.
(Division of Walter Kidde & Company, Inc.)
9 Brighton Road
Clifton, New Jersey 07012

Binishells
Via Fatebenefratelli 20
Milano, Italy

B. M. V.
Esso Research and Engineering Company
P. O. Box 243
Elizabeth, New Jersey 07209

BXL Plastics Materials Group Ltd.
12-18 Grosvenor Gardens
London, S. W. 1, England

CPTB
Clay Products Technical Bureau
Drayton House, Gordon Street
London, W. C. 1, England

Crane
4100 South Kedzie Avenue
Chicago, Illinois

Dow Chemical (Spirit Generator)
Plastics Department
Midland, Michigan

Masonry Research
2600 Wilshire Boulevard
Los Angeles, California 90057

Modumatic Building Units
6710 9th Street North
St. Petersburg, Florida 33702

Onduline (Steel Skin)
Omnium Francais Industriel et Commercial
38 Rue Saint Ferdinand
Paris, France

Pico Safe Stairs
1226 West Olive Avenue
Burbank, California 91506

American Plywood Association
1119 A Street
Tacoma, Washington 98401

Rubberoid Industrialized Roofs
The Rubberoid Company Ltd.
1 New Oxford Street
London, W. C. 1, England

Slant/Fin Corporation
100 Forest Drive
Greenval, New York

SCPI
Structural Clay Products Institute
1750 Old Meadow Road
McLean, Virginia 22101

Thermalite Ytong Ltd.
Hams Hall, Lea Martson, Sutton Coldfield
Warwickshire, England

Threadline
Intercon Research Inc.
1100 East 52nd Street
Indianapolis, Indiana

Tri/Posite System
P. O. Box 1021
Skokie, Illinois 60076

Universal-Rundle Corporation
New Castle, Pennsylvania

Universal Atlas Cement
United States Steel Corporation
P. O. Box 2969
Pittsburgh, Pennsylvania

Yel Fac Facade System
Velux A/S Maskinvej 4
Søborg, Denmark

Vin-Lox Corporation
1352 Northwest 29th Street
Miami, Florida

Whalen System
The Whalen Company
Brock Bridge Road
Laurel, Maryland 20810

INDEX